IntechOpen

Supporting open minds since 2005

Advanced Supercritical Fluids Technologies
http://dx.doi.org/10.5772/intechopen.83197
Edited by Igor Pioro

Contributors
Molla Tadesse Abate, Ada Ferri, Guan Jinping, Guoqiang Chen, Vincent Nierstrasz, Igor Leonardovich Pioro, Ludmila Otilia Cinteza, Daniela Bala, Bofeng Bai, Fangbo Li, Binbin Pei, Huixiong Li, Xiangfei Kong, Qian Zhang, Changjiang Liao, Xianliang Lei, Lingtong Gou, Dongfeng Sun, Masaru Watanabe, Yuya Hiraga, Masayoshi Wagatsuma, Keisuke Suzuki, Takuma Kato, Yasuto Goto, Yukihiro Kanaguri, Shuzhong Wang, Yanhui Li, Igor Novosselov, Brian Pinkard, John Kramlich, Per Reinhall, Antonio Montes, Clara Pereyra, Enrique Martinez De La Ossa

First published in London, United Kingdom, 2020 by IntechOpen
IntechOpen is the global imprint of INTECHOPEN LIMITED, registered in England and Wales, registration number: 11086078, 7th floor, 10 Lower Thames Street, London, EC3R 6AF, United Kingdom
Printed in Croatia

British Library Cataloguing-in-Publication Data
A catalogue record for this book is available from the British Library

Additional hard and PDF copies can be obtained from orders@intechopen.com

Advanced Supercritical Fluids Technologies
Edited by Igor Pioro
p. cm.
Print ISBN 978-1-83880-708-5
Online ISBN 978-1-83880-709-2
eBook (PDF) ISBN 978-1-83880-710-8

For EU product safety concerns:
IN TECH d.o.o., Prolaz Marije Krucifikse Kozulić 3, 51000 Rijeka, Croatia, info@intechopen.com or visit our website at intechopen.com.

Meet the editor

Professor Igor Pioro – Ph.D. (1983); Doctor of Technical Sciences (1992); Professional Engineer (Ontario, Canada) (2008); Fellow of ASME (2012), Canadian SME (2015), and Engineering Institute of Canada (EIC) (2013); member of ANS (2004), and Canadian NS (2010); is an internationally recognized scientist within the areas of nuclear engineering and thermal sciences/engineering (https://nuclear.ontariotechu.ca/people/faculty/dr-igor-pioro.php). He is an author/co-author of about 500 publications. Dr Pioro is a Founding Editor of the ASME Journal of Nuclear Engineering and Radiation Science. He was a Chair of the Executive Committee of the Nuclear Engineering Division of the ASME (2011-2012) and a Chair of the International Conference on Nuclear Engineering (ICONE-20) (2011-2012). Professor Pioro has received many international and national awards and certificates of appreciation.

Contents

Preface

IntechOpen presents a new technical book, *Advanced Supercritical Fluids Technologies*, which covers modern and advanced applications of supercritical fluids in various industries. The editorial team consists of top international experts, mainly from academia, but with engineering and technological experience, and representing the following countries: 1) Canada; 2) China; 3) Italy; 4) Japan; 5) Romania; 6) Spain; 7) Sweden; and 8) USA.

The need for this book is based on the following rationale.

Using SuperCritical Fluids (SCFs) in various processes is not new, because Mother Nature has been processing minerals in aqueous solutions at critical and supercritical pressures for billions of years. In the late 1800s, scientists started to use this natural process in their laboratories to create various crystals. The first studies dedicated to the investigation of various properties and heat transfer in SCFs started possibly as early as the 1930s. It was found that heat transfer near a critical point was quite high, and this discovery was used in single-phase thermosyphons with an intermediate working fluid at the near-critical state. Therefore, solid knowledge of specifics of thermophysical properties and heat transfer is important for many applications of SCFs in industry.

In the 1950s, the idea of using supercritical water was rather attractive for coal-fired thermal power plants. At supercritical pressures, there is no liquid-vapour phase transition; therefore, there is no such phenomenon as critical heat flux or dry-out. Only within a certain range of parameters may a deterioration of heat transfer occur. The objective of operating "steam" generators at supercritical pressures was to increase the total efficiency of a power plant. Currently, the use of supercritical water in power-plant "steam" generators is the largest application of a fluid at supercritical pressures in industry.

Also, at the end of the 1950s and the beginning of the 1960s, some studies were conducted to investigate the possibility of using SCFs in nuclear reactors. Several designs of nuclear reactors using water as the reactor coolant at supercritical pressures were developed in the USA and the former USSR. This idea was abandoned for almost 30 years but regained support in the 1990s.

Currently, six concepts of nuclear-power reactors/nuclear power plants of the next generation – Generation-IV – were proposed, which will have thermal efficiencies comparable with those of modern thermal power plants. The SuperCritical Water-cooled Reactor (SCWR) is one of these six concepts under development in a number of countries. Analysis of Generation-IV concepts shows that SCFs, such as helium and water, will be used as reactor coolants, and SCFs such as helium, nitrogen (or mixture of nitrogen (80%) and helium (20%)), carbon dioxide, and water will be used as Working Fluids (WFs) in power Brayton and Rankine cycles.

However, there are other areas where SCFs are used or will be implemented in the near future. The latest developments within these areas focus on:

- Production of liquid fuels from biomass in sub and supercritical water and carbohydrate up-conversion in ionic liquid and SCFs mixtures;

- Supercritical-water gasification for an emerging technology with synergistic applications in renewable energy and waste processing;

- Supercritical-water oxidation (SCWO), which is a promising, environment-friendly technology to efficiently deal with a wide variety of organic wastes such as wastewaters, industrial and municipal sludge, etc.;

- Fabrication of inorganic and organic nanoparticles in supercritical carbon dioxide with particular emphasis on the metallic and polymeric nanoparticles used in biomedicine;

- Impregnation of materials in supercritical carbon dioxide to impart various functionalities;

- One-step process of foaming and impregnation using supercritical carbon dioxide.

This book on *Advanced Supercritical Fluids Technologies* provides engineers and specialists in various industries dealing with SCFs as well as researchers, scientists, and students of the corresponding departments with a comprehensive overview of the current status, latest trends and developments of these technologies.

Igor Pioro, Ph.D., Dr. Tech. Sc., P.Eng., Fellow ASME, CSME & EIC
Professor,
Founding Editor of the ASME Journal of Nuclear Engineering and Radiation Science,
Faculty of Energy Systems and Nuclear Science,
University of Ontario Institute of Technology,
Canada

Chapter 1

Supercritical-Fluids Thermophysical Properties and Heat Transfer in Power-Engineering Applications

Igor L. Pioro

Abstract

Researches on specifics of thermophysical properties and heat transfer at supercritical pressures (SCPs) started as early as the 1930s with the study on free-convection heat transfer to fluids at a near-critical point. In the 1950s, the concept of using SC "steam" to increase thermal efficiency of coal-fired thermal power plants became an attractive option. Germany, USA, the former USSR, and some other countries extensively studied heat transfer to SC fluids (SCFs) during the 1950s till the 1980s. This research was primarily focused on bare circular tubes cooled with SC water (SCW). However, some studies were performed with modeling fluids such as SC carbon dioxide and refrigerants instead of SCW. Currently, the use of SC "steam" in coal-fired thermal power plants is the largest industrial application of fluids at SCPs. Near the end of the 1950s and at the beginning of the 1960s, several studies were conducted to investigate a possibility of using SCW as a coolant in nuclear reactors with the objective to increase thermal efficiency of nuclear power plants (NPPs) equipped with water-cooled reactors. However, these research activities were abandoned for some time and regained momentum in the 1990s. In support of the development of SCW-cooled nuclear-power reactor (SCWR) concepts, first experiments have been started in annular and various bundle flow geometries. At the same time, more numerical and CFD studies have been performed in support of our limited knowledge on specifics of heat transfer at SCPs in various flow geometries. As the first step in this process, heat transfer to SCW in vertical bare tubes can be investigated as a conservative approach (in general, heat transfer in fuel bundles will be enhanced with various types of appendages, that is, grids, end plates, spacers, bearing pads, fins, ribs, etc.). New experiments in the 1990–2000s were triggered by several reasons: (1) thermophysical properties of SCW and other SCFs have been updated from the 1950s–1970s, for example, a peak in thermal conductivity in the critical/pseudocritical points was "officially" introduced in 1990s; (2) experimental techniques have been improved; (3) in SCWRs, various bundle flow geometries will be used instead of bare-tube geometry; (4) in SC "steam" generators of thermal power plants, larger diameter tubes/pipes (20–40 mm) are used, however in SCWRs hydraulic-equivalent diameters of proposed bundles will be within 5–12 mm; (5) with Research and Development (R&D) of next-generation or Generation-IV nuclear-power-reactor concepts, new areas of application for SCFs have appeared—for example, SCP helium was proposed to be used as a reactor coolant, SCP Brayton and Rankine cycles with SC carbon dioxide as a working fluid are being developed, etc. A comparison of thermophysical properties of SCFs with those of subcritical-pressure fluids showed that SCFs as single-phase fluids have unique properties, which are close to "liquid-like" behavior below critical or pseudocritical points and are quite similar to the behavior of "gas-like" substances above these points. A comparison of selected SCW heat transfer correlations has shown that their results may differ from one to another by more than 200%. Based on these comparisons, it became evident that there is a need for reliable, accurate, and wide-range SCW heat transfer correlation(s) to be developed and verified. Therefore, the objective of this chapter is to summarize in concise form specifics of supercritical-fluids thermophysical properties and heat transfer in power-engineering applications.

Keywords: supercritical water, carbon dioxide, refrigerant, forced convective heat transfer

1. Introduction

1.1 Historical note on using supercritical fluids (SCFs)

The use of supercritical fluids (SCFs) in various processes is not new and, actually, is not a human invention. Nature has been processing minerals in aqueous solutions at near or above the critical point of water for billions of years. In the late 1800s, scientists started to use this natural process in their labs for creating various crystals. During the last 50–60 years, this process, called hydrothermal processing (operating parameters: water pressure from 20 to 200 MPa and temperatures from 300 to 500°C), has been widely used in the industrial production of high-quality single crystals (mainly gem stones) such as sapphire, tourmaline, quartz, titanium oxide, zircon and others [1].

Also, compressed water, that is, water at a supercritical pressure (SCP), but at a temperature below $T_{cr} \approx 374°C$, exists in oceans at the depth of ~2.2 km and deeper. If at this depth there is an active underwater volcano with the temperature of a magma above T_{cr} of water, conditions for existence of supercritical water (SCW) can be reached.

The first works devoted to the problem of heat transfer at supercritical pressures (SCPs) started as early as the 1930s. Schmidt et al. [2] investigated free-convection heat transfer to fluids at a near-critical point with the application to a new effective cooling system for turbine blades in jet engines. They found that the free-convection heat transfer coefficient (HTC) at the near-critical state was quite high, and decided to use this advantage in single-phase thermosyphons with an intermediate working fluid at the near-critical point [3].

In the 1950s, the idea of using SC "steam" (actually, SCW) appeared to be rather attractive for the Rankine power cycle. The objective was to increase a thermal efficiency of coal-fired thermal power plants (ThPPs) (see **Table 1**). This change, that is, substantially higher operating pressures in the Rankine cycle from subcritical ones, and, correspondingly to that, higher inlet-turbine temperature up to 625°C, has allowed increasing of thermal efficiencies from 40–43% to 50–55% (gross) (in total by 7–15%). Currently, SCP coal-fired thermal power plants (world electricity generation with coal 38%—the largest source for electricity generation; in India—77%; China—65%; Germany—37%; and in USA—30%) are the second ones by thermal efficiencies after gas-fired combined-cycle ThPPs (world electricity generation with natural gas 23%—second largest source for electricity generation; in

No.	Power plant	Gross thermal efficiency
1	Combined-cycle ThPP (combination of Brayton gas-turbine cycle (fuel—natural gas or LNG); combustion-products parameters at gas turbine: $P_{in} \approx 2.3$ MPa and $T_{in} \approx 1650°C$) and Rankine cycle steam-turbine parameters: $P_{in} \approx 12.5$ MPa and $T_{in} \approx 585°C$ (T_{cr} = 374°C)	Up to 62%
2	SCP coal-fired ThPP (Rankine cycle "steam"-turbine parameters (see **Figure 1**): $P_{in} \approx 23.5$–38 MPa (P_{cr} = 22.064 MPa), $T_{in} \approx 540$-625°C (T_{cr} = 374°C) and steam reheat at: $P \approx 0.25{\cdot}P_{in}$ and $T_{reheat} \approx 540$-625°C)	Up to 55%
3	Subcritical-pressure coal-fired ThPP (older plants; Rankine cycle steam-turbine parameters (see **Figure 2**): P_{in} = 17 MPa (T_{sat} = 352°C), T_{in} = 540°C (T_{cr} = 374°C), and steam reheat at: $P \approx 0.25{\cdot}P_{in}$ and T_{reheat} = 540°C)	Up to 43%
4	Carbon dioxide-cooled reactor (advanced gas-cooled reactor (AGR)) NPP (Generation-III) (reactor coolant (carbon dioxide): P = 4 MPa and T = 290–650°C; Rankine cycle steam-turbine parameters (see **Figure 2**): P = 17 MPa (T_{sat} = 352°C); T_{in} = 540°C (T_{cr} = 374°C), and steam reheat at: $P \approx 0.25{\cdot}P_{in}$ and T_{in} = 540°C)	Up to 42%
5	Sodium-cooled fast reactor (SFR) (BN-600; BN-800) NPP (reactor coolant (sodium): $P \approx 0.1$ MPa (above sodium level) and T_{max} = 550°C; Rankine cycle steam-turbine parameters (see **Figure 3**): P = 14 MPa (T_{sat} = 337°C); T_{in} = 505°C (T_{cr} = 374°C) and steam reheat at: $P \approx 0.25{\cdot}P_{in}$ and T_{in} = 505°C)	Up to 40%
6	Pressurized water reactor (PWR) NPP (Generation-III+, new reactors) (reactor coolant (light water): P = 15.5 MPa (T_{sat} = 345°C) and T = 280-322°C; Rankine cycle steam-turbine parameters (see **Figure 4**): P = 7.8 MPa and $T_{in} = T_{sat}$ = 293°C and steam reheat at $P_{in} \approx 1$ MPa and $T_{in} \approx 273°C$)	Up to 36-38%
7	Pressurized water reactor (PWR) NPP (Generation-III, current fleet) (reactor coolant: P = 15.5 MPa (T_{sat} = 345°C) and T = 292–329°C; Rankine cycle steam-turbine parameters (see **Figure 4**): P = 6.9 MPa and $T_{in} = T_{sat}$ = 285°C and steam reheat at $P_{in} \approx 1$ MPa and $T_{in} \approx 265°C$)	Up to 34-36%
8	Boiling-water-reactor (BWR) or advanced BWR NPP (Generation-III and III+, current fleet) (P_{in} = 7.2 MPa and $T_{in} = T_{sat}$=288°C (direct cycle) and steam reheat at $P_{in} \approx 1$ MPa and $T_{in} \approx 268°C$ (see **Figure 4**))	Up to 34%
9	Pressurized heavy water reactor (PHWR) NPP (Generation-III, current fleet) (reactor coolant: P_{out} = 10 MPa (T_{sat} = 311°C) and T = 260–310°C; Rankine cycle steam-turbine parameters: P = 4.6 MPa and $T_{in} = T_{sat}$ = 259°C and steam reheat at l $P_{in} \approx 1$ MPa and $T_{in} \approx 240°C$)	Up to 32%

Table 1.
Typical ranges of thermal efficiencies (gross) of modern thermal and nuclear power plants (NPPs) [4, 5] (for details including schematics and T-s diagrams, see Handbook [6] and Dragunov et al. [7]).

Russia—59%; UK—44%; Italy—42%; and in USA—34%) [4, 5]. More details on ThPPs can be found in Pioro and Kirillov [8] and many other sources.

Also, at SCPs there is no liquid-vapor-phase transition; therefore, there is no such phenomenon as critical heat flux (CHF) or dryout. It is only within a certain range of parameters a deteriorated heat transfer (DHT) regime may occur. Work in this area was mainly performed in Germany, USA, former USSR, and some other countries in the 1950–1980s [9].

1.2 Future applications of SCFs in next-generation nuclear-power reactors and NPPs

At the end of the 1950s and the beginning of the 1960s, early studies were conducted to investigate a possibility of using SCW in nuclear reactors. Several concepts of nuclear reactors using SCW were developed in Great Britain, France,

USA, and former USSR. However, this idea was abandoned for almost 30 years with the emergence of light water reactors (LWRs), but regained interest in the 1990s following LWRs maturation ([6, 9–13]).

This interest was triggered by economical considerations, because nuclear power plants (NPPs) with LWRs (and, especially, with PHWRs) have relatively low thermal efficiencies within the range of 30–36% for Generation-III reactors and up to 37% (38%) for advanced reactors of Generation-III+ (see **Table 1**) compared to those of modern ThPPs (up to 62% for combined-cycle plants and up to 55% for SCP Rankine cycle plants (see **Table 1**)) [6]. Therefore, NPPs with various designs of water-cooled reactors at subcritical pressures cannot compete with modern advanced ThPPs. Also, it should be noted that currently, water-cooled reactors are the vast majority of nuclear-power reactors in the world [14, 15]: (1) PWRs—

No.	Nuclear power plant	Gross eff., %
1	Very high-temperature reactor (VHTR) NPP (reactor coolant—helium (SCF): P = 7 MPa and T_{in}/T_{out} = 640/1000°C; primary power cycle—direct SCP Brayton helium-gas-turbine cycle; possible back-up—indirect Brayton or combined cycles (see **Figures** 5 and **6**))	≥55
2	Gas-cooled fast reactor (GFR) or high-temperature reactor (HTR) NPP (reactor coolant—helium (SCF): P = 9 MPa and T_{in}/T_{out} = 490/850°C; primary power cycle—direct SCP Brayton helium-gas-turbine cycle (see **Figure** 7); possible back-up—indirect SCP Brayton or combined cycles (see **Figures 8** and **9**))	≥50
3	Supercritical water-cooled reactor (SCWR) NPP (one of Canadian concepts; reactor coolant—SC light water: P = 25 MPa and T_{in}/T_{out} = 350/625°C (T_{cr} = 374°C); direct cycle; SCP Rankine cycle with high-temperature secondary-steam superheat: T_{out} = 625°C; possible back-up–indirect SCP Rankine "steam"-turbine cycle with high-temperature secondary-steam superheat) (for details of SCP Rankine cycle, see **Table 1** Item No. 2 and **Figure 1**)	45–50
4	Molten salt reactor (MSR) NPP (reactor coolant—sodium-fluoride salt with dissolved uranium fuel: T_{in}/T_{out} = 700/800°C; primary power cycle—indirect SCP carbon dioxide Brayton gas-turbine cycle; possible back-up—indirect Rankine steam-turbine cycle)	~50
5	Lead-cooled fast reactor (LFR) NPP (Russian design BREST-OD-300*: reactor coolant—liquid lead: $P \approx$ 0.1 MPa and T_{in}/T_{out} = 420/540°C; primary power cycle—indirect subcritical-pressure Rankine steam cycle: $P_{in} \approx$ 17 MPa (P_{cr} = 22.064 MPa) and T_{in}/T_{out} = 340/505°C (T_{cr} = 374°C); high-temperature secondary-steam superheat (in one of the previous designs of BREST-300 NPP primary power cycle was indirect SCP Rankine "steam" cycle: $P_{in} \approx$ 24.5 MPa (P_{cr} = 22.064 MPa) and T_{in}/T_{out} = 340/520°C (T_{cr} = 374°C); also, note that power-conversion cycle in a different LFR designs from other countries is based on SCP carbon dioxide Brayton gas-turbine cycle	~41–43
6	Sodium-cooled fast reactor (SFR) NPP (Russian design BN-600: reactor coolant—liquid sodium (primary circuit): $P \approx$ 0.1 MPa and T_{in}/T_{out} = 380/550°C; liquid sodium (secondary circuit): T_{in}/T_{out} = 320/520°C; primary power cycle—indirect Rankine steam-turbine cycle: $P_{in} \approx$ 14.2 MPa ($T_{sat} \approx$ 337°C) and $T_{in\ max}$ = 505°C (T_{cr} = 374°C); secondary-steam superheat: $P \approx$ 2.45 MPa and T_{in}/T_{out} = 246/505°C; possible back-up in some other countries—indirect SCP carbon dioxide Brayton gas-turbine cycle)	~40

**BREST-OD-300 is Fast Reactor with "NATural safety"-Test-Demonstration in Russian abbreviations (БРЕСТ-OD-300—Быстрый Реактор с ЕСТественной безопасностью—Опытно –Демонстрационный).*

Table 2.
Estimated ranges of thermal efficiencies (gross) of Generation-IV NPP concepts (Generation-IV concepts are listed according to thermal-efficiency decrease) [6, 16].

299 units or 68% from the total number of 441 units; (2) BWRs—65 units or 15%; (3) PHWRs—48 units or 11%; (4) light water, graphite-moderated reactors (LGRs)—13 units of 3%.

Therefore, six concepts of nuclear-power reactors/NPPs of next generation, Generation-IV, were proposed (see **Table 2**), which will have thermal efficiencies comparable with those of modern thermal power plants. Supercritical water-cooled reactor (SCWR) is one of these six concepts under development in a number of countries [6, 17]. Analysis of Generation-IV concepts listed in **Table 2** shows that SCFs, such as helium and water, will be used as reactor coolants, and SCFs such as helium, nitrogen (or mixture of nitrogen (80%) and helium (20%)), carbon dioxide, and water will be used as working fluids (WFs) in power Brayton and Rankine cycles (critical parameters of selected SCFs are listed in **Table 3**). However, it should be mentioned that helium as the reactor coolant and as the working fluid in Brayton power cycle will be at supercritical conditions, which are far above by pressure and temperature critical parameters, that is, helium will behave as compressed gas.

Nowadays, the most widely used SCFs are water, carbon dioxide, and refrigerants [9]. Quite often, carbon dioxide and refrigerants are considered as modeling fluids and used instead of SCW due to significantly lower critical pressures and temperatures, which decreases the complexity and costs of thermalhydraulic experiments. However, they can be/will be used as working fluids in new SCP power cycles: Brayton and Rankine ones [6] (for details, see **Table 3**).

Also, other applications of SCFs will be discussed in the following chapters and are listed in Pioro and Duffey [9].

No.	Fluid	Molar mass	T_{cr}	P_{cr}	ρ_{cr}	Application in power engineering at SCPs
		kg/kmol	°C	MPa	kg/m^3	
1	Carbon dioxide,[1] CO_2	44.01	30.978	7.3773	467.6	WF in Brayton and Rankine power cycles (see **Figures 5** and **6**)
2	Ethanol, C_2H_6O	46.068	241.56	6.268	273.19	N/A
3	Helium,[2] He	4.0026				Reactor coolant in VHTR & GFR (see **Figure 7**); WF in Brayton power cycle (see **Figure 7**)
4	Methanol, CH_3OH	32.042	239.45	8.1035	275.56	N/A
5	Nitrogen, N_2	28.013	-146.96	3.3958	313.3	WF in Brayton cycle (also, mixture of N_2 (80%) & He (20%) is proposed (see **Figures 8** and **9**))
6	R-12, CCl_2F_2	120.91	111.97	4.1361	565.0	Modeling fluid in thermalhydraulic tests
7	R-134a, CF_3CH_2F	102.03	101.06	4.0593	511.9	Modeling fluid in thermalhydraulic tests
8	Water[3], H_2O	18.015	373.95	22.064	322.0	WF in Rankine cycle of coal-fired ThPP; reactor coolant in SCWR; WF in Rankine power cycle (see **Figure 1**)

[1,2,3]*Thermodynamics diagrams: P-T and T-s can be found in Handbook [6].*
[3]*Thermodynamics diagrams: P-T and T-s are shown in* **Figure 10**.

Table 3.
Critical parameters of selected fluids and gases (based on NIST [25]).

2. Specifics of thermophysical properties of SCFs

Prior to a general discussion on specifics of forced-convective heat transfer at critical and supercritical pressures, it is important to define special terms and expressions used at these conditions [6, 9]. For a better understanding of these

No.	Literature source	Fluid	*P*, MPa	*T*, °C	Properties
1	Pioro et al. [19]	Properties of selected metals, alloys, and diamond Properties of selected insulating materials Radiative properties of selected materials Properties of selected nuclear fuels Properties of selected gases at atmospheric pressure Properties of selected cryogenic gases Properties of selected fluids on saturation line Properties of selected supercritical fluids Properties of selected liquid alkali metals Thermophysical properties of nuclear-reactor coolants			
2	Handbook [6]	H_2O, CO_2, He	-	-	*T-s* diagrams
		H_2O (BWR, PHWR, PWR)	7, 11, 15	50-375	$\rho, k, \mu, \nu, c_p, H, \mathbf{Pr}, \beta$
		H_2O (SCW)	P_{cr}, 25, 30, 35, 40	350-600	$\rho, k, \mu, \nu, c_p, H, \mathbf{Pr}, \beta$
		CO_2 (SC CO_2)	P_{cr}, 8.4, 10.0, 11.7	0-165	$\rho, k, \mu, \nu, c_p, H, \mathbf{Pr}, \beta$
		He	P_{cr} and other pressures	Range of *T*	k, c_p, β
		Air, Ar, CO_2, He, H_2, Kr (gases)	0.1	0-1000	$\rho, k, \mu, c_p, \mathbf{Pr}, \beta$
		CO_2 (AGR)	4	250-1000	$\rho, k, \mu, c_p, H, \mathbf{Pr}, \beta$
		FLiNaK (MSR)	0.1		
		H_2O/SCW (PWR/SCWR)	15.5/25		
		He (VHTR, GFR)	7, 9		
		Na, Pb, Pb-Bi (SFR, LFR)	0.1		
3	Mann and Pioro [20]	SC R-134a	P_{cr}, 5, 10, 13, 15	-100-175	k, c_p, β
4	Gupta et al. [21]	SCW SC CO_2 SC R-134a (three fluids on same graph)	25.0 8.4* 4.6*	$(0.5-1.6)\frac{T}{T_{cr}}$	$\rho, k, \mu, c_p, H, \mathbf{Pr}$
5	Pioro and Mokry [22]	H_2O	-	-	*T-s* diagram
		H_2O (SCW)	P_{cr}, 25, 30, 35	350-600	$\rho, k, \mu, \nu, c_p, H, \mathbf{Pr}, \beta$
		R-12 (SC R-12)	P_{cr}, 4.65	0-350	$\rho, k, \mu, \nu, c_p, H, \mathbf{Pr}, \beta$
6	Pioro and Duffey [9]	R-134a (SC R-134a)	P_{cr}, 4.6	70-150	$\rho, k, \mu, \nu, c_p, H, \mathbf{Pr}, \beta$

**Pressures for SC carbon dioxide, R-134a, and R-12 are equivalent for SCW pressure of 25 MPa, based on, so-called, reduced-pressure scaling:* $\left(\frac{P}{P_{cr}}\right)_{Fluid} = \left(\frac{P}{P_{cr}}\right)_{SCW}$.

Table 4.
Selected list of literature sources on thermophysical properties of fluids, gases, and other materials.

terms and expressions their definitions are listed in Glossary (see below) (also, see **Figures 10–35**). Specifics of thermophysical properties at SCPs are described in Pioro et al. [23]; Handbook [6]; Mann and Pioro [24]; Gupta et al. [25]; Pioro and Mokry [26]; and Pioro and Duffey [9] (for more details, see **Table 4**).

Glossary

Compressed fluid is the fluid at a pressure above the critical pressure, but at a temperature below the critical temperature (see **Figure 10**).

Critical point (also called a critical state) is the point in which the distinction between the liquid and gas (or vapor) phases disappears (see **Figure 10**), that is, both phases have the same temperature, pressure, and specific volume or density. The critical point is characterized with the phase-state parameters: T_{cr}, P_{cr} and v_{cr} (or ρ_{cr}), which have unique values for each pure substance.

Deteriorated heat transfer (DHT) is characterized with lower values of the HTC compared to those for normal heat transfer (NHT); and hence, has higher values of wall temperature within some part of a heated channel (see **Figures 12, 13a, 24b, 25b, 27, 31**, and **35**) or within the entire heated length (see **Figure 14b**).

Improved heat transfer (IHT) is characterized with higher values of the HTC compared to those for NHT; and hence, lower values of wall temperature within some part of a heated channel (see **Figures 12, 21, 25, 27b, 33**, and **34**) or within the entire heated length. In our opinion, the IHT regime or mode includes peaks or "humps" in the HTC profile near the critical or pseudocritical points.

Normal heat transfer (NHT) can be characterized in general with HTCs similar to those of subcritical convective heat transfer far from the critical or pseudocritical regions, when they are calculated according to the conventional single-phase Dittus-Boelter-type correlations: $\mathbf{Nu} = 0.0243\ \mathbf{Re}^{0.8}\mathbf{Pr}^{0.4}$ (see **Figures 12, 13a, 14a, 21, 24, 25, 27**, and **30–34**).

Overheated vapor is the vapor at pressures below the critical pressure, and at temperatures above the saturation temperature, but below the critical temperature (see **Figure 10**).

Pseudocritical line is the line, which consists of pseudocritical points (see **Figure 10**).

Pseudo-boiling is a physical phenomenon similar to subcritical-pressure nucleate boiling, which may appear at SCPs. Due to heating of an SCF with a bulk-fluid temperature below the pseudocritical temperature (high-density fluid, i.e., "liquid-like") (see **Figures 10, 11, 13b** and **15**), some layers near the heated surface may attain temperatures above the pseudocritical temperature (low-density fluid, i.e., "gas-like"). This low-density "gas-like" fluid leaves the heated surface in a form of variable density volumes (bubbles). During the pseudo-boiling, the HTC usually increases (IHT regime).

Pseudocritical point (characterized with P and T_{pc}) is the point at a pressure above the critical pressure and at a temperature ($T_{pc} > T_{cr}$) corresponding to the maximum value of specific heat at this particular pressure (see **Figures 10, 11**, and **13b**).

Pseudo-film boiling is a physical phenomenon similar to subcritical-pressure film boiling, which may appear at SCPs. At pseudo-film boiling, a low-density fluid (a fluid at temperatures above the pseudocritical temperature, i.e., "gas-like") prevents a high-density fluid (a fluid at temperatures below the pseudocritical temperature, i.e., "liquid-like") from contacting ("rewetting") a heated surface. Pseudo-film boiling leads to the DHT regime.

Supercritical fluid is the fluid at pressures and temperatures that are higher than the critical pressure and critical temperature (see **Figure 10**). However, in the present paper, the term *supercritical fluid* usually includes both terms—*supercritical fluid* and *compressed fluid*.

Supercritical "steam" is actually supercritical water, because at supercritical pressures fluid is considered as a single-phase substance (see **Figure 10**). However, this term is widely (and incorrectly) used in the literature in relation to supercritical-"steam" generators and turbines.

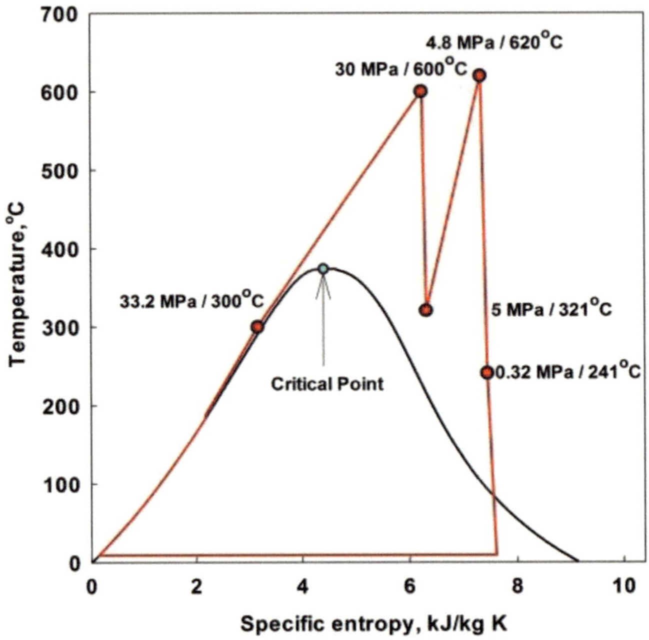

Figure 1.
T-s diagram of generic SCP Rankine "steam"-turbine power cycle (modern advanced coal-fired thermal power plants and future SCWR NPPs) [6, 7].

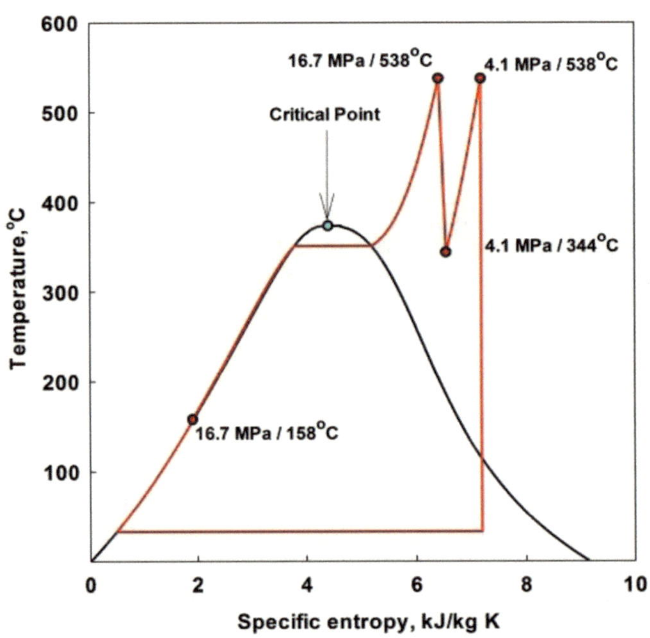

Figure 2.
T-s diagram of generic subcritical-pressure Rankine steam-turbine power cycle (older coal-fired thermal power plants and AGR Torness NPP) [6, 7].

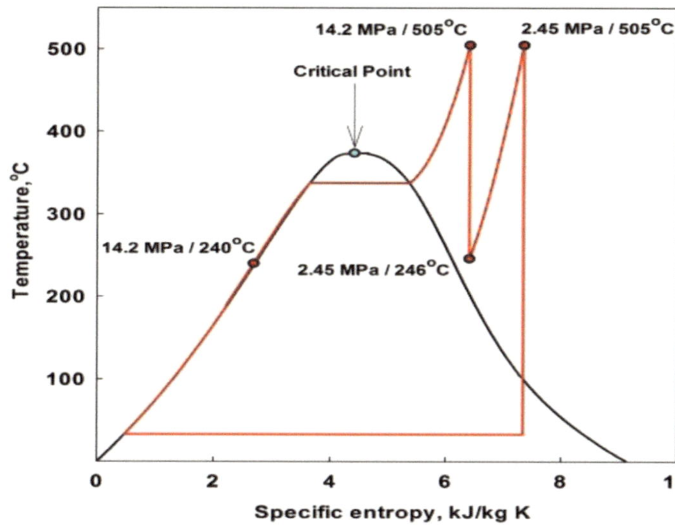

Figure 3.
T-s diagram of generic subcritical-pressure Rankine steam-turbine power cycle (old coal-fired thermal power plants and SFR NPPs) [6, 7].

Superheated steam is the steam at pressures below the critical pressure, but at temperatures above the critical temperature (see **Figure 10**).

Also, profiles of the basic thermophysical properties (density, thermal conductivity, dynamic viscosity, specific heat and specific enthalpy) and Prandtl number for four SCFs: water, ethanol, methanol, and carbon dioxide; at critical and one supercritical pressure, which is 25 MPa for water and the corresponding to that equivalent pressures for all other SCFs vs. reduced temperature (temperature) are shown in **Figures 15–20**.

3. Specifics of forced-convection heat transfer at supercritical pressures

3.1 Vertical bare tubes

Water is the most widely used coolant or working fluid at SCPs. The largest application of SCW is in SC "steam" generators and turbines, which are widely used in the thermal power industry worldwide. Currently, upper limits of pressures and temperatures used in the thermal-power industry are about 30–38 MPa and 600–625°C, respectively (see **Table 1**). A new direction in SCW application in the power industry has been the development of SCWR concepts (see **Table 2**), as part of the Generation-IV International Forum (GIF) [27] initiative (for details, see [6, 9–13, 28–30]; and Proceedings of the International Symposiums on SCWRs (ISSCWR) (selected augmented and revised papers from ISSCWRs have been published in the ASME Journal of Nuclear Engineering and Radiation Science in 2020, Vol. 6 No. 3; in 2018, Vol. 4, No. 1, and 2016, Vol. 2, No. 1).

Experiments at SCPs are very expensive and require sophisticated equipment and measuring techniques. Therefore, some of these studies (e.g., heat transfer in fuel-bundle simulators) are proprietary and, hence, usually are not published in open literature.

The majority of studies deal with heat transfer and hydraulic resistance of working fluids, mainly water, carbon dioxide, refrigerants, and helium, in circular bare tubes [9, 22, 31–34]. A limited number of studies were devoted to heat transfer and pressure drop in annuli and bundles [9, 10, 35–45].

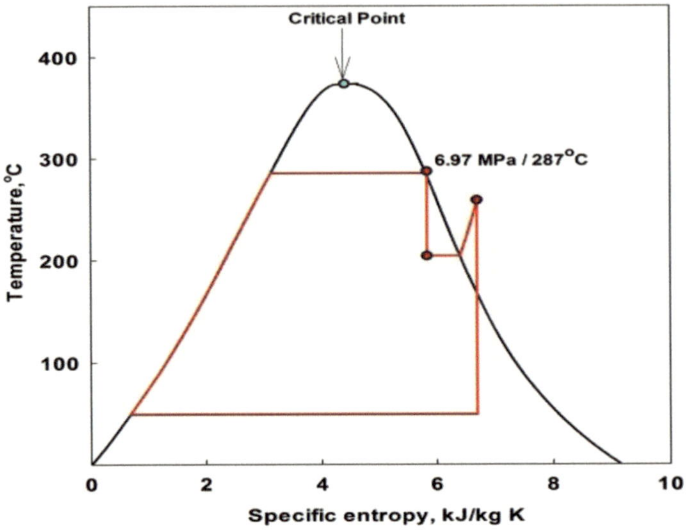

Figure 4.
T-s diagram of generic subcritical-pressure Rankine saturated-steam-turbine power cycle (PWR and BWR NPPs) [6, 7].

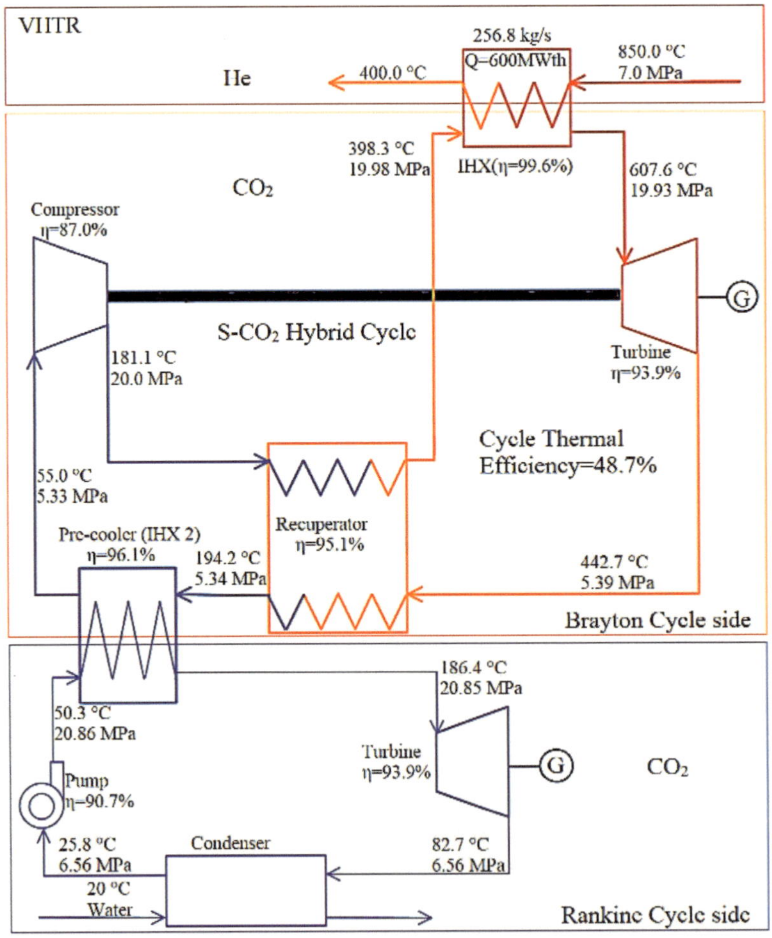

Figure 5.
Layout of 600-MW_{th} VHTR NPP with SC-CO_2 power cycle (based on figure from Bae et al. [17]) [18].

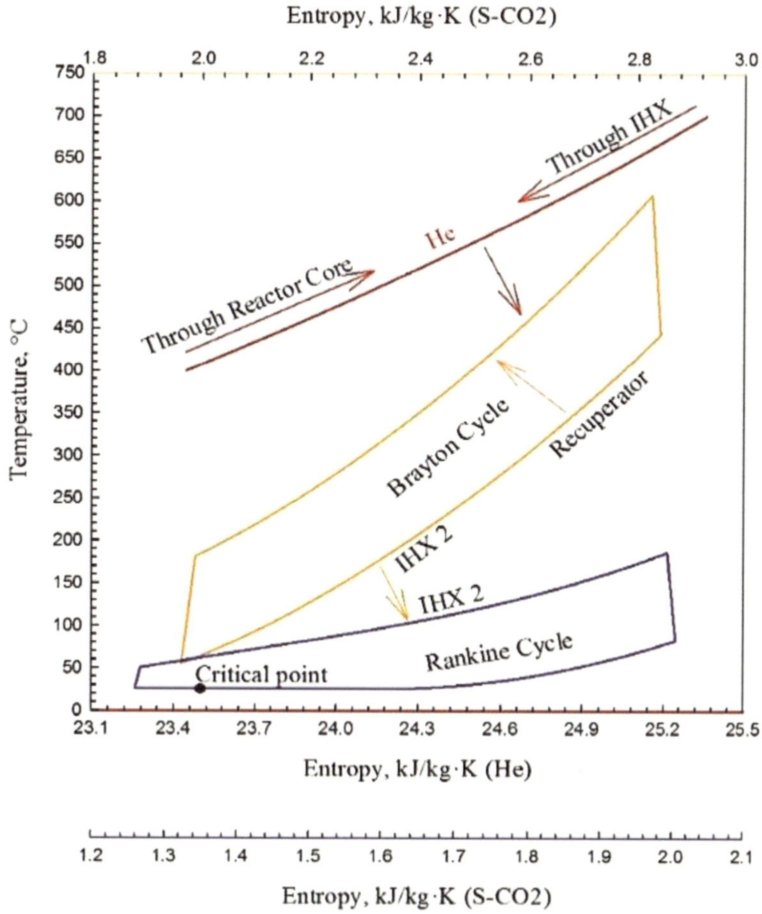

Figure 6.
T-s diagram for 600-MW_{th} VHTR NPP with SC-CO_2 (S-CO_2) power cycle (based on **Figure** *5) [18].*

New experiments in the 1990s–2000s were triggered by several reasons: (1) thermophysical properties of SCW have been updated from the 1950s–1970s, for example, a peak in thermal conductivity in the critical/pseudocritical points was "officially" introduced in the 1990s; (2) experimental techniques have been improved; (3) in SCWRs various bundle flow geometries will be used instead of bare-tube geometry; and (4) in SC "steam" generators of thermal power plants larger diameter tubes/pipes (20–40 mm) are used, however, in SCWRs hydraulic-equivalent diameters of proposed bundles will be within 5–12 mm.

Accounting that SCW, SC carbon dioxide and SC R-12 are the most widely used fluids, specifics of heat transfer, including generalized correlations, will be discussed in this paper. Specifics of heat transfer and pressure drop at other conditions and/or for other fluids are discussed in the book by Pioro and Duffey [9].

All primary sources (i.e., all sources found by the authors from a total of 650 references dated mainly from 1950 till beginning of 2006) of heat transfer experimental data for water and carbon dioxide flowing inside circular tubes at supercritical pressures are listed in the book by Pioro and Duffey [9].

In general, three major heat transfer regimes (for their definitions, see Section 2, Glossary) can be noticed at critical and supercritical pressures (for details, see **Figures 12, 13a, 14, 21, 24, 25, 27, 30–35**):

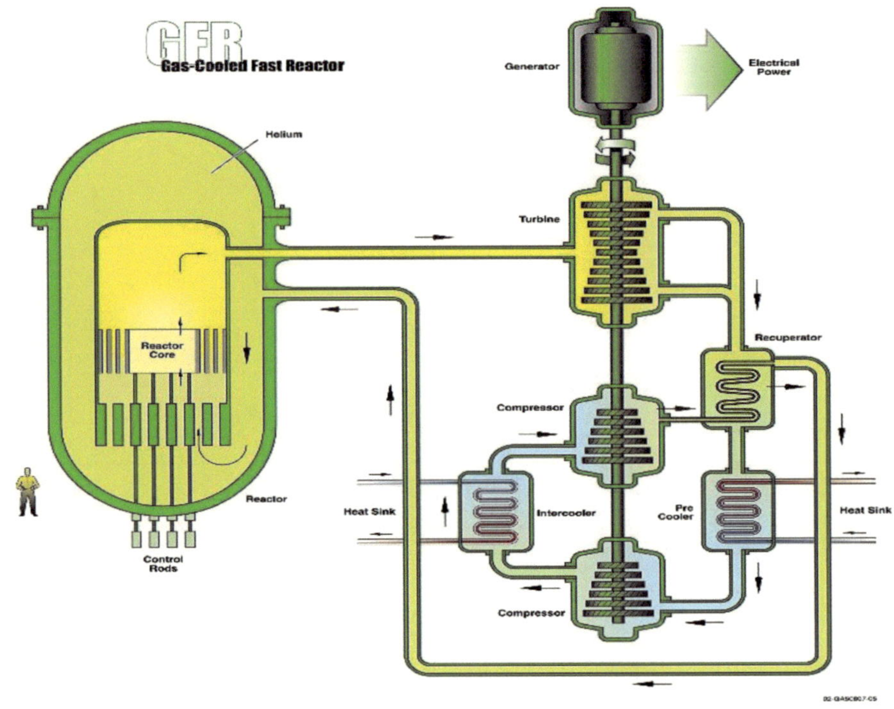

Figure 7.
Schematic of 600-MW$_{th}$ GFR concept considered initially by GIF with direct Brayton helium cycle (Courtesy of GIF) (see also [6]).

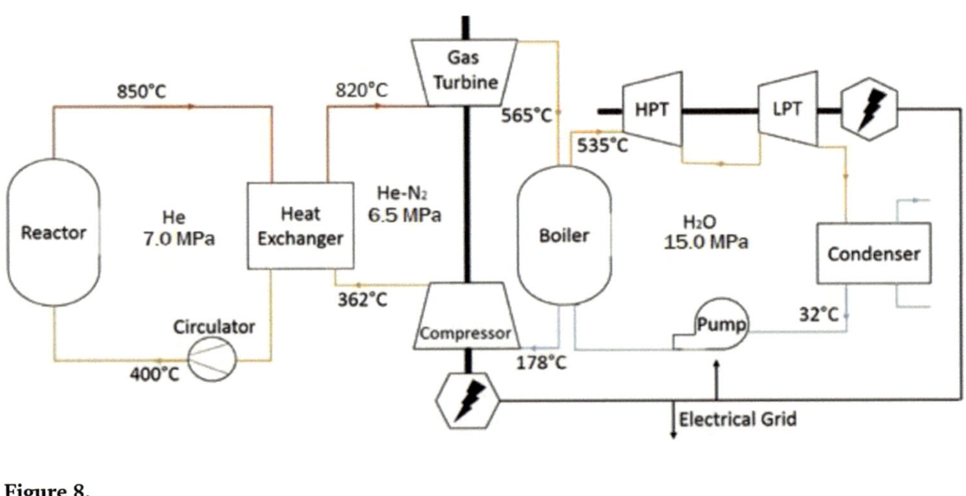

Figure 8.
Layout of 2400-MW$_{th}$ GFR NPP with He-N$_2$ indirect combined power cycle (based on figure from Anzieu [23]) [18].

1. Normal heat transfer;

2. Improved heat transfer; and

3. Deteriorated heat transfer.

Also, two special phenomena (for their definitions, see Section 2, Glossary) may appear along a heated surface: (1) pseudo-boiling; and (2) pseudo-film boiling.

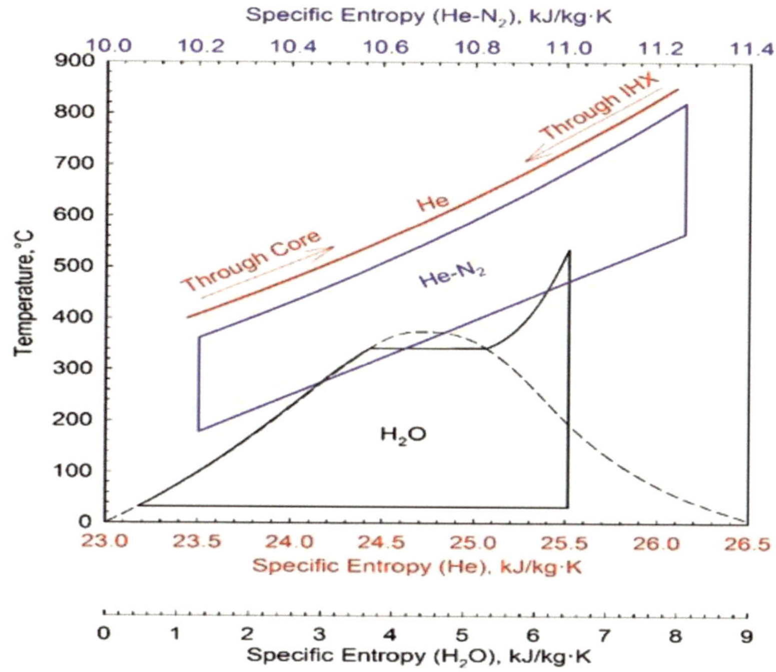

Figure 9.
T-s diagrams of 2400-MW$_{th}$ GFR NPP combined power cycle (based on **Figure 8**) *[18].*

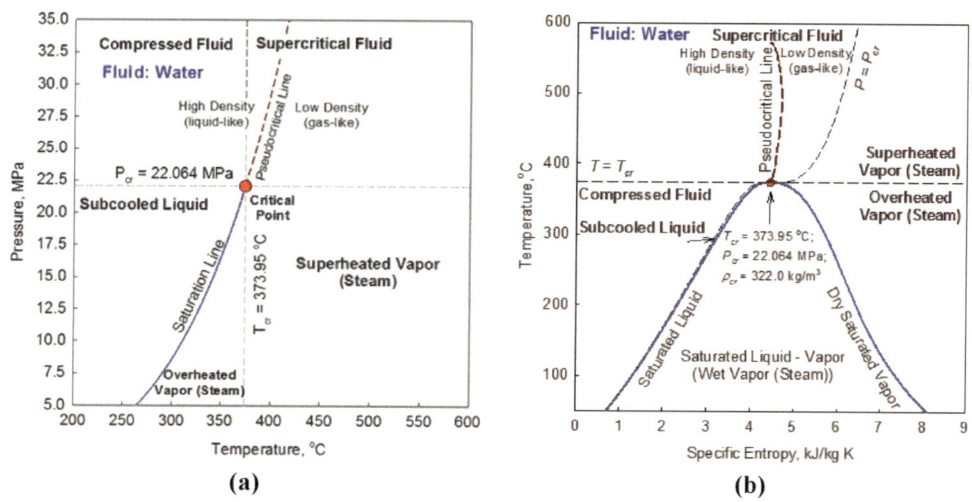

Figure 10.
Thermodynamics diagrams for water: (a) pressure-temperature and (b) temperature-specific entropy (based on NIST [25]).

These heat transfer regimes and special phenomena appear to be due to significant variations of thermophysical properties near the critical and pseudocritical points and due to operating conditions.

Therefore, the following conditions can be distinguished at critical and SCPs:

a. Wall and bulk-fluid temperatures are below a pseudocritical temperature within a part of (see **Figure 12**) or the entire heated channel (see **Figures 14a, 24a,** and **30**);

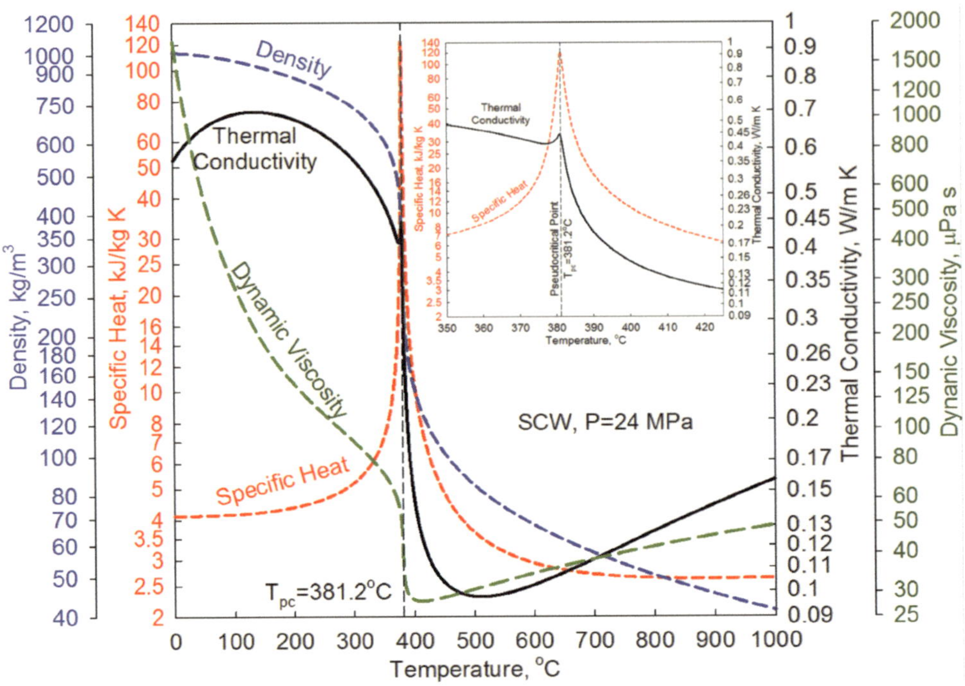

Figure 11.
Profiles of selected thermophysical properties (density, specific heat, thermal conductivity, and dynamic viscosity) vs. temperature for SCW at pressure of 24.0 MPa (based on NIST [25]).

b. Wall temperature is above, and bulk-fluid temperature is below a pseudocritical temperature within a part of (see **Figures 13a, 31, 34,** and **35**) or the entire heated channel (see **Figure 14b**);

c. Wall temperature and bulk-fluid temperature is above a pseudocritical temperature within a part of or the entire heated channel (see **Figures 12, 13a, 21, 31–35**);

d. High heat fluxes (see **Figures 13a**, **24** and **25**);

e. Entrance region (see **Figures 12, 13a, 32,** and **34**);

f. Upward and downward flows;

g. Horizontal flows; and

h. Effect of gravitational forces at lower mass fluxes; etc.

All these conditions can affect SC heat transfer.

Figure 13b shows bulk-fluid-temperature and thermophysical-properties (thermal conductivity, dynamic viscosity, specific heat, and Prandtl number) profiles along the heated length of a vertical bare circular tube (operating conditions in this figure correspond to those in **Figure 13a**).

Some researchers have suggested that variations in thermophysical properties near critical and pseudocritical points result in the maximum value of HTC. Thus, Yamagata et al. [46] found that for SCW flowing in vertical and horizontal tubes, the HTC increases significantly within the pseudocritical region (**Figure 21**). The magnitude of the peak in HTC decreases with increasing heat flux and pressure. The maximum HTC values correspond to a bulk-fluid enthalpy, which is slightly less than the pseudocritical bulk-fluid enthalpy.

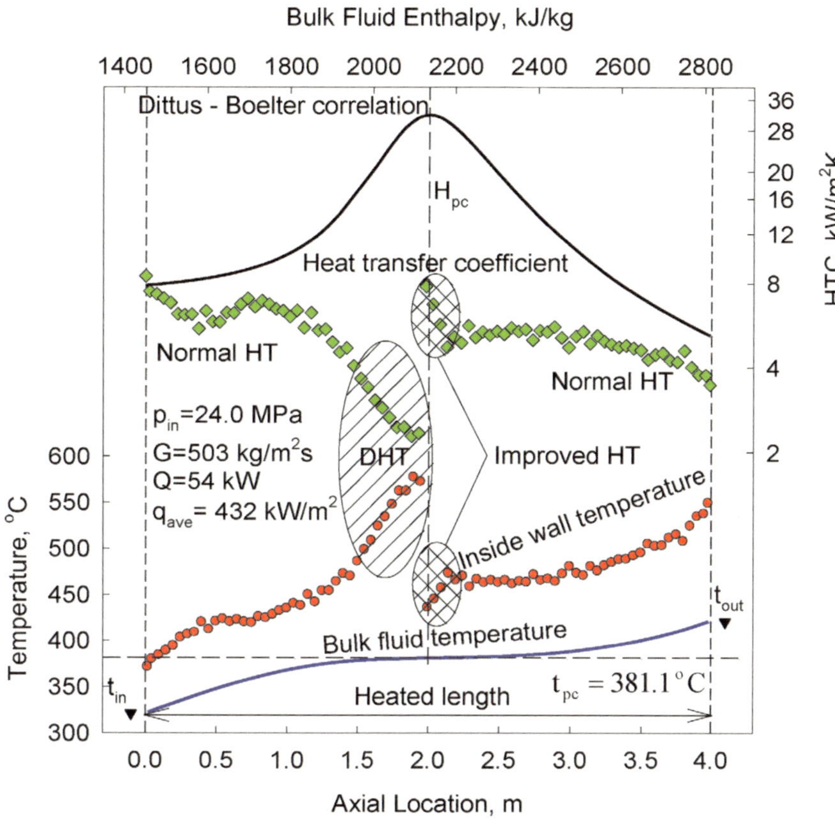

Figure 12.
Temperature and HTC profiles along heated length of vertical bare tube with upward flow of SCW (data by Kirillov et al. [26]): D = 10 mm; L_h = 4 m; q_{dht} = 316 kW/m² at G = 503 kg/m²s; points—experimental data; curves—calculated data; curve for HTC is calculated through Dittus-Boelter correlation (Eq. (1)). Profiles of density, specific heat, thermal conductivity, and dynamic viscosity vs. temperature for SCW at pressure of 24.0 MPa are shown in **Figure 11**. *Uncertainties of primary parameters are listed in* **Table 5**.

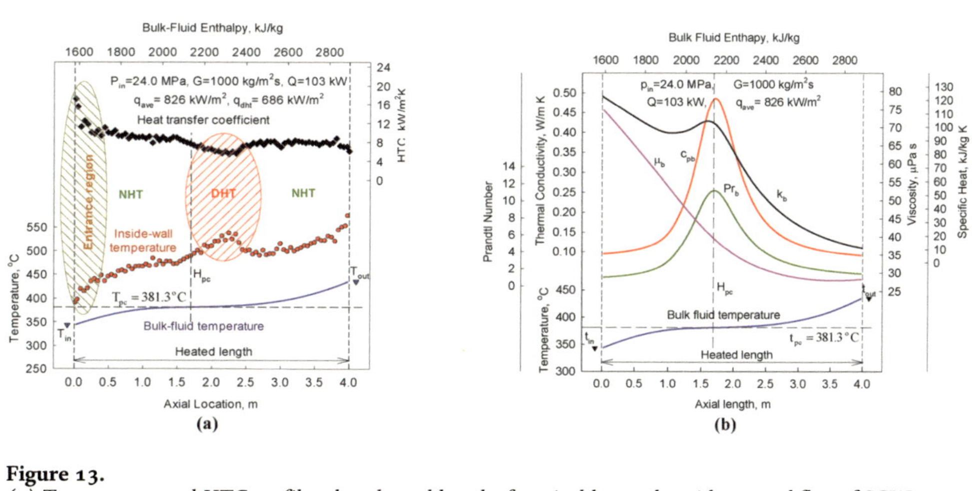

Figure 13.
(a) Temperature and HTC profiles along heated length of vertical bare tube with upward flow of SCW (data by Kirillov et al. [26]): D = 10 mm; L_h = 4 m; points—experimental data; curves—calculated data. Uncertainties of primary parameters are listed in **Table 5**; *and (b) temperature and thermophysical-properties profiles along heated length of vertical tube: operating conditions in this figure correspond to those in (a); and thermophysical properties based on bulk-fluid temperature. Profiles of density, specific heat, thermal conductivity, and dynamic viscosity vs. temperature for SCW at pressure of 24.0 MPa are shown in* **Figure 11**.

3.2 Vertical annular channel, and three- and seven-rod bundles cooled with SCW

In future SCWRs the main flow geometry will be bundles of various designs [6, 10]. Therefore, a limited number of experiments have been performed in simplified bundle simulators cooled with SCW and heated with an electrical current [10, 35–44].

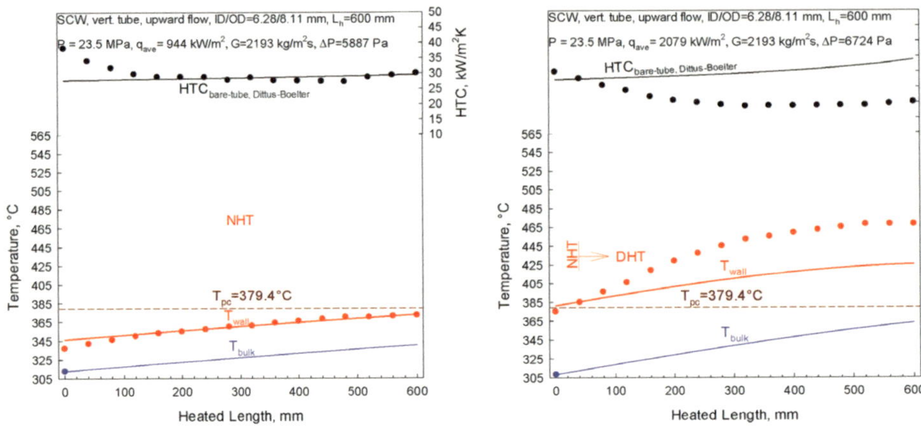

(a) (b)

Figure 14.
Profiles of bulk-fluid and inside-wall temperatures, and HTC along heated length of vertical bare tube with upward flow of SCW at various heat fluxes: (a) $q = 944\ kW/m^2$; $T_{b\ in} = 313°C$ (entrance region can be identified within $L_h = 0–150\ mm$) and (b) $q = 2079\ kW/m^2$; $T_{b\ in} = 308°C$ (data by Razumovskiy et al.). For both graphs, $q_{dht} = 1575\ kW/m^2$ at $G = 2193\ kg/m^2s$ (based on Eq. (5) [51]: $P = 23.5\ MPa$; $G = 2193\ kg/m^2s$; and. Points—experimental data; curves—calculated data; curves for HTC and T_w are calculated through Dittus-Boelter correlation (Eq. (1)). Uncertainties of primary parameters are similar to those listed in **Table 6**.

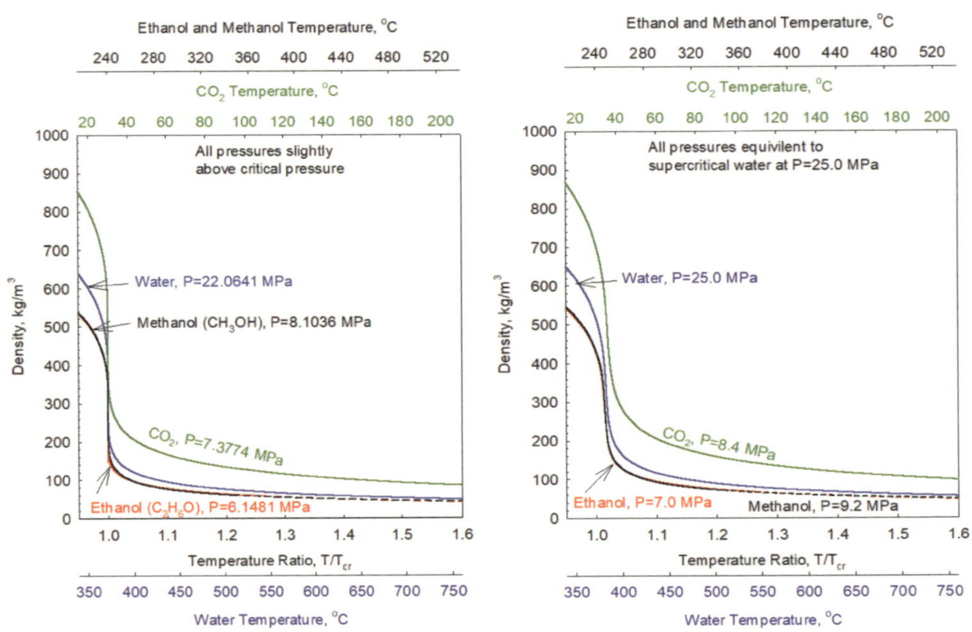

(a) (b)

Figure 15.
Density profiles vs. reduced temperature and temperature for water, carbon dioxide, ethanol, and methanol (based on NIST [25]) (prepared by D. Mann): (a) at critical pressures; and (b) at 25 MPa for water and equivalent pressures for other SCFs (based on reduced-pressure scaling (for details, see **Table 4** *and [21])).*

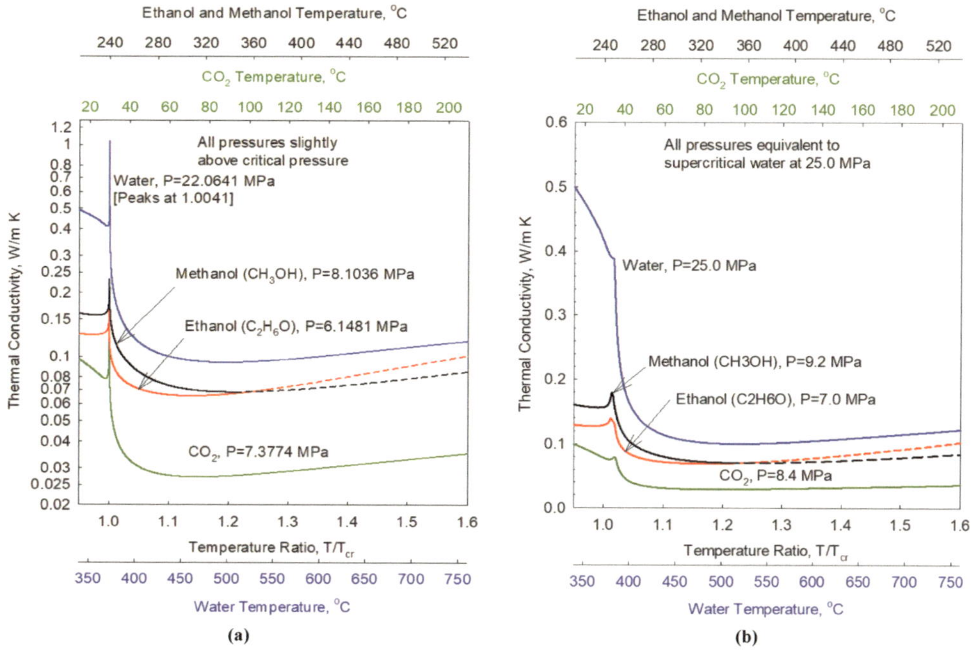

Figure 16.
Thermal-conductivity profiles vs. reduced temperature and temperature for water, carbon dioxide, ethanol, and methanol (based on NIST [25]) (prepared by D. Mann): (a) at critical pressures; and (b) at 25 MPa for water and equivalent pressures for other SCFs (based on reduced-pressure scaling (for details, see **Table 4** *and [21])).*

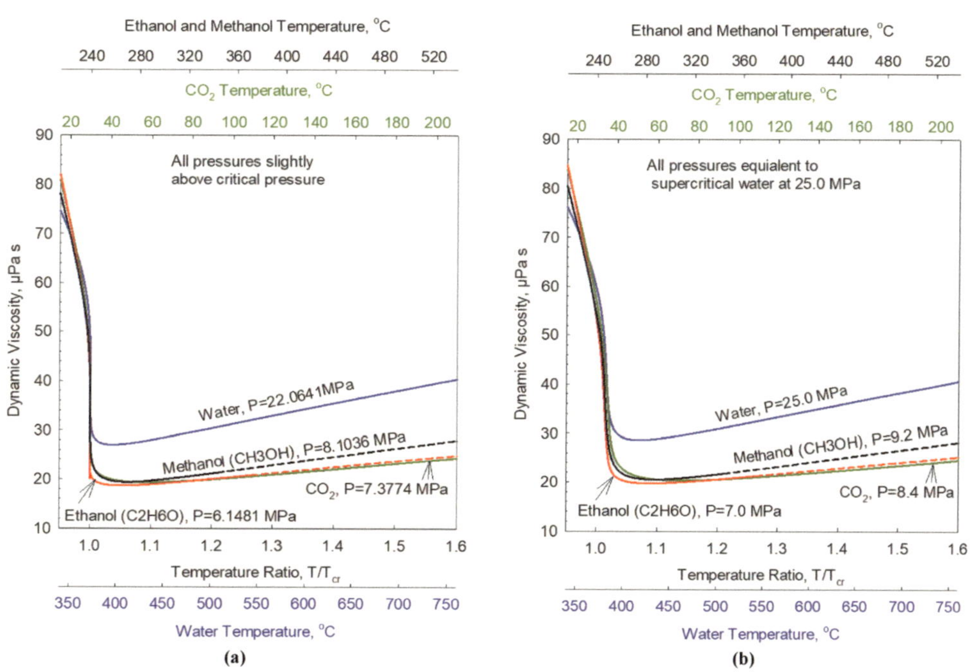

Figure 17.
Dynamic-viscosity profiles vs. reduced temperature and temperature for water, carbon dioxide, ethanol, and methanol (based on NIST [25]) (prepared by D. Mann): (a) at critical pressures; and (b) at 25 MPa for water and equivalent pressures for other SCFs (based on reduced-pressure scaling (for details, see **Table 4** *and [21])).*

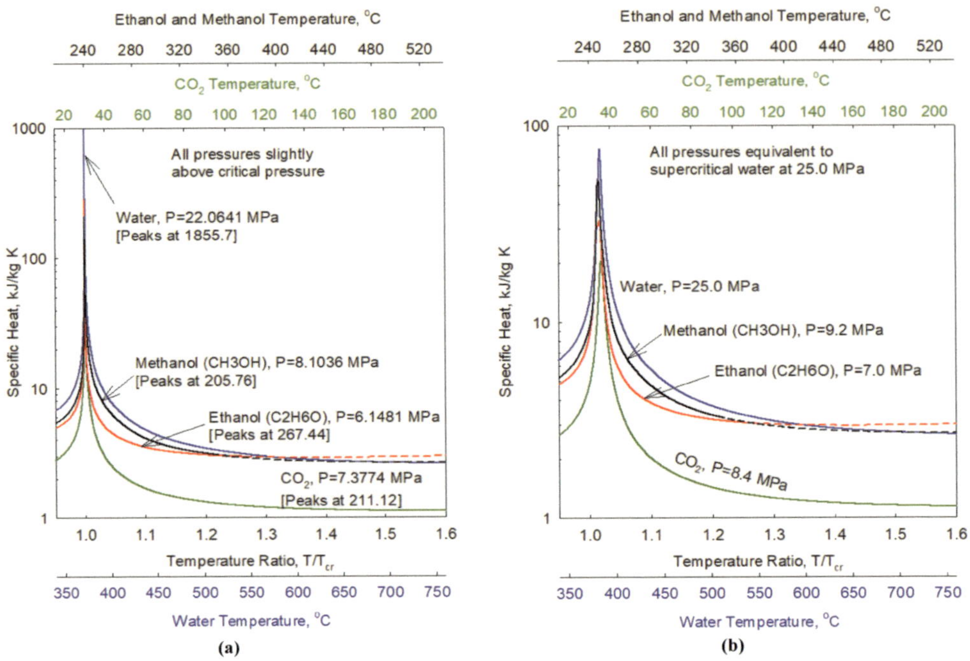

Figure 18.
Specific-heat profiles vs. reduced temperature and temperature for water, carbon dioxide, ethanol, and methanol (based on NIST [25]) (prepared by D. Mann): (a) at critical pressures; and (b) at 25 MPa for water and equivalent pressures for other SCFs (based on reduced-pressure scaling (for details, see **Table 4** *and [21])).*

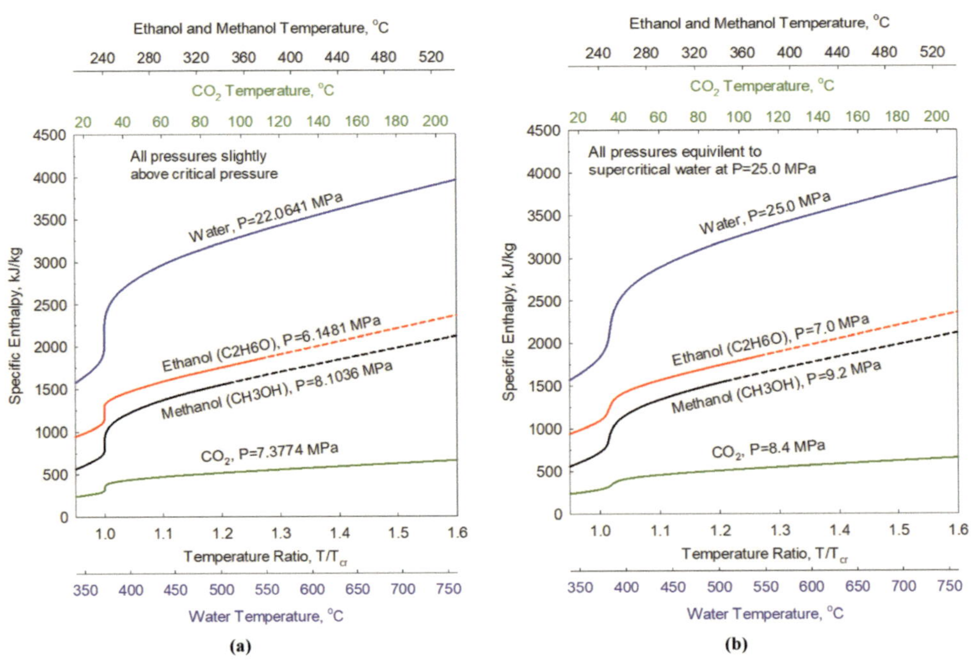

Figure 19.
Specific-enthalpy profiles vs. reduced temperature and temperature for water, carbon dioxide, ethanol, and methanol (based on NIST [25]) (prepared by D. Mann): (a) at critical pressures; and (b) at 25 MPa for water and equivalent pressures for other SCFs (based on reduced-pressure scaling (for details, see **Table 4** *and [21])).*

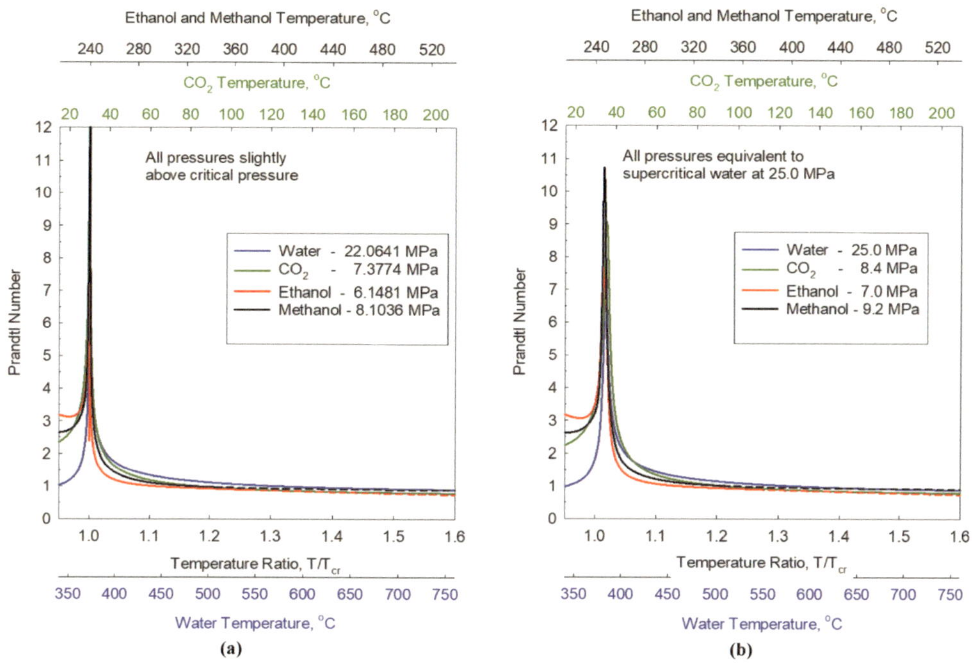

Figure 20.
Prandtl-Number profiles vs. reduced temperature and temperature for water, carbon dioxide, ethanol, and methanol (based on NIST [25]) (prepared by D. Mann): (a) at critical pressures; and (b) at 25 MPa for water and equivalent pressures for other SCFs (based on reduced-pressure scaling (for details, see **Table 4** *and [21])).*

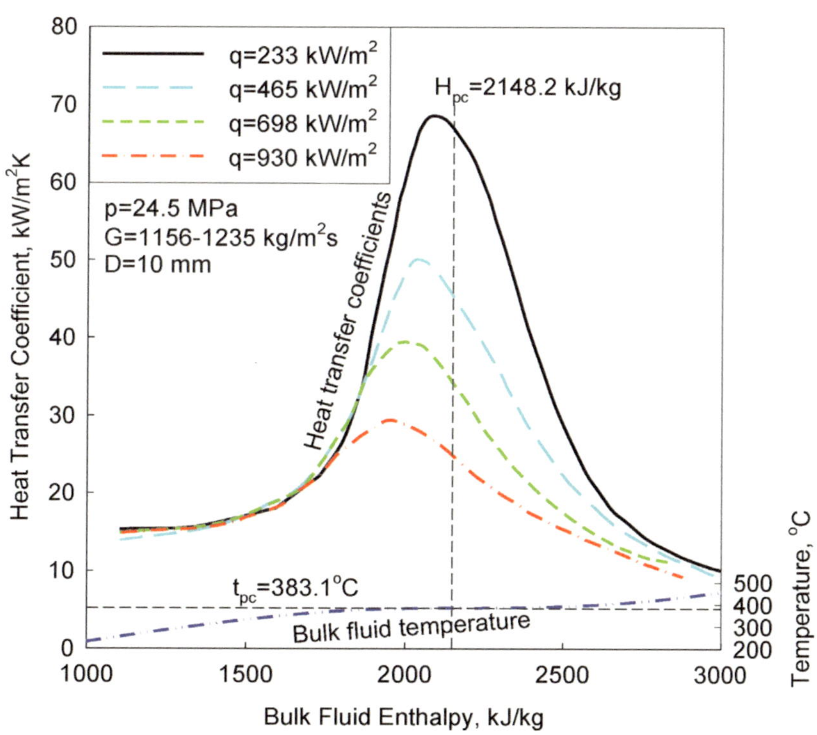

Figure 21.
Heat transfer coefficient vs. bulk-fluid enthalpy in vertical tube with upward flow of SCW at various heat fluxes (data from Yamagata et al. [46]).

An annulus or a one-rod (single-rod) bundle is the simplest bundle geometry (see **Figures 22a** and **23**), and **Figure 24** shows profiles of bulk-fluid and wall temperatures, and HTC along heated length of vertical annular channel (one-rod bundle). **Figures 22b** and **23** show three-rod-bundle flow geometry, and **Figure 25** shows profiles of bulk-fluid and wall temperatures, and HTC along heated length of vertical three-rod bundle. **Figure 26** shows seven-rod-bundle flow geometry, and **Figure 27** shows profiles of bulk-fluid and wall temperatures, and HTC along heated length of the vertical seven-rod bundle.

Analysis of data in **Figures 25b** and **27b** shows that all three HT regimes, which were noticed in bare circular tubes, are also possible in annuli and bundle flow geometries. **Figures 24** and **25** show a comparison between the HTC experimental data obtained in annulus and three-rod bundle with those calculated through the Dittus-Boelter correlation (Eq. (1)). The comparison showed that, in general, there is no significant difference between calculated HTC values and experimental ones. This finding means that in spite of the presence of rod(s) with four helical ribs in SCW flow, which can be considered as an HT enhancement surface(s), there is no significant increase in HTC. However, when q_{dht} values reached in SCW-cooled annulus and 3- and seven-rod bundles were compared to those obtained in bare tubes, it was found that q_{dht} in bare tubes were 1.6–1.8 times lower (see **Table** 7).

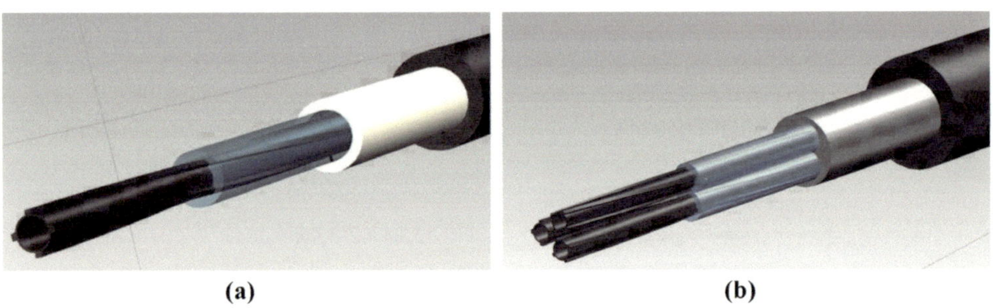

Figure 22.
3-D image of vertical annular channel (a) and three-rod bundle (b) cooled with upward flow of SCW (for other details, see **Figure 23***) [35]: heated rods equipped with four helical ribs.*

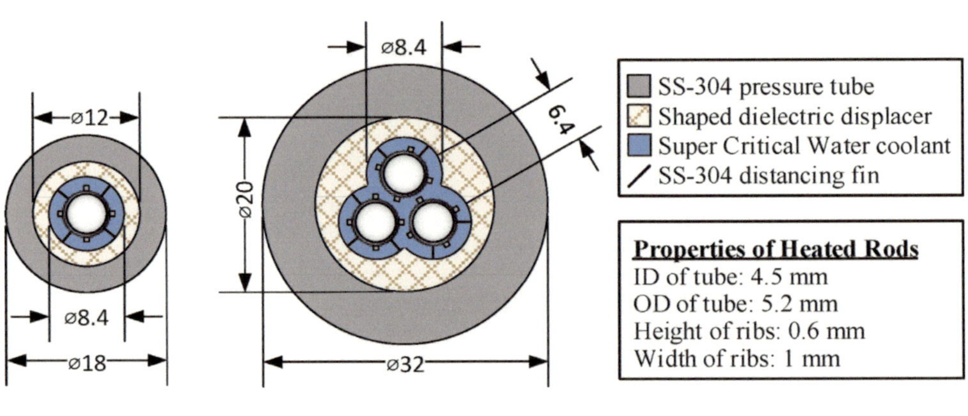

Figure 23.
Radial cross-sections of annular channel (single rod) and three-rod bundle (for other details, see **Figure 22***) [35]: heated rods equipped with four helical ribs; all dimensions in mm; and Ukrainian stainless steel has been used for heated rods, by content and other parameters, this steel is very close to those of SS-304.*

3.3 Vertical seven-rod bundle cooled with SC R-12

Figures 28 and **29** show a seven-rod bundle test section, which can be considered as a bare bundle, and **Figures 30** and **31** show profiles of bulk-fluid and wall temperatures, and HTC vs. heated length of the central rod at three circumferential locations. Analysis of **Figures 30** and **31** shows that we also have here all three HT

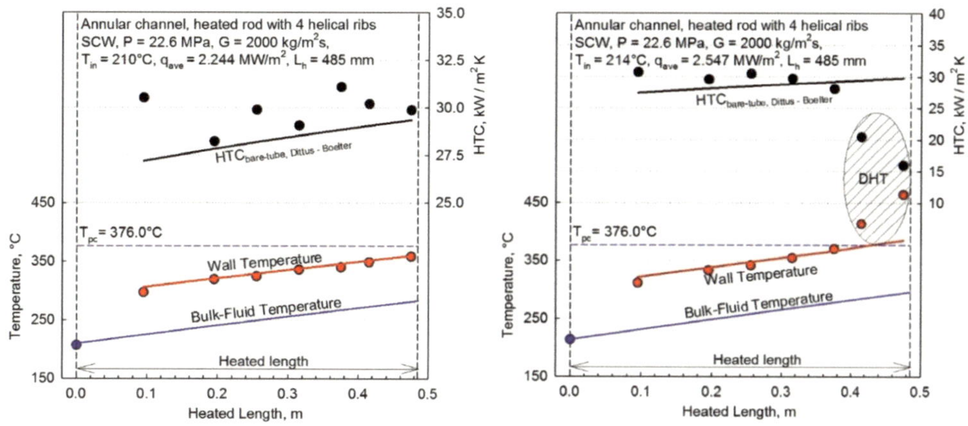

(a) (b)

Figure 24.
Profiles of bulk-fluid and wall temperatures, and HTC along heated length of vertical annular channel (one-rod bundle; rod with four helical ribs) cooled with upward flow of SCW ([36])—P = 22.6 MPa and G = 2000 kg/m²s (bare tube q_{dht} = 1431 kW/m² (based on Eq. (5)): (a) q_{ave} = 2.244 MW/m² and T_{in} = 210°C; and (b) q_{ave} = 2.547 MW/m² and T_{in} = 214°C). For details of test section, see **Figure 23***. Points are experimental data; curves are calculated data; curves for HTC and T_w are calculated through Dittus-Boelter correlation (Eq. (1)). Uncertainties of primary parameters are listed in* **Table 6***.*

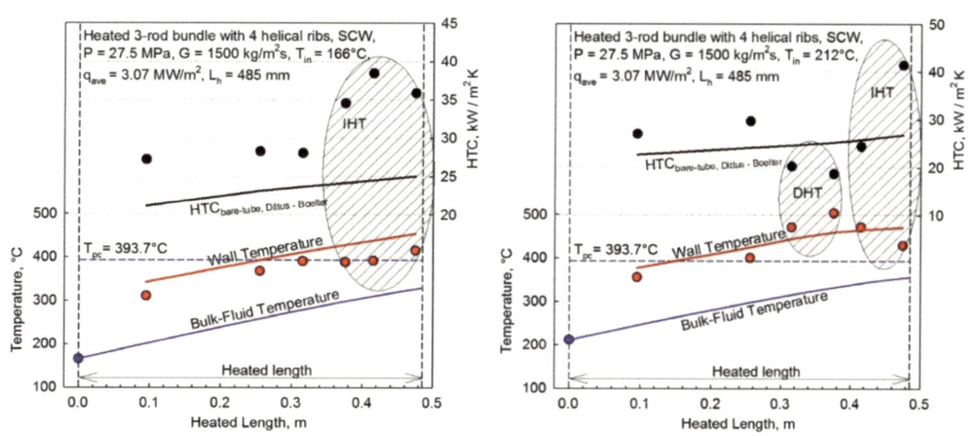

(a) (b)

Figure 25.
Profiles of bulk-fluid and wall temperatures, and HTC along heated length of vertical annular channel (three-rod bundle; each rods with 4 helical ribs) cooled with upward flow of SCW ([36])—P = 27.5 MPa; q_{ave} = 3.07 MW/m²; G = 1500 kg/m²s (bare tube q_{dht} = 1059 kW/m² (based on Eq. (5)): (a) T_{in} = 166°C and (b) T_{in} = 212°C. Bare tube q_{dht} = 1431 kW/m² at G = 2000 kg/m²s (based on Eq. (5)); for details of test section, see **Figure 23***). Points are experimental data; curves are calculated data; curves for HTC and T_w are calculated through Dittus-Boelter correlation (Eq. (5)). Uncertainties of primary parameters are listed in* **Table 6***.*

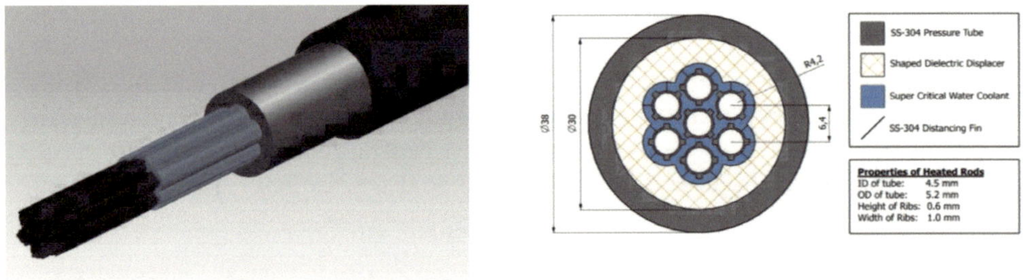

Figure 26.
3-D view (a) and cross-sectional view of vertical seven-rod bundle (b) cooled with upward flow of SCW [41, 42]: heated rods equipped with four helical ribs; all dimensions in mm; and Ukrainian stainless steel has been used for heated rods, by content and other parameters this steel is very close to those of SS-304.

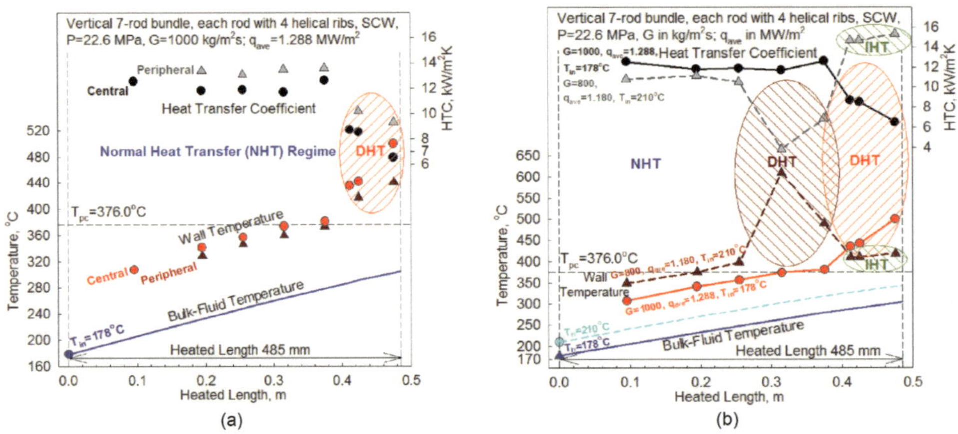

Figure 27.
Profiles of bulk-fluid and wall temperatures, and HTC vs. heated length; vertical seven-rod bundle (see **Figure 26***) cooled with upward flow of SCW [42]: P = 22.6 MPa. Uncertainties of primary parameters are listed in* **Table 6***. (a) G = 1000 kg/m²s; q_{ave} = 1.29 MW/m² (bare tube q_{dht} = 0.69 MW/m²); T_{in} = 178°C; and central and peripheral rods; (b) G = 1000; q_{ave} = 1.29 MW/m² (bare tube q_{dht} = 0.69 MW/m²); T_{in} = 178°C; and G = 800 kg/m²s; q_{ave} = 1.18 MW/m² (bare tube q_{dht} = 0.54 MW/m²); T_{in} = 210°C; and central rod.*

regimes plus sometimes quite significant differences in local HTC values and wall temperatures around the central rod circumference.

4. Practical prediction methods for forced-convection heat transfer at supercritical pressures

4.1 Supercritical water (SCW)

Unfortunately, satisfactory analytical methods for practical prediction of forced-convection heat transfer at SCPs have not yet been developed due to the difficulty in dealing with steep property variations, especially, in turbulent flows and at high heat fluxes [10, 48]. Therefore, generalized correlations based on experimental data are used for HTC calculations at SCPs.

There are numerous correlations for convective heat transfer in circular tubes at SCPs (for details, see in Pioro and Duffey [9]). However, an analysis of these correlations has shown that they are more or less accurate only within the particular

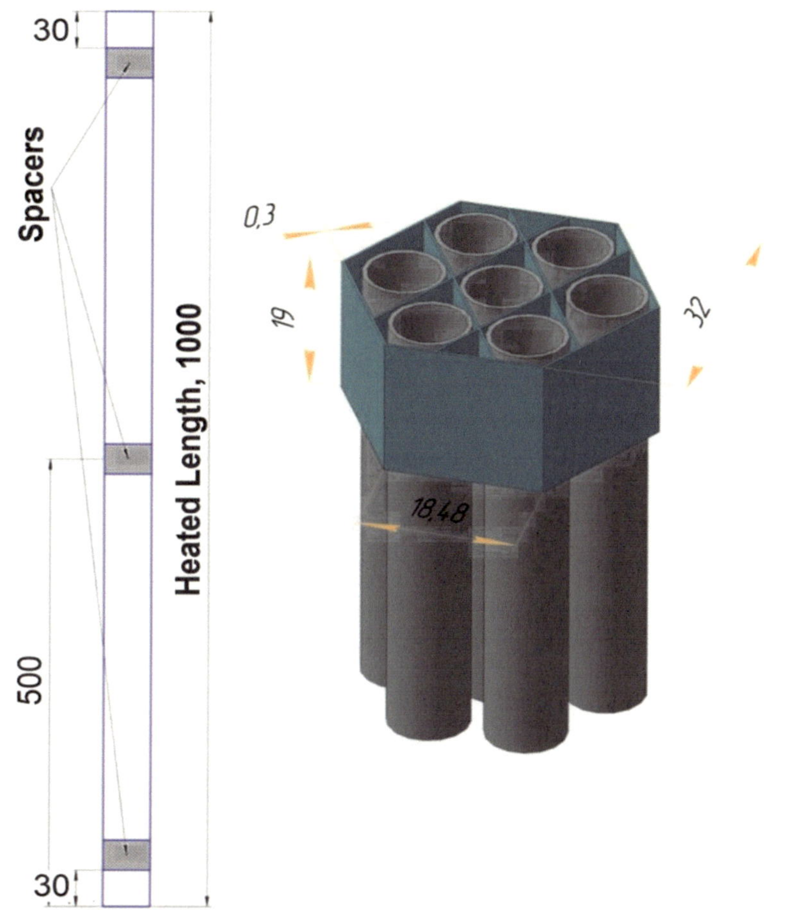

Figure 28.
Spacer grid locations and dimensions (all dimensions are in mm) [43].

dataset, which was used to derive the correlation, but show a significant deviation in predicting other experimental data. Therefore, only selected correlations are considered below.

In general, many of these correlations are based on the conventional Dittus-Boelter-type correlation (see Eq. (1)) in which the "regular" specific heat (i.e., based on bulk-fluid temperature) is replaced with the cross-sectional averaged specific heat within the range of $(T_w - T_b)$; $\left(\frac{H_w - H_b}{T_w - T_b}\right)$, J/kg K. Also, additional terms, such as: $\left(\frac{k_b}{k_w}\right)^k$; $\left(\frac{\mu_b}{\mu_w}\right)^m$; $\left(\frac{\rho_b}{\rho_w}\right)^n$; etc., can be added into correlations to account for significant variations in thermophysical properties within a cross-section due to a nonuniform temperature profile, that is, due to heat flux.

It should be noted that usually generalized correlations, which contain fluid properties at a wall temperature, require iterations to be solved, because there are two unknowns: (1) HTC and (2) the corresponding wall temperature. Therefore, the initial wall temperature value at which fluid properties will be estimated should be "guessed" to start iterations.

The most widely used heat transfer correlation at subcritical pressures for forced convection is the Dittus-Boelter [49] correlation. In 1942, McAdams [50] proposed to use the Dittus-Boelter correlation in the following form, for forced-convective heat transfer in turbulent flows:

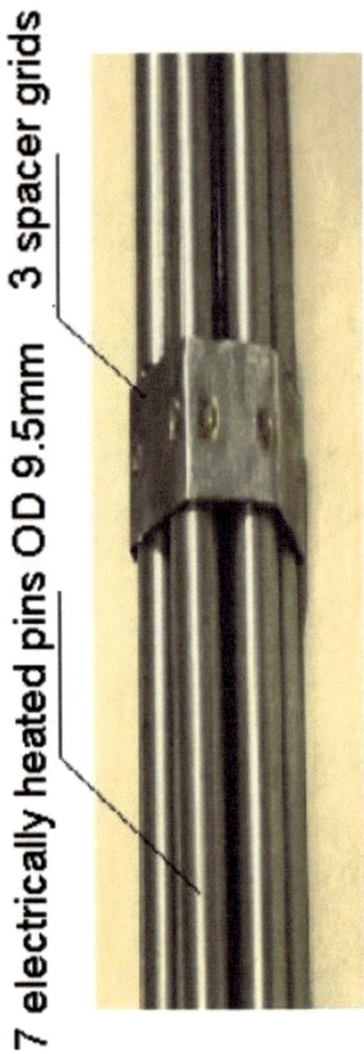

Figure 29.
Photo of central part of 7-element bundle with spacer grid [43].

$$\mathrm{Nu_b} = 0.0243\ \mathrm{Re}_b^{0.8}\mathrm{Pr}_b^{0.4} \tag{1}$$

However, it was noted that Eq. (1) might produce unrealistic results at SCPs within some flow conditions (see **Figure 12**), especially, near the critical and pseudocritical points, because it is very sensitive to properties variations.

In general, experimental HTC values show just a moderate increase within the pseudocritical region. This increase depends on mass flux and heat flux: higher heat flux—less increase. Thus, the bulk-fluid temperature might not be the best characteristic temperature at which all thermophysical properties should be evaluated. Therefore, the cross-sectional averaged Prandtl number, which accounts for thermophysical-properties variations within a cross-section due to heat flux, was proposed to be used in many SC HT correlations instead of the regular Prandtl number. Nevertheless, this classical correlation (Eq. (1)) was used extensively as a basis for various SC HT correlations [9].

The majority of empirical correlations were proposed in the 1960s–1970s [9], when experimental techniques were not at the same level (i.e., advanced level) as they are today. Also, thermophysical properties of SCW have been updated since

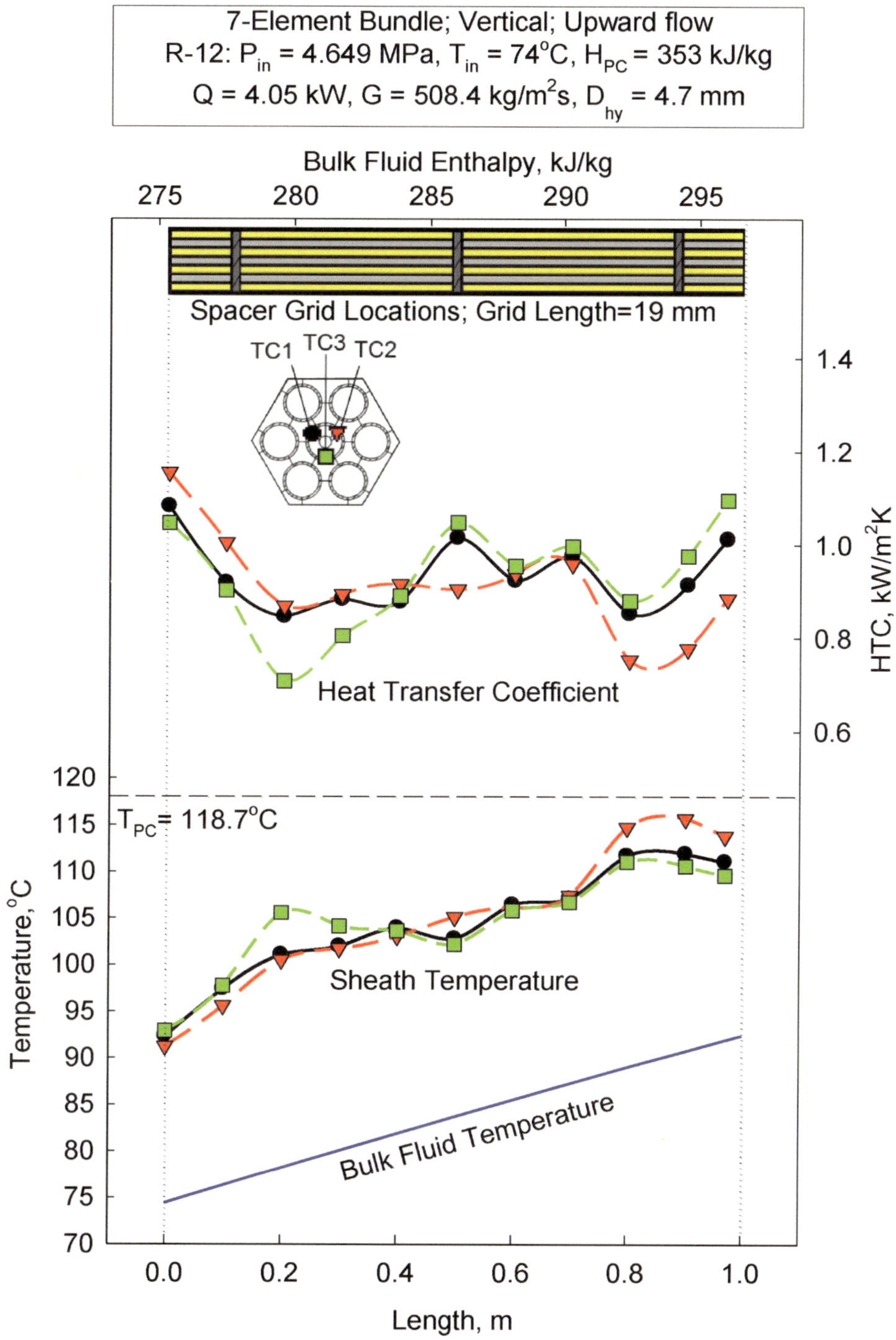

Figure 30.
Bulk-fluid and wall temperatures, and HTC profiles along heated length of vertical bare 7-element bundle (D_{hy} = 4.7 mm) cooled with upward flow of SC R-12 [43, 44]: Run 3: P_{in} = 4.65 MPa; G = 508 kg/m²s; q_{ave} = 19.4 kW/m², and T_{in} = 74°C.

that time (for example, a peak in thermal conductivity in critical and pseudocritical points within a range of pressures from 22.1 to 25 MPa for water was not officially recognized until the 1990s).

Therefore, new correlations within the SCWRs operating range, were developed and evaluated by I. Pioro with his students (mainly, by S. Mokry et al. (bulk-fluid-temperature approach) and S. Gupta et al. (wall temperature approach)) using the

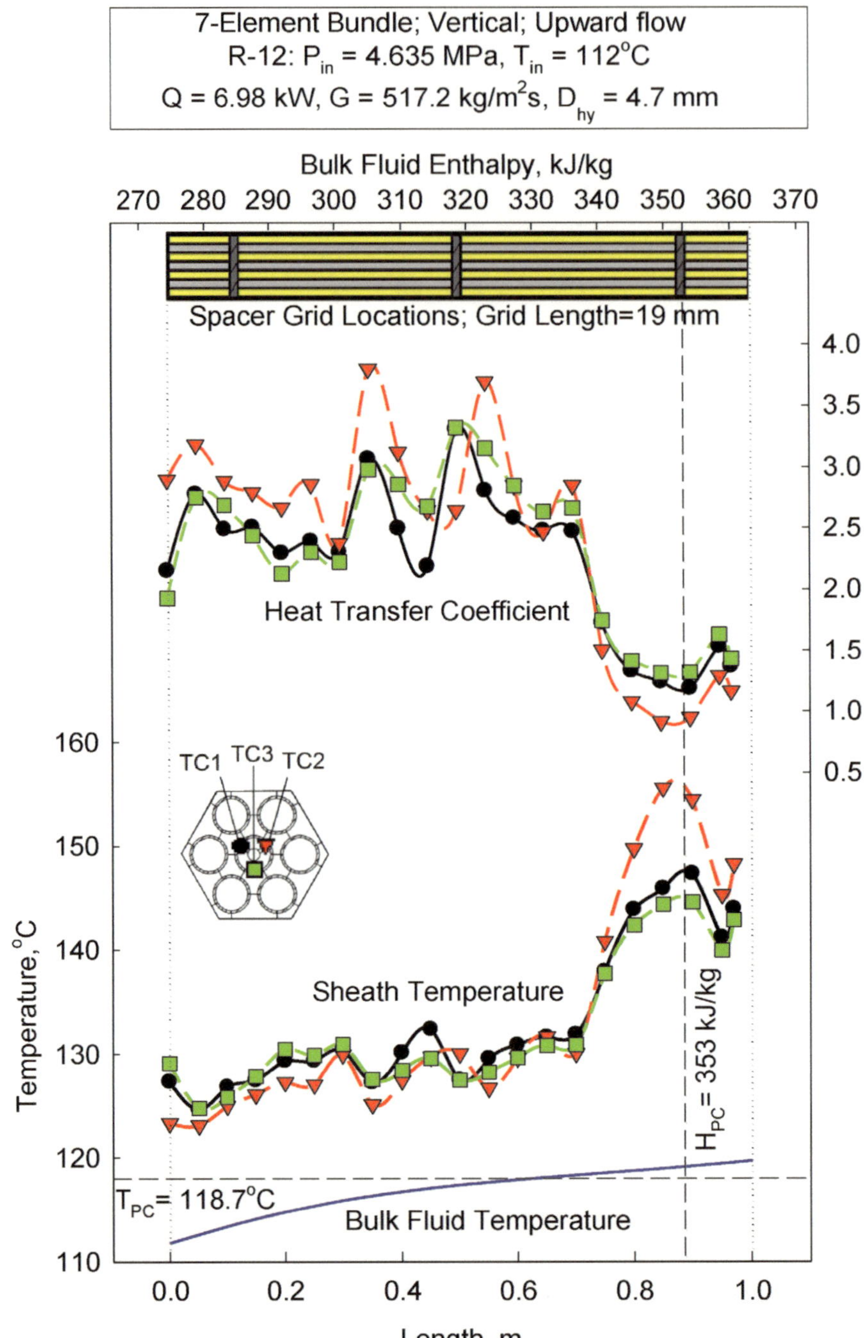

Figure 31.
Bulk-fluid and wall temperatures, and HTC profiles along heated length of vertical bare 7-element bundle (D_{hy} = 4.7 mm) cooled with upward glow of SC R-12 [43, 44]: Run 7: P_{in} = 4.64 MPa; G = 517 kg/m²s; q_{ave} = 33.4 kW/m², and T_{in} = 112°C.

best SCW dataset by P.L. Kirillov and his co-workers and adding smaller datasets by other researchers:

1. Pioro-Mokry correlation (bulk-fluid-temperature approach) [51, 52]:

$$\mathrm{Nu_b} = 0.0061 \quad \mathrm{Re}_\mathrm{b}^{0.904}\overline{\mathrm{Pr}_\mathrm{b}}^{0.684}\left(\frac{\rho_w}{\rho_b}\right)^{0.564}. \tag{2}$$

The Pioro-Mokry correlation (Eq. (2)) was verified within the following operating conditions (only for NHT and IHT regimes (see **Figures 32** and **33**), but not for the DHT regime): SCW, upward flow, vertical bare circular tubes with inside diameters of 3–38 mm, pressure—22.8–29.4 MPa, mass flux—200–3000 kg/m²s, and heat flux—70–1250 kW/m². All thermophysical properties of SCW were calculated according to NIST REFPROP software [25]. This correlation has accuracy of ±25% for HTCs and ±15 for wall temperatures (**Figure 34**). Eventually, this nondimensional correlation can be also used for other SCFs. However, its accuracy can be less or even significantly less in these cases.

2. Pioro-Gupta correlation (wall temperature approach) [53]:

$$\mathbf{Nu_w} = 0.0033\,\mathbf{Re_w^{0.941}}\overline{\mathbf{Pr_w}}^{0.764}\left(\frac{\mu_w}{\mu_b}\right)^{0.398}\left(\frac{\rho_w}{\rho_b}\right)^{0.156} \tag{3}$$

Eq. (3) has an uncertainty of about ±25% for HTC values and about ±15% for calculated wall temperatures within the same ranges as those for Eq. (2). Also, it was decided to add an entrance effect to make this correlation even more accurate. This entrance effect was modeled by an exponentially-decreasing term as shown below:

$$\mathbf{Nu_w} = \mathbf{Nu_w}\left[1 + exp\left(-\frac{x}{24D}\right)\right]^{0.3}, \tag{4}$$

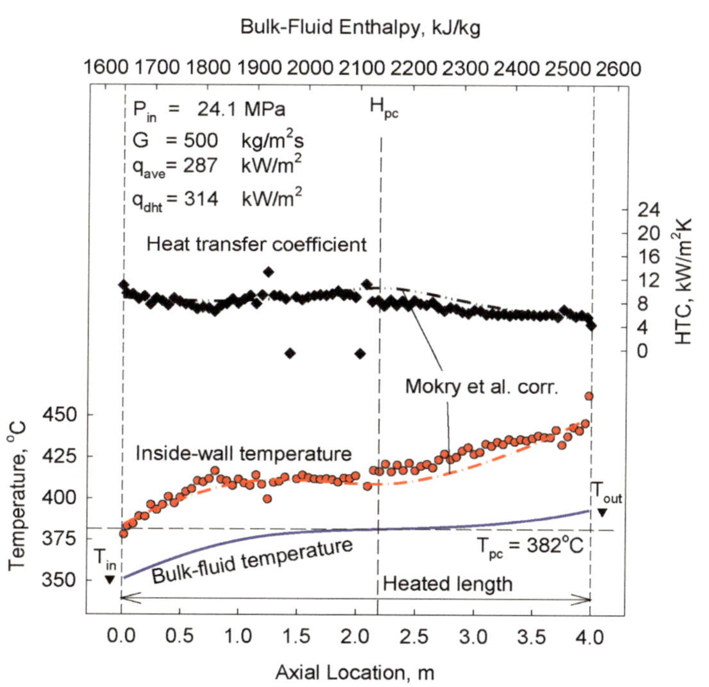

Figure 32.
Temperature and HTC profiles along 4-m circular tube (D = 10 mm) with upward flow of SCW (data by Kirillov et al. [26]) [54]: $P_{in} \approx 24$ MPa, G = 500 kg/m²s; q_{ave} = 287 kW/m²; comparison of calculated HTC values through the "proposed correlation"—Eq. (2) with experimental data within Normal Heat Transfer (NHT) regime.

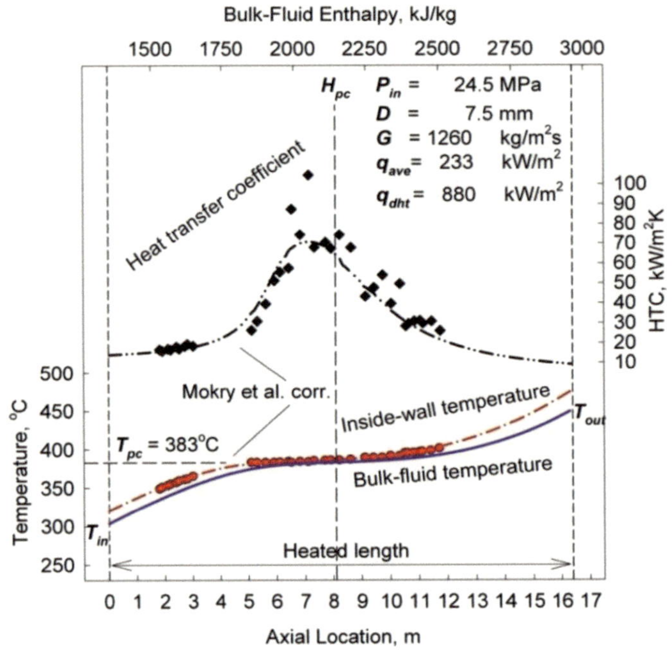

Figure 33.
Temperature and HTC profiles along circular tube (D = 7.5 mm) with upward flow of SCW (data by Yamagata et al. [46]) [54]: P_{in} = 24.5 MPa; G = 1260 kg/m²s; q_{ave} = 233 kW/m²; comparison of calculated HTC values through the "proposed correlation"—Eq. (2) with experimental data within normal and improved heat transfer (NHT and IHT) regimes.

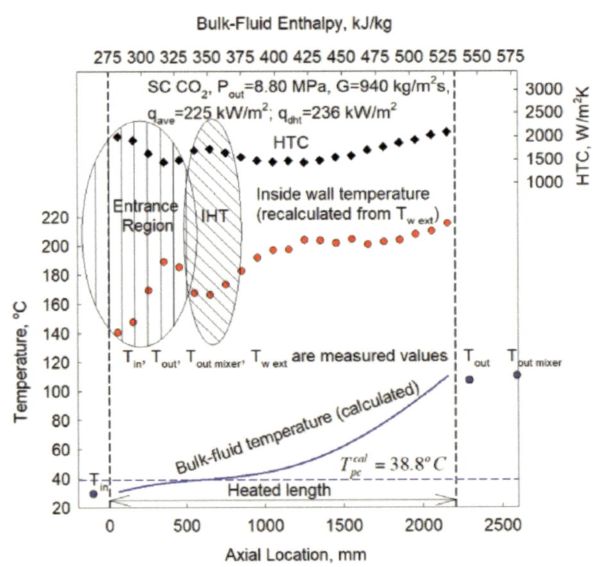

Figure 34.
Wall temperature and HTC profiles along vertical circular tube (D = 8 mm and L = 2.208 m) with upward flow of SC CO_2 (data by I. Pioro): P = 8.8 MPa; G = 940 kg/m²s; q = 225 kW/m², and T_{in} = 30°C.

where, $\mathbf{Nu}_w$ is calculated using Eq. (3). It should be noted that this HT correlation is also intended only for NHT and IHT regimes.

The following empirical correlation was proposed by I. Pioro and S. Mokry for calculating the minimum heat flux at which the DHT regime appears in vertical bare circular tubes:

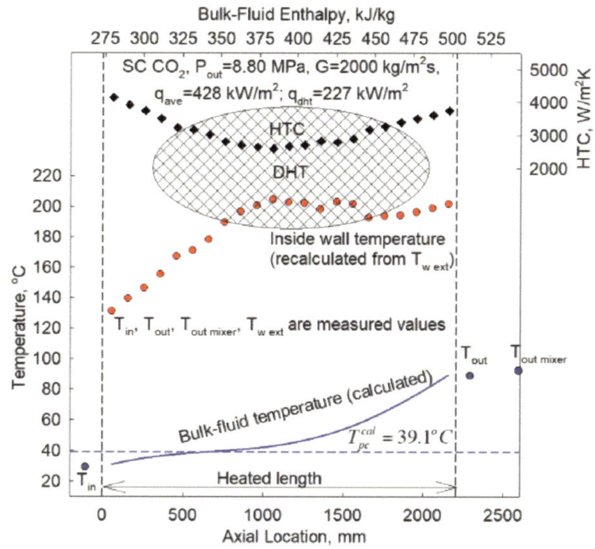

Figure 35.
Wall temperature and HTC profiles along vertical circular tube (D = 8 mm and L = 2.208 m) with upward flow of SC CO_2 (data by I. Pioro): P = 8.8 MPa; G = 2000 kg/m²s; q = 428 kW/m², and T_{in} = 29°C.

Pioro-Mokry correlation for q_{dht} [51]:

$$q_{\mathrm{dht}} = -58.97 + 0.745 \cdot G, \quad \mathrm{kW/m^2}. \qquad (5)$$

Correlation (Eq. (5)) is valid within the following range of experimental parameters: SCW, upward flow, vertical bare tube with inside diameter 10 mm, pressure 24 MPa, mass flux 200–1500 kg/m²s, and bulk-fluid inlet temperature 320–350°C. Uncertainty is about ±15% for the DHT heat flux.

Wang et al. [33] have evaluated 15 q_{dht} correlations for SCW, and they have concluded that Pioro-Mokry correlation (Eq. (5)) "may be used for preliminary estimations."

A recent study was conducted by Zahlan et al. [55, 56] in order to develop a heat transfer look-up table for the critical/SCPs. An extensive literature review was conducted, which included 28 datasets and 6663 trans-critical heat transfer data (**Figure 35**). **Tables 8** and **9** list results from this study in the form of the overall-weighted average and root-mean-square (RMS) errors: (a) within three SC sub-regions; and (b) for subcritical liquid and superheated steam. Many of the correlations listed in these tables can be found in Zahlan et al. [55, 56] and Pioro and Duffey [9]. In their conclusions, Zahlan et al. [55, 56] determined that within the SC region, the latest correlation by Pioro-Mokry [51] (Eq. (2)) showed the best

Parameters	Uncertainty
Test-section power	±1.0%
Inlet pressure	±0.25%
Wall temperature	±3.0%
Mass-flow rate	±1.5%
Heat loss	≤3.0%

Table 5.
Uncertainties of primary parameters [51].

Parameters		Maximum uncertainty
Measured	Inlet pressure	±0.2%
	Bulk-fluid temperature	±3.4%
	Wall temperature	±3.2%
Calculated	Mass-flow rate	±2.3%
	Heat flux	±3.5%
	HTC	±12.7%
	Heat loss	≤3.4%

Table 6.
Maximum uncertainties of measured and calculated parameters [35–40].

No.	Test section	Operating conditions	q_{dht}, MW/m²	Increase in q_{dht} value compared to that of bare tube
1	Bare tube	P = 24.1 MPa and G = 2000 kg/m²s	1.43	1.8
2	Annulus	P = 22.6 MPa and G = 2000 kg/m²s	2.55	
3	Bare tube	P = 24.1 MPa and G = 2700 kg/m²s	1.95	1.6
4	Three-rod bundle	P = 22.6 MPa and G = 2700 kg/m²s	3.20	
5	Bare tube	P = 24.5 MPa and G = 800 kg/m²s	0.54	1.8
6	Seven-rod bundle	P = 24.5 MPa and G = 800 kg/m²s	0.96	

Table 7.
Comparison of DHT values in bare-tube, annular channel (one-rod), and three-rod and seven-rod bundles [35, 42].

prediction for the data within all three sub-regions investigated (based on RMS error) (see **Table 8**). Also, the Pioro-Mokry correlation showed quite good predictions for subcritical-pressure water and superheated steam compared to other several correlations (see **Table 9**). Also, it was concluded that Pioro-Gupta correlation (Eq. (3)) was quite close by RMS errors to the Pioro-Mokry correlation.

Chen et al. [57] has also concluded that the Pioro-Mokry correlation for SCW HT "performs best" compared to other 14 correlations.

4.2 Supercritical carbon dioxide

The following correlation was proposed by S. Gupta (an MASc student of I. Pioro) [21] for SC carbon dioxide flowing inside vertical bare tubes:

$$\mathbf{Nu_w} = 0.0038\mathbf{Re_w}^{0.957}\overline{\mathbf{Pr}}_\mathbf{w}^{-0.14}\left(\frac{\rho_w}{\rho_b}\right)^{0.84}\left(\frac{k_w}{k_b}\right)^{-0.75}\left(\frac{\mu_w}{\mu_b}\right)^{-0.22} \quad (6)$$

Uncertainties associated with this correlation are ±30% for HTC values and ± 20% for calculated wall temperatures (see **Figures 36** and **37**). Ranges of parameters for the dataset used to develop Eq. (6) are listed in **Table 10**.

Table 11 list mean and root-mean square (RMS) errors in HTC and T_w for proposed correlations using equations shown below:

It was also decided to develop the q_{dht} correlation for SC carbon dioxide based on the dataset obtained by I. Pioro in vertical bare tube with upward flow, which ranges are listed in **Table 10** [58]. Therefore, based on the identified 41 cases of

No.	Correlation	Regions					
		Liquid-like		Gas-like		Critical or pseudocritical	
		Errors, %					
		Ave.	RMS	Ave.	RMS	Ave.	RMS
1	Dittus-Boelter [49]	24	44	90	127	-	-
2	Sieder and Tate [59]	46	65	97	132	-	-
3	Bishop et al. [60]	5	28	5	20	23	31
4	Swenson et al. [61]	**1**	31	-16	21	4	23
5	Krasnoshchekov et al. [62]	18	40	-30	32	24	65
6	Hadaller and Banerjee [63]	34	53	14	24	-	-
7	Gnielinski [64]	10	36	99	139	-	-
8	Watts and Chou [65], NHT	6	30	-6	21	11	28
9	Watts and Chou [65], DHT	2	**26**	9	24	17	30
10	Griem [66]	2	28	11	28	9	35
11	Koshizuka and Oka [67]	26	47	27	54	39	83
12	Jackson [68]	15	36	15	32	30	49
13	Mokry et al. [51, 52]	-5	**26**	-9	**18**	**-1**	**17**
14	Kuang et al. [69]	-6	27	10	24	-3	26
15	Cheng et al. [70]	4	30	**2**	28	21	85
16	Gupta et al. [53]	-26	33	-12	20	**-1**	18

In bold—minimum values.

Table 8.
Overall-weighted average and RMS errors within three supercritical sub-regions (correlations are listed according to the year of publication, that is, from early ones to the latest ones) [55, 56].

No.	Correlation	Subcritical liquid		Superheated steam	
		Error, %			
		Ave.	RMS	Ave.	RMS
1	Dittus and Boelter [49]	10	23	75	127
2	Sieder and Tate [59]	28	37	84	138
3	Hadaller and Banerjee [63]	27	36	19	34
4	Gnielinski [64]	-4	**18**	80	130
5	Mokry et al. [51]	**-1**	19	**-5**	**20**

In bold—minimum values.

Table 9.
Overall average and RMS error within subcritical region [55, 56].

P, MPa	T_{in}, °C	T_{out}, °C	T_w, °C	q, kW/m^2	G, kg/m^{2}s
7.57-8.8	20-40	29-136	29-224	9.3-616.6	706-3169

Table 10.
Ranges of parameters of dataset used to develop Eq. (6).

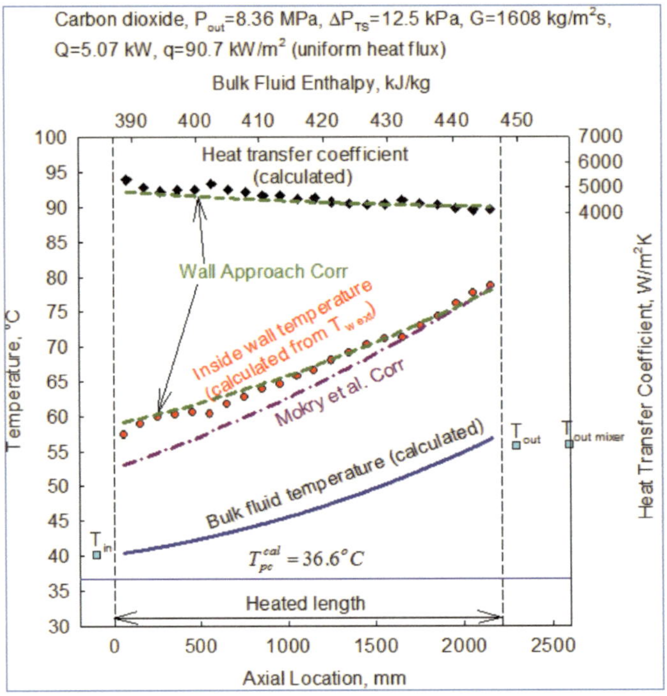

Figure 36.
HTC and T_w *variations along* L = 2.208 *m circular tube* (D = 8 *mm*)*:* q = 90.7 *kW/m*² P = 8.4 *MPa, and* G = 1608 *kg/m²s. Wall Approach Corr. is Eq. (6) and Mokry et al. Corr. – Eq. (2).*

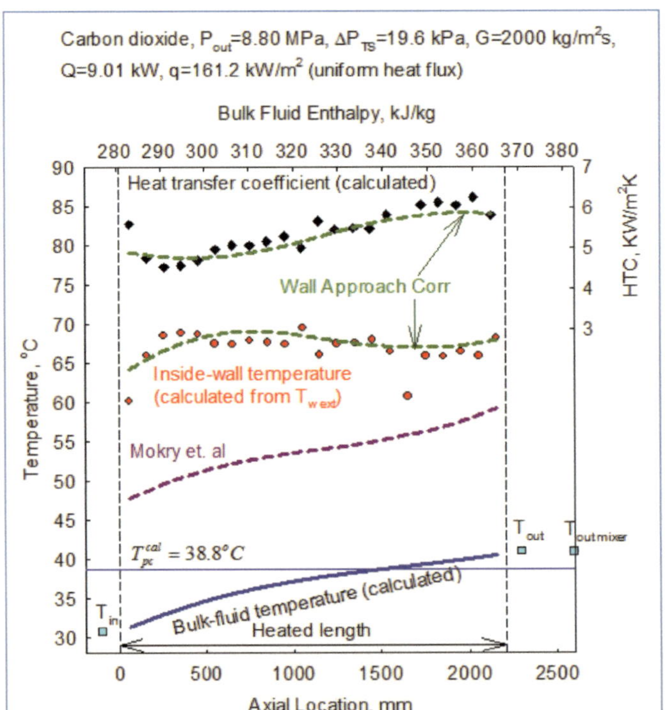

Figure 37.
HTC and T_w *variations along* L = 2.208 *m circular tube* (D = 8 *mm*)*:* q = 161.2 *kW/m*² P = 8.8 *MPa, and* G = 2000 *kg/m²s. Wall Approach Corr. is Eq. (6) and Mokry et al. Corr. – Eq. (2).*

Errors in HTC (for the reference dataset), %		
	Mean Error	**RMS**
Proposed new correlation (T_b approach)	0.9%	22.4%
Proposed new correlation (T_{film} approach)	**0.2%**	21.7%
Proposed new correlation (T_w approach—Eq. (6))	0.8%	**20.3%**
Swenson et al. [61] correlation	89%	132%
Mokry et al. [51] correlation for SCW	68%	123%
Gupta et al. [53] correlation for SCW	78%	130%

Table 11.
Mean and RMS errors for HTC values of proposed correlations (values in bold represent minimum errors) [21].

DHT within the SC carbon dioxide dataset, the following correlation for the minimal heat flux at which deterioration occurs was proposed:

$$q_{min} = 66.81 + 0.18 \cdot G \quad (7)$$

In general, the total pressure drop for forced convection inside a channel can be calculated according to expressions listed in Pioro and Duffey [9] and Pioro et al. [71].

5. Conclusions

Supercritical fluids are used quite intensively in various industries. Therefore, understanding specifics of thermophysical properties, heat transfer, and pressure drop in various flow geometries at supercritical pressures is an important task.

In general, three major heat transfer regimes were noticed at critical and supercritical pressures in various flow geometries (vertical bare tubes, annulus, three- and seven-rod bundles) and several SCFs (SCW, SC carbon dioxide, and SC R-12): (1) normal heat transfer; (2) improved heat transfer; and (3) deteriorated heat transfer. Also, two special phenomena may appear along a heated channel: (1) pseudo-boiling; and (2) pseudo-film boiling. These heat transfer regimes and special phenomena appear to be due to significant variations of thermophysical properties near the critical and pseudocritical points and due to operating conditions.

Comparison of heat transfer-coefficient values obtained in bare circular tubes with those obtained in annulus (one-rod bundle)/three-rod bundle (rod(s) equipped with four helical ribs) shows that there are almost no differences between these values. However, the minimal heat flux at which deterioration occurs (q_{dht}) in annulus, and three- and seven-rod bundles are in 1.6–1.8 times higher compared to that recorded in bare tubes.

The current analysis of a number of well-known heat transfer correlations for supercritical fluids showed that the Dittus-Boelter correlation [49] significantly overestimates experimental HTC values within the pseudocritical range. The Bishop et al. [60] and Jackson [68] correlations tend also to deviate substantially from the experimental data within the pseudocritical range. The Swenson et al. [61] correlation provided a better fit for the experimental data than the previous three correlations within some flow conditions, but does not follow up closely the experimental data within others.

Therefore, new correlations were developed by Pioro with his students Mokry et al. [51] (bulk-fluid-temperature approach) and Gupta et al. [21] (wall temperature approach), which showed the best fit for the experimental data within a wide range of operating conditions. These correlations have uncertainties of about ±25% for HTC values and about ±15% for calculated wall temperature. Also, based on an independent study performed by Zahlan et al. [55, 56], Pioro-Mokry correlation (given as Eq. (2)) is the best for superheated steam compared to other well-known correlations. Also, this correlation showed quite good predictions for subcritical-pressure fluids.

Acknowledgements

The author would like to express his appreciation to his former and current students, S. Clark, A. Dragunov, S. Gupta, M. Mahdi, D. Mann, S. Mokry, R. Popov, G. Richards, Eu. Saltanov, H. Sidawi, E. Tamimi, and A. Zvorykin, for their assistance in the preparation of figures and developing of correlations.

Nomenclature

A	area, m^2
c_p	specific heat at constant pressure, J/kg K
$\bar{c}_p$	averaged specific heat within the range of $(T_w - T_b)$; $\left(\frac{H_w-H_b}{T_w-T_b}\right)$, J/kg K
D	inside diameter, m
G	mass flux, kg/m^2s; $\left(\frac{m}{A_{fl}}\right)$
H	specific enthalpy, J/kg
h	heat transfer coefficient, W/m^2K
k	thermal conductivity, W/m K
L	heated length, m
m	mass-flow rate, kg/s; $(\rho \cdot V)$
P, p	pressure, Pa
Q	heat transfer rate, W
q	heat flux, W/m^2; $\left(\frac{Q}{A_h}\right)$
s	specific entropy, J/kg K
T, t	temperature, °C
T_{film}	film temperature, °C; $\left(\frac{T_w+T_b}{2}\right)$
V	volume-flow rate, m^3/s
v	specific volume, m^3/kg
x	axial coordinate, m

Greek letters

α	thermal diffusivity, m^2/s; $\left(\frac{k}{c_p \cdot \rho}\right)$
β	volumetric expansion coefficient, 1/K
Δ	difference
η	efficiency, %
μ	dynamic viscosity, Pa·s
ρ	density, kg/m^3
υ	kinematic viscosity, m^2/s; $\left(\frac{\mu}{\rho}\right)$

Non-dimensional numbers

Nu	Nusselt number; $\left(\frac{h \cdot D}{k}\right)$
Pr	Prandtl number; $\left(\frac{\mu \cdot c_p}{k}\right) = \left(\frac{\upsilon}{\alpha}\right)$

$\overline{\mathrm{Pr}}$	cross-sectional average Prandtl number within the range of $(T_w - T_b)$; $\left(\frac{\mu \cdot \bar{c}_p}{k}\right)$
Re	Reynolds number; $\left(\frac{G \cdot D}{\mu}\right)$

Subscripts or superscripts

ave.	average
b	bulk
cal	calculated
corr.	correlation
cr	critical
dht	deteriorated heat transfer
fl	flow
h	heated
hy	hydraulic-equivalent
in	inlet
max	maximum
min	minimum
out	outlet
pc	pseudocritical
sat	saturation
th	thermal
w	wall

Abbreviations and acronyms

AECL	Atomic Energy of Canada Limited
AGR	advanced gas-cooled reactor
ASME	American Society of Mechanical Engineers
Ave.	average
BN	fast sodium (reactor; in Russian abbreviations)
BWR	boiling water reactor
CHF	critical heat flux
CFD	computational fluid dynamics
corr.	correlation
CRL	Chalk River Laboratotries (AECL)
DHT	deteriorated heat transfer
GFR	Gas-cooled fast reactor
GIF	Generation-IV International Forum
HT	heat transfer
HTC	heat transfer coefficient
HTR	high-temperature reactor
HPT	high-pressure turbine
IAEA	International Atomic Energy Agency
ID	inside diameter
IHT	improved heat transfer
IHX	intermediate heat exchanger
LFR	lead-cooled fast reactor
LGR	light-water-cooled graphite-moderated reactor
LNG	liquified natural gas
LPT	low-pressure turbine

LWR	light water reactor
MSR	molten salt reactor
N/A	not applicable
NIST	National Institute of Standards and Technology (USA)
NHT	normal heat transfer
NPP	nuclear power plant
OD	outside diameter
PHWR	pressurized heavy water reactor
PWR	pressurized water reactor
REFPROP	REFerence PROPerties
RMS	root-mean square (error)
S-CO_2 (SC-CO_2)	supercritical carbon dioxide
SC	supercritical
SCF	supercritical fluid
SCP	supercritical pressure
SCW	supercritical water
SCWR	supercritical water-cooled reactor
SFR	sodium-cooled fast reactor
SS	stainless steel
TC	thermocouple
TECDOC	TEChnical DOCument
ThPP	thermal power plant
UK	United Kingdom
USA	United States of America
USSR	Union of Soviet Socialist Republics
VHTR	very high temperature reactor
WF	working fluid

Author details

Igor L. Pioro
Institute of Technology, University of Ontario, Oshawa, Ontario, Canada

*Address all correspondence to: igor.pioro@uoit.ca

References

[1] Levelt Sengers JMHL. Supercritical fluids: Their properties and applications, Chapter 1. In: Kiran E et al., editors. Supercritical Fluids. Vol. 366. Netherlands: NATO Advanced Study Institute on Supercritical Fluids—Fundamentals and Application, NATO Science Series, Series E, Applied Sciences, Kluwer Academic Publishers; 2000. pp. 1-29

[2] Schmidt E, Eckert E, Grigull V. Heat transfer by liquids near the critical state. AFF Translation, No. 527, Air Materials Command, Wright Field, Dayton, OH, USA, April; 1946

[3] Pioro LS, Pioro IL. Industrial Two-Phase Thermosyphons. New York, NY, USA: Begell House, Inc.; 1997. 288 pp

[4] Pioro I, Duffey RB, Kirillov PL, Pioro R, Zvorykin A, Machrafi R. Current status and future developments in nuclear-power industry of the world. ASME Journal of Nuclear Engineering and Radiation Science. 2019;**5**(1):27. Available from: http://nuclearengineering.asmedigitalcollection.asme.org/article.aspx?articleID=2718229

[5] Pioro I, Kirillov P. Current status of electricity generation at nuclear power plants. In: Méndez-Vilas A, editor. Materials and Processes for Energy: Communicating Current Research and Technological Developments, Energy Book Series #1. Spain: Formatex Research Center; 2013. pp. 806-817. Available from: http://www.formatex.info/energymaterialsbook/book/806-817.pdf

[6] Handbook of Generation IV Nuclear Reactors. Pioro IL, editor. Duxford, UK: Elsevier, Woodhead Publishing (WP); 2016. pp. 940. Available from: https://www.gen-4.org/gif/jcms/c_9373/publications

[7] Dragunov A, Saltanov E, Pioro I, Kirillov P, Duffey R. Power cycles of Generation III and III$^+$ nuclear power plants. ASME Journal of Nuclear Engineering and Radiation Science. 2015;**1**(2):10

[8] Pioro I, Kirillov P. Current status of electricity generation at thermal power plants. In: Méndez-Vilas A, editor. Materials and Processes for Energy: Communicating Current Research and Technological Developments, Energy Book Series #1. Spain: Formatex research Center; 2013. pp. 796-805. Available from: http://www.formatex.info/energymaterialsbook/book/796-805.pdf

[9] Pioro IL, Duffey RB. Heat Transfer and Hydraulic Resistance at Supercritical Pressures in Power Engineering Applications. New York, NY, USA: ASME Press; 2007. p. 328

[10] IAEA-TECDOC-1900, 2020. Understanding and Prediction of Thermohydraulic Phenomena Relevant to Supercritical Water Cooled Reactors (SCWRs), Final Report of a Coordinated Research Project, IAEA TECDOC Series, Vienna, Austria, 544 pages. Free download from: https://www.iaea.org/publications/13636/understanding-and-prediction-of-thermohydraulic-phenomena-relevant-to-supercritical-water-cooled-reactors-scwrs

[11] Oka Y, Koshizuka S, Ishiwatari Y, Yamaji A. Super Light Water Reactors and Super Fast Reactors. New York, NY, USA: Springer; 2010. p. 416

[12] Pioro I. The potential use of supercritical water-cooling in nuclear reactors. In: Krivit SB, Lehr JH, Kingery TB, editors. Nuclear Energy Encyclopedia: Science, Technology, and Applications. Hoboken, NJ, USA: J. Wiley & Sons; 2011. pp. 309-347

[13] Schulenberg T, Starflinger J, editors. High performance light water reactor.

In: Design and Analyses. Germany: Scientific Publishing, Karlsruher Institut für Technologie (KIT); 2012. p. 241

[14] Duffey R, Pioro I. Ensuring the future of nuclear power, Mechanical Engineering Magazine. ASME. 2019; **141**(11):30-35

[15] Pioro I. Current status and future developments in nuclear-power industry of the world. In: Proceedings of the international conference on climate change and the role of nuclear power, IAEA, Vienna, Austria, October 7–11, Paper No. CN275-42. 2019. p. 4

[16] Pioro I, Duffey R. Current and future nuclear-power reactors and plants, Chapter 4. In: Letcher T, editor. Managing global warming, an Interface of technology and human Issues. London, UK: Elsevier, Academic Press; 2018. p. 820

[17] Bae SJ, Ahn Y, Lee JI. Preliminary study of the supercritical CO_2 hybrid cycle for the HTGR application. In: Transactions of the Korean Nuclear Society, Spring Meeting, Jeju, Korea, May 29–30. 2014

[18] Mahdi M, Popov R, Pioro I. Research on Thermal efficiencies of various power cycles for GFRs and VHTRs. Proceedings of the 26th International Conference on Nuclear Engineering (ICONE-26), July 22–26, London, England, Paper #81618. p. 9

[19] Pioro I, Mahdi M, Popov R. Heat transfer media and their properties, Chapter 33. In: Kulacki FA, editor. Handbook of thermal science and engineering. Cham, Switzerland: Springer; 2018. pp. 1353-1446

[20] Mann D, Pioro I. Study on specifics of thermophysical properties of supercritical fluids in power engineering applications. In: Proceedings of the 23rd International Conference on Nuclear Engineering (ICONE-23), May 17–21, Chiba, Japan, Paper #1730. 2015. p. 11

[21] Gupta S, Saltanov E, Mokry SJ, Pioro I, Trevani L, McGillivray D. Developing empirical heat-transfer correlations for supercritical CO_2 flowing in vertical bare tubes. Nuclear Engineering and Design. 2013;**261**: 116-131

[22] Pioro I, Mokry S. Heat transfer to fluids at supercritical pressures, chapter. In: Belmiloudi A, editor. Heat Transfer. Theoretical Analysis, Experimental Investigations and Industrial Systems. Rijeka, Croatia: IntechOpen; 2011. pp. 481-504. Available from: http://www.intechopen.com/books/heat-transfer-theoretical-analysis-experimental-investigations-and-industrial-systems/heat-transfer-to-supercritical-fluids

[23] Anzieu P. Gas-cooled Gen-IV systems VHTR & GFR. In: Presentation for CEA Meeting, Demanova, Slovakia, February. 2010. 34 Slides

[24] Pioro I, Kirillov P. Generation IV nuclear reactors as a basis for future electricity production in the world, chapter. In: Méndez-Vilas A, editor. Materials and Processes for Energy: Communicating Current Research and Technological Developments, Energy Book Series #1. Spain: Formatex Research Center; 2013. pp. 818-830. Available from: http://www.formatex.info/energymaterialsbook/book/818-830.pdf

[25] NIST REFPROP. Reference fluid thermodynamic and transport properties. In: Lemmon EW, Bell IH, Huber ML, McLinden MO, editors. NIST Standard Reference Database 23, Version 10.0, Developers. Boulder, CO, 80305, USA: Applied Chemicals and Materials Division, National Institute of Standards and Technology (NIST); 2018

[26] Kirillov PL, Lozhkin VV, Smirnov AM. Investigation of Borders

of Deteriorated Regimes of a Channel at Supercritical Pressures (in Russian). Obninsk, Russia: State Scientific Center of Russian Federation Institute of Physics and Power Engineering by the name of A.I. Leypunskiy, FEI-2988; 2003. p. 20

[27] Generation IV International Forum (GIF). 2019. Available from: https://www.gen-4.org/gif/jcms/c_9260/public

[28] Kirillov PL, Bogoslovskaya GP. Generation IV supercritical water-cooled nuclear reactors: Realistic prospects and research program. Nuclear Energy and Technology. 2019; **5**(1):67-74

[29] Leung LKH, Nava-Dominguez A. Thermal-hydraulics program in support of Canadian SCWR concept development. ASME Journal of Nuclear Engineering and Radiation Science. 2018;**4**(1):8

[30] Peiman W, Pioro I, Gabriel K. Thermal-hydraulic and neutronic analysis of a re-entrant fuel channel design for pressure-channel supercritical water-cooled reactors. ASME Journal of Nuclear Engineering and Radiation Science. 2015;**1**(2):10

[31] Pioro IL. Current status of research on heat transfer in forced convection of fluids at supercritical pressures. Nuclear Engineering and Design. 2019;**354**:14

[32] Schatte GA, Kohlhepp A, Wieland C, Spliethoff H. Development of a new empirical correlation for the prediction of the onset of the deterioration of heat transfer to supercritical water in vertical tubes. International Journal of Heat and Mass Transfer. 2016;**102**:133-141

[33] Wang H, Leung LKH, Wang W, Bi Q. A review on recent heat transfer studies to supercritical pressure water in channels. Applied Thermal Engineering. 2018;**142**:573-596

[34] Zahlan HAM, Leung LKH, Huang Y-P, Liu G-X. Assessment of convective heat transfer correlations against an expanded database for different fluids at supercritical pressures. ASME Journal of Nuclear Engineering and Radiation Science. 2018;**4**(1):14

[35] Razumovskiy VG, Pis'mennyi EN, Sidawi K, Pioro IL, Koloskov AE. Experimental heat transfer in an annular channel and 3-rod bundle cooled with upward flow of supercritical water. ASME Journal of Nuclear Engineering and Radiation Science. 2016;**2**(1):8

[36] Razumovskiy VG, Pis'mennyi EN, Sidawi K, Pioro IL, Maevskiy EM, Koloskov AE. Specifics of heat transfer to supercritical water flowing upward in annular channel and 3-rod bundle. In: Proceedings of the 7th International Symposium on SCWR (ISSCWR-7), March 15–18, Helsinki, Finland, Paper #2095. 2015. p. 11

[37] Razumovskiy VG, Mayevskiy EM, Koloskov AE, Pis'mennyi EN, Pioro IL. Heat transfer to water at supercritical parameters in vertical tubes, annular channels, 3- and 7-rod bundles. In: Proceedings of the 21st International Conference on Nuclear Engineering (ICONE-21), July 29–August 2, Chengdu, China, Paper #16442. 2013. p. 8

[38] Razumovskiy VG, Pis'mennyy EN, Koloskov AE, Pioro IL. Heat transfer to supercritical water in vertical annular channel and 3-rod bundle. In: Proceedings of the 17th International Conference on Nuclear Engineering (ICONE-17), Brussels, Belgium, July 12–16, Paper #75212. 2009. p. 6

[39] Razumovskiy VG, Pis'menny EN, Koloskov AE, Pioro IL. Heat transfer to supercritical water in vertical 7-rod bundle. In: Proceedings of the 16th International Conference on Nuclear Engineering (ICONE-16), Orlando, FL,

USA, May 11–15, Paper #48954. 2008. p. 6

[40] Sidawi K, Pioro I, Razumovskiy VG, Pis'mennyi EN, Koloskov AE. HTC correlation applications to supercritical water flowing upward in a vertical annular channel and 3-rod bundle. In: Proceedings of the 23rd International Conference on Nuclear Engineering (ICONE-23), May 17–21, Chiba, Japan, Paper #1743. 2015. p. 11

[41] Clark S, Pioro R, Zvorykin A, Fialko NM, Pioro IL. Comparison of experimental and calculated HTC values for short vertical 7-rod bundle cooled with SCW. In: Proceedings of the 39th Annual Conference of the Canadian Nuclear Society and 43rd Annual CNS/CNA Student Conference, Ottawa, ON, Canada, June 23–26, Paper #41. 2019. p. 7

[42] Zvorykin A, Pioro R, Fialko NM, Pioro IL. Specifics of heat transfer to supercritical water in vertical 7-element bundle. In: Proceedings of the 27th International Conference on Nuclear Engineering (ICONE-27), May 19–24, Tsukuba, Ibaraki, Japan, Paper #1916. 2019. p. 10

[43] Richards G, Harvel GD, Pioro IL, Shelegov AS, Kirillov PL. Heat transfer profiles of a vertical bare, 7-element bundle cooled with supercritical freon R-12. Nuclear Engineering and Design. 2013;**264**:246-256

[44] Richards G. Study of heat transfer in a 7-element bundle cooled with the upward flow of supercritical Freon-12 [M.A.Sc. thesis]. Oshawa, ON, Canada: Faculty of Energy Systems and Nuclear Science, University of Ontario Institute of Technology; 2012. p. 177

[45] Leung LKH, Rao Y, Podila K. Assessment of computational tools in support of heat-transfer correlation development for fuel assembly of Canadian supercritical water-cooled reactor. ASME Journal of Nuclear Engineering and Radiation Science. 2016;**2**(1):9

[46] Yamagata K, Nishikawa K, Hasegawa S, et al. Forced convective heat transfer to supercritical water flowing in tubes. International Journal of Heat & Mass Transfer. 1972;**15**(12): 2575-2593

[47] Zvorykin A, Mahdi M, Popov R, Barati Far K, Pioro I. Heat transfer to supercritical water (liquid-like state) flowing in a short vertical bare tube with upward flow. In: Proceedings of the 26th International Conference on Nuclear Engineering (ICONE-26), July 22–26, London, England, Paper #81608. 2018. p. 14

[48] Farah A, Harvel G, Pioro I. Analysis of CFD code FLUENT capabilities for supercritical water heat transfer applications in vertical bare tubes. ASME Journal of Nuclear Engineering and Radiation Science. 2016;**2**(3):12

[49] Dittus FW, Boelter LMK. Heat Transfer in Automobile Radiators of the Tubular Type. Vol. 2(13). Berkeley: University of California; 1930. pp. 443-461 Publications in Engineering (or International Communications in Heat and Mass Transfer, 1985, Vol. 12, pp. 3–22)

[50] McAdams WH. Heat Transmission. 2nd ed. New York, NY, USA: McGraw-Hill; 1942. p. 459

[51] Mokry S, Pioro IL, Farah A, King K, Gupta S, Peiman W, et al. Development of supercritical water heat-transfer correlation for vertical bare tubes. Nuclear Engineering and Design. 2011; **241**:1126-1136

[52] Mokry S, Naidin M, Baig F, Gospodinov Y, Zirn U, Bakan K, et al. Conceptual thermal-design options for pressure-tube SCWRs with thermochemical co-generation of hydrogen. In: Proceedings of the 16th

International Conference On Nuclear Engineering (ICONE-16), Orlando, FL, USA, May 11–15, Paper #48313. 2008. p. 13

[53] Gupta S, Mokry S, Pioro I. Developing a heat-transfer correlation for supercritical-water flowing in vertical tubes and its application in SCWR. In: Proceedings of the 19th International Conference on Nuclear Engineering (ICONE-19), Makuhari, Japan, May 16–19, Paper #43503. 2011. p. 11

[54] Mokry S. Development of a heat-transfer correlation for supercritical water in SCWR applications [M.A.Sc. thesis]. Oshawa, ON, Canada: Faculty of Energy Systems and Nuclear Science, University of Ontario Institute of Technology; 2009. p. 190

[55] Zahlan H, Groeneveld D, Tavoularis S. Look-up table for trans-critical heat transfer. In: Proc. 2nd Canada-China Joint Workshop on Supercritical Water-Cooled Reactors (CCSC-2010), Toronto, Ontario, Canada: Canadian Nuclear Society (CNS), April 25-28. 2010

[56] Zahlan H, Groeneveld DC, Tavoularis S, Mokry S, Pioro I. Assessment of supercritical heat transfer prediction methods. In: Proceedings of the 5th International Symposium on SCWR (ISSCWR-5), Vancouver, BC, Canada, March 13–16, Paper P08. 2011. p. 20

[57] Chen W, Fang X, Xu Y, Su X. An assessment of correlations of forced convection heat transfer to water at supercritical pressure. Annals of Nuclear Energy. 2015;**76**:451-460

[58] Saltanov E, Pioro I, Mann D, Gupta S, Mokry S, Harvel G. Study on specifics of forced-convective heat transfer in supercritical carbon dioxide. ASME Journal of Nuclear Engineering and Radiation Science. 2015;**1**(1):8

[59] Sieder NM, Tate GE. Heat transfer and pressure drop of liquids in tubes. Industrial and Engineering Chemistry. 1936;**28**(12):1429-1435

[60] Bishop AA, Sandberg RO, Tong LS. Forced convection heat transfer to water at near-critical temperatures and supercritical pressures. A.I.Ch.E.-I. Chem.E. Symposium Series. 1965;**2**: 77-85

[61] Swenson HS, Carver JR, Kakarala CR. Heat transfer to supercritical water in smooth-bore tubes. Journal of Heat Transfer, Transactions of the ASME Series C. 1965;**87**(4):477-484

[62] Krasnoshchekov EA, Protopopov VS, Van F, Kuraeva IV. Experimental investigation of heat transfer for carbon dioxide in the supercritical region. In: Gazley C Jr, Hartnett JP, Ecker ERC, editors. Proceedings of the 2nd All-Soviet Union Conference on Heat and Mass Transfer, Minsk, Belarus', May, 1964, Published as Rand Report R-451-PR. Vol. 1. 1967. pp. 26-35

[63] Hadaller G, Banerjee S. Heat transfer to superheated steam in round tubes. In: AECL Report. 1969

[64] Gnielinski V. New equation for heat and mass transfer in turbulent pipe and channel flow. International Journal of Chemical Engineering. 1976;**16**(2): 359-366

[65] Watts MJ, Chou C-T. Mixed convection heat transfer to supercritical pressure water. In: Proceedings of the 7th International Heat Transfer Conference (IHTC). Germany: Munich; 1982. pp. 495-500

[66] Griem H. A new procedure for the at near-and supercritical prediction pressure of forced convection heat transfer. International Journal of Heat Mass Transfer. 1996;**3**:301-305

[67] Koshizuka S, Oka Y. Computational analysis of deterioration phenomena and thermal-hydraulic design of SCR. In: Proceedings of the 1st International Symposium on Supercritical Water-Cooled Reactor Design and Technology (SCR-2000), Paper No. 302. 2000

[68] Jackson JD. Consideration of the heat transfer properties of supercritical pressure water in connection with the cooling of advanced nuclear reactors. In: Proceedings of the 13th Pacific Basin Nuclear Conference (PBNC-13), Shenzhen City, China, October 21–25. 2002

[69] Kuang B, Zhang Y, Cheng X. A new, wide-ranged heat transfer correlation of water at supercritical pressures in vertical upward ducts. In: Proceedings of the 7th International Topical Meeting on NUclear Reactor Thermal-Hydraulics, Operations and Safety (NUTHOS-7), Seoul, Korea, October 5–9. 2008

[70] Cheng X, Yang YH, Huang SF. A simple heat transfer correlation for SC fluid flow in circular tubes. In: Proceedings of the 13th International Topical Meeting on NUclear REactor THermalhydraulics (NURETH-13), Kanazawa City, Ishikawa Prefecture, Japan, September 27–October 2. 2009

[71] Pioro I, Duffey R, Dumouchel T. Hydraulic resistance of fluids flowing in channels at supercritical pressures (survey). Nuclear Engineering and Design. 2004;**231**(2):187-197

Chapter 2

Heat Transfer Correlations for Supercritical Water in Vertically Upward Tubes

Huixiong Li, Xiangfei Kong, Xianliang Lei, Qian Zhang, Changjiang Liao, Lingtong Gou and Dongfeng Sun

Abstract

Supercritical pressure water (SCW) has been widely used in many engineering fields and industries, such as fossil fuel-fired power plants, newly developed Gen-IV nuclear power plants and so forth. Heat transfer characteristics of SCW are of great importance for both design and safe operation of the related systems. Many heat transfer correlations have been developed in the history for predicting the heat transfer characteristics of SCW. However, the prediction accuracy of the existing correlations is less than satisfactory, especially in the cases with deteriorated heat transfer (DHT) because of the severe and quick variation in thermal physical properties of SCW in the vicinity of the fluids' pseudo critical point. It is very necessary to develop new correlations for the heat transfer of SCW to meet the engineering requirements for satisfactory prediction of the heat transfer behavior of SCW. In this chapter, experimental data on heat transfer of SCW are extensively collected from published literatures, and the performance of the existing heat transfer correlations for SCW are reviewed and quantitatively evaluated against the collected experimental data, and then a new heat transfer correlation for SCW with high prediction accuracy is proposed.

Keywords: supercritical water, enhanced heat transfer, deteriorated heat transfer, correlation, smooth tube

1. Introduction

Supercritical pressure fluids (SCFs) have been widely used in many engineering fields and industries, such as fossil fuel-fired power plants, newly developed Gen-IV nuclear power plants, chemical processes, and so forth. Heat transfer characteristics of SCFs are of great importance for both design and safe operation of the related systems. As one of the most popular SCFs, supercritical pressure water (SCW) has been selected as the ideal working fluid in nuclear power plants and fossil-fired power stations, and the heat transfer behavior of SCW has attracted more and more attention from both scholars and engineers in these fields [1, 2].

As well known, most of the newly designed thermal power plants are operated at supercritical pressures, and the operating parameters of future fossil-fired power plants have been suggested to raise to even higher levels to gain higher thermal efficiency, and at the same time, to obtain effective gains in reduction of the

IntechOpen

pollution products emission. For example, the U.S. Department of Energy has ever proposed a research program (AD 760), in which the operating parameters of the future fossil-fired power plants are suggested to raise to 37.9 MPa/732°C/760°C with an efficiency of up to 47%. A program similar to AD 760, i.e., the AD 700 thermal program, has also been proposed in Europe, and planned to build the ultra-supercritical steam condition at 720°C and 35 MPa in the future thermal power plants so as to obtain a cycle efficiency up to 50% [2]. There is no doubt that the heat transfer characteristics of SCW is an important issue for the thermal power plants with high level parameters. The higher the operating parameters, the higher the thermal efficiency, and as well, accordingly, the more important is the precise prediction of heat transfer characteristics of SCW. It should be noted that deteriorated heat transfer (DHT) is one kind of special heat transfer phenomena of SCW that might happen in the heat transfer tube and even lead to the burst of heat transfer tube or other failure accidents in the plants. On the other hand, the supercritical water-cooled reactors (SCWRs) has been selected as one of the six most promising reactors concepts presented at the Generation IV International Forum due to its high thermal efficiency and other features [3]. Similar to the situation in supercritical pressure thermal power plants, the heat transfer of SCW is an important issue for SCWRs as well. Furthermore, due to the possible risks in release of radioactive medium under conditions with tube or vessel failures, the heat transfer of SCW is of much more concern for the safe operation of SCWRs than that of thermal power stations, and as a result, precise knowledge about the heat transfer characteristics of SCW is one of the major tasks for the design and operation of SCWRs [3, 4]. Besides, heat transfer of SCW is of significance for other applications with SCW as the working fluid, such as SCW gasification system [5].

Two methods have been proposed by different scholars to predict the heat transfer characteristics of SCW: the look-up Table [6, 7] and the heat transfer correlations [8]. The look-up table for heat transfer of SCW is based on real experimental data. The prediction accuracy of such look-up table is pretty high; however, the applicable ranges of the operation parameters (e.g., pressures, mass fluxes, and heat fluxes) of the look-up table are limited by the experimental conditions. In the development of ultra-supercritical pressure boilers or other heat exchangers with high operation parameters, the applicability of the look-up table seems to be limited. Unlike the look-up table, heat transfer correlations have been developed on the basis on experimental data, dimensionless analysis, and theoretical analysis of the heat transfer phenomena. The applicability of heat transfer correlations are much flexible than that of look-up tables, and as a result, heat transfer correlations for SCW have found wide applications in related industries and engineering.

In the last few decades, many heat transfer correlations for SCW were proposed (see **Table 3** in the later section of this chapter). Generally, the existing heat transfer correlations could be divided into three categories [9], as listed in **Table 1**. The first type heat transfer correlation for SCW was built on the basis of the

Types	Forms	Remarks
I	$\mathrm{Nu} = C_0 \times \mathrm{Re}^{C_1} \times \mathrm{Pr}^{C_2}$	C_0, C_1, C_2, C_3, C_4 are constants; F represents thermophysical properties correction terms; **Gr*** represents the buoyancy correction terms; q^+ represents the thermal acceleration correlation terms
II	$\mathrm{Nu} = C_0 \times \mathrm{Re}^{C_1} \times \mathrm{Pr}^{C_2} \times F$	
III	$\mathrm{Nu} = C_0 \times \mathrm{Re}^{C_1} \times \mathrm{Pr}^{C_2} \times F \times (\mathrm{Gr}^*)^{C_3} \times (q^+)^{C_4}$	

Table 1.
Three types of heat transfer correlations for SCW.

Dittus-Boelter et al.'s correlation [10]. Under supercritical pressure conditions, there exists no phase change of the fluid, but the thermophysical properties of the SCW experience dramatic change in the vicinity of pseudocritical point which might have great impact on the heat transfer behavior of the SCW. In view of this fact, a few correction terms consisting of the fluid thermophysical properties, such as ρ_w/ρ_b, μ_w/μ_b, λ_w/λ_b among the others, were introduced to the first type correlations, resulting in the emerge of second type heat transfer correlations for SCW. With the deepening research on heat transfer phenomena of SCW, buoyancy effect and thermal acceleration were considered to be the main reasons for DHT of SCW and then two special correction terms, i.e., $\mathbf{Gr^*}$ and q^+ proposed by Jackson [11] and were added to the second type correlations, yielding the third type heat transfer correlations for SCW.

Pioro et al. [8, 12] have conducted a comprehensive review of the existing heat transfer correlations for SCW and argued that none of the existing correlations could predict the heat transfer of SCW accurately under all heat transfer regimes. Lei et al. [2] ever put emphasis on the deterioration heat transfer of SCW by analyzing the physical mechanism of special heat transfer behavior of the SCW and tried to build new correlation for SCW, and found that most of the existing correlations for heat transfer of SCW focused mainly on the enhanced heat transfer (EHT) regimes of SCW, and these correlations were not capable of providing satisfactory prediction accuracy when it was applied to the DHT regime of SCW, and he thus suggested that excluding DHT data in the development of heat correlations should be responsible for the distinct deviations between the calculated heat transfer values and the experimental data of SCW in the DHT regimes. Another reason for the unsatisfied prediction accuracies of the existing correlations for heat transfer of SCW might be, to some extent, attributed to old thermophysical properties data adopted in the development of some of the existing heat transfer correlations [13]. Therefore, a new and accurate correlation for heat transfer of SCW is expected to be effectively accurate under both the DHT and EHT regimes of SCW, and special emphasis should be directed onto the DHT phenomena, which is the most dangerous situation to not only thermal power plants but also to the Gen-IV SCWRs and other heat transfer equipment with SCW as the working fluid.

In this chapter, experimental data on heat transfer of SCW are collected, including both the EHT regimes and the DHT regimes. Existing heat transfer correlations for SCW are well reviewed. Detailed assessments of prediction accuracy of the existing correlations are also conducted against the collected experimental data. On the basis of the evaluation of the prediction behavior of the existing heat transfer correlations for SCW, a new correlation is proposed for the heat transfer of SCW to cover both the EHT regime and the DHT regime. The prediction capability of the new correlation is also assessed against the collected experimental data.

2. Experimental data and existing heat transfer correlations

Since 1960s, a series of experimental investigations on heat transfer of SCW have been conducted, and a large amount of experimental data have been accumulated. A comprehensive survey of experimental investigations on heat transfer correlations of SCW has been provided by Pioro et al. [4, 8, 14] and showed that most of previous experimental data for heat transfer of SCW focused primarily on the flow of SCW in vertical circular tubes [2]. In the present study, approximately 12,704 data points and 250 experimental cases about the flow and heat transfer of SCW in vertical circular tubes are retrieved and collected from the published literatures.

Table 2 outlines the main information about the experimental data collected in this study on the heat transfer of SCW flowing in vertically upward smooth tubes. It is seen that the experimental data used in the present study covers the parameters of pressures from 2.25×10^7 to 3.103×10^7 Pa, mass fluxes from 200 kg/(m^2·s) to 2500 kg/(m^2·s), heat fluxes from 1.48×10^5 to 2×10^6 W/m^2, and the tube

Author	Year	$P \times 10^6$ [Pa]	G [kg/(m^2·s)]	$q \times 10^3$ [W/m^2]	$d \times 10^{-3}$ [m]
Shitsman et al. [15]	1963	23.3/25.3	430/449	210–386	8
Bishop et al. [16]	1964	24.1/24.2	543/678	252/606	5/38
Swenson et al. [17]	1965	23/31	2150	789	9.42
Vikrev et al. [18, 19]	1964 1967	26.5	495–1400	507–1160	20.4
Styrikovich et al. [20]	1967	24	700	348–872	22
Herkenrath et al. [21]	1967	22.5/24/25	700/1000/1500	300–1410	10/20
Kondratev et al. [22]	1969	25.3	700	581	18
Ackerman et al. [23]	1970	22.75/31.03	542.52–1220	472.9–1261	9.42
Ornatsky et al. [24]	1970	25.5	1500	1810	3
Yamagata et al. [25]	1972	22.6/24.5/29.4	1120–1260	233–930	7.5/10
Lee et al. [26]	1974	24.1/24.5	376/543	252–379	16/38
Polyakov et al. [27]	1991	24.5/29.4	595/675	500/570	8
Griem et al. [28]	1996	25	500/1000	300	14
Koshizuka et al. [29]	2000	31	540/680	473	9.4
Yoshida et al. [30]	2000	24.5	376/410/1180	329/350/698	10/16
Hu et al. [31]	2001	23/26/30	600/900/1200	200–500	26
Xu et al. [32]	2004	23/25/27/30	800/1000/1200	200–600	12
Kirillov et al. [33]	2005	24–24.9	200/1500	227–884	10
Zhu et al. [34]	2009	30	600	250	26
Wang et al. [35, 36]	2010 2011	25–29	600/1200	350/660	17/19.8/26
Mokry et al. [3, 37]	2010 2011	24.0/24.1/24.2	201–500	148–335	10/38
Pan et al. [38]	2011	22.5/27/28/30	1009–1626	216–649	17
Li et al. [39, 40]	2011 2018	23/25/26	459.8–1497.5	192–1326.5	7.6
Wang et al. [41]	2012	23/25/26	449–1520.6	192–1154.3	7.6/10
Wang et al. [42]	2013	23/24/25/26	450–1500	450–1250	10
Huang et al. [43]	2013	23–25.01	631–1263	420–1102	6
Li et al. [44]	2013	23/24/25	607.5–1263	466–1102	6
Zhao et al. [45]	2014	25	1000	570/760	7.6
Gu et al. [46]	2015	23/25/26	780/1000	700/900	10
Shen et al. [47, 48]	2016	28.5	1536	468	19
Qu et al. [49]	2018	24/24.8/30	404/407/420/690	205/250/284/315	19

Table 2.
Experimental data of heat transfer of SCW collected from literatures.

diameters from 3 to 38 mm. Detailed analysis of the experimental data shows that among the 250 experimental cases collected, including both in this study, 134 cases are in the EHT regimes, while 116 cases of them are in the DHT regimes.

As well known, many heat transfer correlations have been proposed for predicting the heat transfer characteristics of SCW, and **Table 3** listed 34 of them as examples, for SCW in vertically upward tubes. It is seen from **Table 3** that many of the correlations were proposed 40 years ago and old thermophysical property data

Author	Year	Correlation
Dittus-Boelter et al. [10]	1985	$\mathrm{Nu} = 0.023\,\mathrm{Re}_b^{0.8}\mathrm{Pr}_b^{1/3}$
Shitsman et al. [50]	1957	$\mathrm{Nu} = 0.023\,\mathrm{Re}_b^{0.8}\mathrm{Pr}_{min}^{0.8}$ $\mathrm{Pr}_{min} = \mathrm{Pr}_b$ for $\mathrm{Pr}_b < \mathrm{Pr}_w$, $\mathrm{Pr}_{min} = \mathrm{Pr}_w$ for $\mathrm{Pr}_w < \mathrm{Pr}_b$
Petukhov et al. [8]	1963	$\mathrm{Nu} = \dfrac{\left[f\,\mathrm{Re}_b\mathrm{Pr}_b\left(\frac{\mu}{\mu_w}\right)^{0.11}\left(\frac{\lambda_w}{\lambda_b}\right)^{0.33}\left(\frac{\overline{C_p}}{C_{p,b}}\right)^{0.35}/8\right]}{\left[12.7\sqrt{\frac{f}{8}}\left(\mathrm{Pr}_b^{2/3}-1\right)+1.07\right]}$ $f = \left[1.82\log_{10}(\mathrm{Re}_b) - 1.64\right]^{-2}$
Domin et al. [51]	1963	$\mathrm{Nu} = \begin{cases} 0.1\,\mathrm{Re}_b^{0.66}\overline{\mathrm{Pr}}_b^{1.2}, T_w \geq 623.15\mathrm{K} \\ 0.036\,\mathrm{Re}_b^{0.8}\overline{\mathrm{Pr}}_b^{0.4}\left(\frac{\mu_w}{\mu_b}\right), 523.15\mathrm{K} < T_w < 623.15\mathrm{K} \end{cases}$
Swenson et al. [17]	1965	$\mathrm{Nu} = 0.00459\,\mathrm{Re}_w^{0.923}\overline{\mathrm{Pr}}_w^{0.613}(\rho_w/\rho_b)^{0.231}$
Krasnoschekov et al. [52]	1967	$\mathrm{Nu} = \mathrm{Nu}_0\left(\frac{\rho_w}{\rho_b}\right)^{0.3}\left(\frac{\overline{C_p}}{C_{p,b}}\right)^{n}$ $\mathrm{Nu}_0 = \dfrac{(\xi/8)\,\mathrm{Re}_b\overline{\mathrm{Pr}}}{12.7\sqrt{\xi/8}\left(\overline{\mathrm{Pr}}^{2/3}-1\right)+1.07}$ $n = \begin{cases} 0.4 & T_b < T_w < T_{pc} \text{ or } 1.2T_{pc} < T_b < T_w \\ n_1 = 0.22 + 0.18(T_w/T_{pc}) & 1 \leq (T_w/T_{pc}) \leq 2.5 \\ n_1 + (5n_1 - 2)(1 - (T_b/T_{pc})) & T_b < T_w \text{ and } 1 \leq (T_b/T_{pc}) \leq 1.2 \end{cases}$
Kondratev et al. [22]	1969	$\mathrm{Nu} = 0.02\,\mathrm{Re}_b^{0.8}$
Ornatsky et al. [24]	1970	$\mathrm{Nu} = 0.023\,\mathrm{Re}_b^{0.8}\mathrm{Pr}_{min}^{0.8}\left(\frac{\rho_w}{\rho_b}\right)^{0.3}$ $\mathrm{Pr}_{min} = \mathrm{Pr}_b$ for $\mathrm{Pr}_b < \mathrm{Pr}_w$, $\mathrm{Pr}_{min} = \mathrm{Pr}_w$ for $\mathrm{Pr}_w < \mathrm{Pr}_b$
Grass et al. [53]	1971	$\mathrm{Nu} = \dfrac{(f/8)\,\mathrm{Re}_b\mathrm{Pr}_b}{1.07 + 12.7\sqrt{f/8}\left(\mathrm{Pr}_G^{2/3}C_{p,b}/C_{p,w} - 1\right)}$ $f = \left[1.82\log_{10}(\mathrm{Re}_b) - 1.64\right]^{-2}$ $\mathrm{Pr}_G = \mathrm{Pr}_b$ for $\mathrm{Pr}_b < 0.5\mathrm{Pr}_w$, $\mathrm{Pr}_G = \mathrm{Pr}_w$ for $\mathrm{Pr}_b > 0.5\mathrm{Pr}_w$
Yamagata et al. [25]	1972	$\mathrm{Nu} = 0.0135\,\mathrm{Re}_b^{0.85}\overline{\mathrm{Pr}}_b^{0.8}F_C$ $F_C = \begin{cases} 1.0 & (T_{pc} - T_b)/(T_w - T_b) \geq 1 \\ 0.67\mathrm{Pr}_{pc}^{-0.05}\left(\overline{C_p}/C_{p,b}\right)^{n_1} & 0 \leq (T_{pc} - T_b)/(T_w - T_b) \leq 1 \\ \left(\overline{C_p}/C_{p,b}\right)^{n_1} & (T_{pc} - T_b)/(T_w - T_b) \leq 0 \end{cases}$ $n_1 = -0.77\left(1 + \frac{1}{\mathrm{Pr}_{pc}}\right) + 1.49,\ n_2 = 1.44\left(1 + \frac{1}{\mathrm{Pr}_{pc}}\right) - 0.53$
Yeroshenko et al. [54]	1981	$\mathrm{Nu} = 0.023\,\mathrm{Re}_b^{0.8}\mathrm{Pr}_b^{0.4}\left[\dfrac{2}{(0.8\psi + 0.2)^{0.5} + 1}\right]^2 F$ $\psi = 1 + \beta(T_w - T_b)$ $F = \begin{cases} \left(\frac{\overline{C_p}}{C_{p,b}}\right)^{0.28} & \text{for } \overline{C_p} > C_{p,b} \\ 1 & \text{for } \overline{C_p} \leq C_{p,b} \end{cases}$

Author	Year	Correlation
Watts-Chou et al. [55]	1982	$Nu = 0.021\,Re_b^{0.8}\overline{Pr_b}^{-0.55}\left(\frac{\rho_w}{\rho_b}\right)^{0.35}\phi$ $\phi = \begin{cases} 1 & \text{for} \quad Bu \le 10^{-5} \\ (1-3000Bu)^{0.295} & \text{for} \quad 10^{-5} < Bu \le 10^{-4} \\ (7000Bu)^{0.295} & \text{for} \quad Bu \ge 10^{-4} \end{cases}$ $Bu = \frac{\overline{Gr}}{Re^{2.7}\overline{Pr}^{0.5}}$
Petukhov et al. [56]	1983	$Nu = Re_b\overline{Pr_b}(f/8)/\left[1 + 900/Re_b + 12.7(f/8)^{0.5}\left(Pr^{2/3}-1\right)\right]$ $f = f_0(\rho_w/\rho_b)^{0.4}(\mu_w/\mu_b)^{0.2}$ $f_0 = [1.82\log_{10}(Re_b/8)]^{-2}(\rho_w/\rho_b)^{0.4}$
Kirillov et al. [41]	1990	$Nu = Nu_0\left(\frac{\rho_w}{\rho_b}\right)^{0.4}\left(\frac{\overline{C_p}}{C_{p,b}}\right)^{n}\varphi(k^*)$ $Nu_0 = Re_b\overline{Pr_b}(f/8)/\left(1 + 900/Re_b + 12.7(f/8)^{0.5}\left(\overline{Pr_b^{2/3}}-1\right)\right)$ $f = (1.82\log_{10}Re_b)^{-2}, k^* = \left(1-\frac{\rho_w}{\rho_b}\right)\frac{Gr_b}{Re_b^2}, Gr_b = \frac{g(1-\rho_w/\rho_b)D^3}{\nu_b^2}$ $n = \begin{cases} 0.4 \text{ for } \frac{T_w}{T_{pc}} < 1.0 \text{ and } \frac{T_b}{T_{pc}} > 1.2 \\ 0.22 + 0.18\frac{T_w}{T_{pc}} \text{ for } \frac{T_w}{T_{pc}} < 1.0 \text{ and } \frac{T_b}{T_{pc}} < 1.0 \\ 0.9\frac{T_b}{T_{pc}}\left(1-\frac{T_w}{T_{pc}}\right) + 1.08\frac{T_w}{T_{pc}} - 0.68 \text{ for } \frac{T_w}{T_{pc}} < 1.0 \text{ and } \frac{T_b}{T_{pc}} < 1.2 \\ 0.7 \text{ for } \overline{C_p}/C_{p,b} \ge 1.0 \quad \text{other conditions} \end{cases}$ $\varphi(k^*) = \begin{cases} 0.8 - 1.65\ln k^* - 2.75(\ln k^*)^2 - 1.74(\ln k^*)^3 \\ \quad -0.55(\ln k^*)^4 - 0.09(\ln k^*)^5 - 0.0056(\ln k^*)^6 \\ \qquad \text{for } k^* \le 0.4 \\ 1.4(k^*)^{0.37} \text{ for } k^* > 0.4 \end{cases}$
Gorban et al. [8]	1990	$Nu = 0.0059\,Re_b^{0.9}Pr_b^{-0.12}$
Razumovskiy et al. [41]	1990	$Nu = Re_b Pr_b(f/8)\left(\frac{\overline{C_p}}{C_{p,b}}\right)^{0.65}/\left[1.07 + 12.7 \times \left(\frac{f}{8}\right)^{0.5} \times \left(Pr_b^{2/3}-1\right)\right]$ $f = [1.82 \times \log_{10}(Re_b/8)]^{-2}((\rho_w/\rho_b) \times (\mu_w/\mu_b))^{0.18}$
Griem et al. [28]	1996	$Nu = 0.0169\,Re_b^{0.84}Pr_{sel}^{0.432}\omega$ $Pr_{sel} = \mu_b C_{p,sel}/\lambda_m, \ \lambda_m = 0.5(\lambda_w + \lambda_b)$ $\omega = \begin{cases} 0.82 & \text{for } h_b < 1540\text{kJ/kg} \\ 1 & \text{for } h_b > 1740\text{kJ/kg} \\ 0.82 + 9 \times 10^{-4} & \text{for } 1540\text{kJ/kg} \le h_b \le 1740\text{kJ/kg} \end{cases}$ For detailed of $C_{p,sel}$, please see [64]
Hu et al. [31]	2001	$Nu = 0.0068 \times Re_b^{0.9} \times \overline{Pr_b}^{0.63} \times (\rho_w/\rho_b)^{0.17} \times (\lambda_w/\lambda_b)^{0.29}$
Kitoh et al. [57]	2001	$Nu = 0.015\,Re_b^{0.85}Pr_b^{m}, m = 0.69 - 81000/200G^{1.2} + f_c q$ $f_c = 2.9 \times 10^{-8} + 0.11/200G^{1.2}$ for $0 \le h_b \le 1.5 \times 10^6$J/kg $f_c = -8.7 \times 10^{-8} - 0.65/200G^{1.2}$ for $1.5 \times 10^6\text{J/kg} \le h_b \le 3.3 \times 10^6$J/kg $f_c = -9.7 \times 10^{-7} - 1.3/200G^{1.2}$ for $3.3 \times 10^6\text{J/kg} \le h_b \le 4 \times 10^6$J/kg

Author	Year	Correlation
Xu et al. [32]	2001	$Nu = 0.02269\,Re_b^{0.8079}\overline{Pr}_b^{0.9213}\left(\frac{\mu_w}{\mu_b}\right)^{0.8687}\left(\frac{\rho_w}{\rho_b}\right)^{0.6638}$
Jackson [58]	2002	$Nu = 0.0183\,Re_b^{0.82}Pr_b^{0.5}\left(\frac{\rho_w}{\rho_b}\right)^{0.3}\left(\frac{\overline{C_p}}{C_{p,b}}\right)^{n}$ $n = 0.4$ for $T_b < T_w < T_{pc}, 1.2T_{pc} < T_b < T_w$ $n = 0.4 + 0.2(T_w/T_{pc} - 1)$ for $T_b < T_{pc} < T_w$ $n = 0.4 + 0.2(T_w/T_{pc} - 1)[1 - 5(T_b/T_{pc} - 1)]$ for $T_{pc} < T_b < 1.2T_{pc}$ and $T_b < T_w$
Fewster et al. [59]	2004	$Nu = 0.0183\,Re_b^{0.82}\overline{Pr}_b^{0.5}\left(\frac{\rho_w}{\rho_b}\right)^{0.3}$
Kuang et al. [60]	2008	$Nu = 0.0239\,Re_b^{0.759}Pr_b^{0.833}\left(\frac{\lambda_w}{\lambda_b}\right)^{0.0863}\left(\frac{\rho_w}{\rho_b}\right)^{0.31}\left(\frac{\mu_w}{\mu_b}\right)^{0.1}$ $\cdot(Gr^*)^{0.014}(q^*)^{-0.021}$ $Gr^* = g\beta d^4 q/\lambda\nu^2,\ q^* = \frac{q\beta}{GC_p}$
Cheng et al. [61]	2009	$Nu = 0.023\,Re_b^{0.8}Pr_b^{1/3}F,\ F = \min(F_1, F_2)$ $F_1 = 0.85 + 0.776(\pi_A \cdot 10^3)^{2.4}$ $F_2 = \frac{0.48}{(\pi_{A,pc} \cdot 10^3)^{1.55}} + 1.21(1 - \pi_A/\pi_{A,pc})$ $\pi_A = \frac{\beta_b q_w}{GC_p}$
Yu et al. [62]	2009	$Nu = 0.01378\,Re_b^{0.9078}\overline{Pr}_b^{0.6171}\left(\frac{\rho_w}{\rho_b}\right)^{0.4356}(Gr^*)^{-0.012}(q^*)^{0.0605}$ $Gr^* = g\beta d^4 q/\lambda\nu^2,\ q^* = \frac{q\beta}{GC_p}$
Bringer et al. [63]	2010	$Nu = 0.0266\,Re_X^{0.77}Pr_W^{0.55}$ $T_X = T_b\ for\ (T_{pc} - T_b)/(T_w - T_b) < 0$ $T_X = T_{pc}\ for\ 0 \le (T_{pc} - T_b)/(T_w - T_b) \le 1$ $T_X = T_w\ for\ (T_{pc} - T_b)/(T_w - T_b) > 1$
Gupta et al. [64]	2010	$Nu = 0.004\,Re_w^{0.923}\overline{Pr}_w^{0.773}\left(\frac{\rho_w}{\rho_b}\right)^{0.186}\left(\frac{\mu_w}{\mu_b}\right)^{0.366}$
Mokry et al. [3]	2010	$Nu = 0.0061\,Re_b^{0.904}\overline{Pr}_b^{0.684}\left(\frac{\rho_w}{\rho_b}\right)^{0.564}$
Wang et al. [65]	2012	$Nu = 0.0122\,Re_b^{0.752}\overline{Pr}_b^{0.8775}\left(\frac{\lambda_w}{\lambda_b}\right)^{0.0746}\left(\frac{\rho_w}{\rho_b}\right)^{0.0402}$ $\cdot(Gr^*)^{0.0157}K_V^{-0.031}$ $Gr^* = g\beta d^4 q/\lambda\nu^2, K_V = 4q_w D\beta/(Re^2\mu C_p)$
Liu et al. [66]	2012	$Nu = 0.01\,Re_b^{0.889}\overline{Pr}_b^{0.73}\left(\frac{\lambda_w}{\lambda_b}\right)^{0.24}\left(\frac{\mu_w}{\mu_b}\right)^{0.153}\left(\frac{\rho_w}{\rho_b}\right)^{0.401}\left(\frac{\overline{C_p}}{C_{p,b}}\right)^{0.014}$ $\cdot(Gr^*)^{0.007}(q^*)^{0.041}$ $Gr^* = g\beta d^4 q/\lambda\nu^2,\ q^* = \frac{q\beta}{GC_p}$
Wang et al. [42]	2013	$Nu = 0.0182\,Re_b^{0.8}Pr_b^{0.38}\left(\frac{\lambda_w}{\lambda_b}\right)^{0.39}\left(\frac{\mu_w}{\mu_b}\right)^{1.47}\left(\frac{\rho_w}{\rho_b}\right)^{0.54}$
Liao et al. [9]	2014	$Nu = 0.032089\,Re_b^{0.88832}\overline{Pr}_b^{0.71399}\left(\frac{\lambda_w}{\lambda_b}\right)^{0.03339}\left(\frac{\rho_w}{\rho_b}\right)^{0.68707}$ $\cdot(Gr^*)^{0.00417}(q^*)^{0.18212}$ $Gr^* = g\beta d^4 q/\lambda\nu^2,\ q^* = \frac{q\beta}{GC_p}$
Wang and Li et al. [67]	2014	$Nu = 0.00684\,Re_b^{0.9}\overline{Pr}_b^{0.69}\left(\frac{\lambda_w}{\lambda_b}\right)^{0.26}\left(\frac{\rho_w}{\rho_b}\right)^{0.31}$

Author	Year	Correlation
Zhao et al. [45]	2014	$\mathrm{Nu} = 0.023\,\mathrm{Re}_b^{0.8}\mathrm{Pr}_b^{1/3}F, F = \min(F_1, F_2)$ $F_1 = 0.62 + 0.06\ln(\pi_B)$ $F_2 = 11.46\ln(\pi_B)^{-1.04}$ $\pi_B = \frac{\beta_b q_w D}{\lambda_b}$

Table 3.
Existing heat transfer correlations for SCW in vertical upward tubes.

might have been adopted in the development of those correlations. Kurganov et al. [13] have pointed out that some correlations developed on the basis of the old thermophysical property standard for SCW have become quite impractical when the thermophysical property shifts to IAPWS-97 Standard. From this point of view, accurate correlations for heat transfer of SCW must be developed by using the updated properties database.

3. Assessment of the prediction performance of the existing heat transfer correlations

3.1 Assessment method

As seen in **Table 3**, the heat transfer correlations for SCW have been proposed in different years and might be developed on different basis of experimental data. As a result, the applicability of each correlation might be different. As reported by Pioro et al. [8] and Lei et al. [2], there exist distinct discrepancies between the results predicted by different correlations. It is necessary to quantitatively evaluate the prediction performance of the existing correlations.

In this part, the prediction performance of the existing heat transfer correlations are quantitatively estimated by introducing four parameters, i.e., σ_1 (mean relative deviation, MRD), σ_2 (mean absolute deviation, MAD), σ_3 (standard deviation, SD), and ρ_{xy} (correlation coefficient between the predicted values and experimental values), as defined by Eq. (1) through Eq. (5).

$$\sigma_1 = \sum_{i=1}^{n} e_i/N \tag{1}$$

$$\sigma_2 = \sum_{i=1}^{n} |e_i|/N \tag{2}$$

$$\sigma_3 = \sqrt{\sum\nolimits_{i=1}^{n}(e_i - \sigma_1)/(N-1)} \tag{3}$$

where e_i is

$$e_i = \left[\mathrm{Nu}_{\mathrm{cal}}\text{-}\mathrm{Nu}_{\mathrm{exp}}\right]/\mathrm{Nu}_{\mathrm{exp}} \tag{4}$$

$$\rho_{xy} = \frac{Cov(X,Y)}{\sqrt{D(X)}\sqrt{D(Y)}} \tag{5}$$

where $D(X)$ refers to the variance of the experimental data X, $D(Y)$ refers to the variance of the calculated results Y, and $Cov(X, Y)$ is the covariation of X and Y [9]. The closer the ρ_{xy} is to 1.0, the better the correlation is [9].

3.2 Comparison of the predicted values by using existing correlations to the experimental data

As described in the preceding section, the existing heat transfer correlations listed in **Table 3** could generally be divided into three categories (see **Table 1**). For simplicity, two representative correlations are selected from each category of the correlations.

Figure 1(a) shows the comparison of the results calculated by the first type of correlations for heat transfer of SCW to the experimental data. This type of correlations is represented by the Dittus-Boelter et al.'s correlation [10] and Gorban

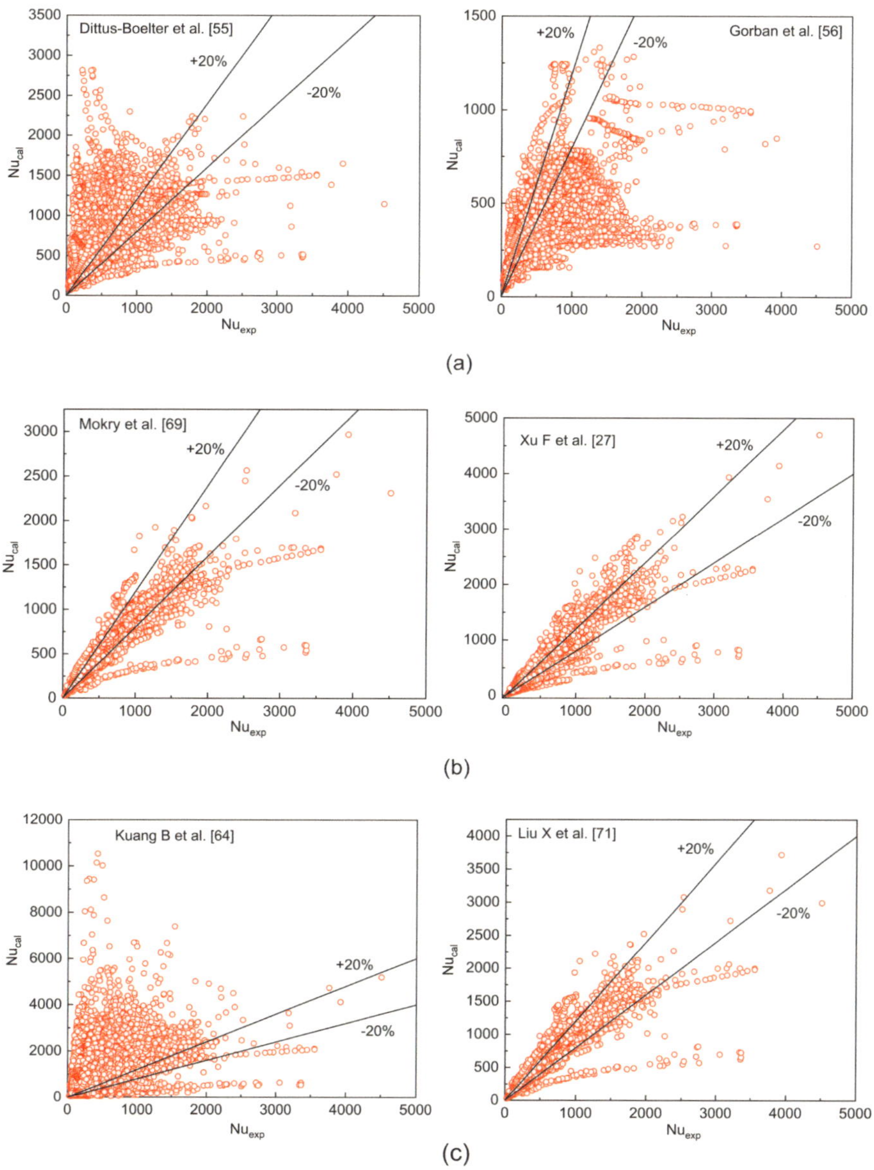

Figure 1.
Comparison of predicted values of correlations with all experimental data. (a) First type of correlations, (b) second type of correlations, (c) third type of correlations.

et al.'s correlation [8] here. It is seen from **Figure 1(a)** that most of the **Nu** values predicted by the correlations are out of ±20% error band, indicating a generally low prediction accuracy of this type of correlations. As explained in the preceding sections, no consideration in these correlations of the dramatic change of thermophysical properties of SCW in the vicinity of its pseudocritical point might be responsible for the low prediction accuracy of the correlations.

Figure 1(b) gives the representative comparison of the results calculated by the second type of correlations for SCW to the experimental data. This type of correlations is represented by the Mokry et al. [3] and Xu [32] correlations here. It can be seen from **Figure 1(b)** that most of predicted results by the correlations concentrate around the line of $\mathbf{Nu_{cal}} = \mathbf{Nu_{exp}}$, suggesting a remarkable improvement in the prediction accuracy of the correlations in comparison to that of the first type of the correlations. This improvement in the prediction accuracy might due to introducing thermophysical properties correction terms into the correlations.

Comparison of the results calculated by the third type correlations to the experimental data are illustrated in **Figure 1(c)**. Here, the third type correlations are represented by the Kuang et al.'s correlation [60] and Liu's correlation [66].

It is seen in **Figure 1(b)**, (c) that the prediction accuracy of the Mokry et al.'s correlation [3], Xu's correlation [32], and Liu's correlation [66] are in roughly the same level. It is surprising that although other two more correction factors (i.e., $\mathbf{Gr}^*$ and q^+) are introduced into the correlations, the prediction accuracy of Kuang et al.'s correlation [60] is unexpectedly worse than that of the Mokry et al.'s correlation [3] and Xu's correlation [32] in the present study. Another two correlations, i.e., Yu et al.'s correlation [62] and Liao's correlation [9], which are of forms similar to that of Kuang et al.'s correlation [60], provide similar prediction performance to that of Kuang et al.'s correlation [60] in this study. This result indicates that adding more correction factors in the correlations does not always produce better accuracy, and the correction terms added to the correlations should be selected carefully.

Figures 2, **3** depict the prediction performance of the 34 existing correlations listed in **Table 3**, under the EHT condition (experimental data from Yamagata et al. [25] is used) and the DHT condition (experimental data from Herkenrath et al. [21] is used), respectively.

It is seen from **Figure 2** that under the EHT conditions, most of existing correlations could provide relatively good prediction accuracy in the enthalpy region lower than 1600 kJ/kg and the region higher than 2800 kJ/kg of SCW (i.e., in the regions far away from the pseudocritical point). However, in the enthalpy region of 1600–2300 kJ/kg (a region around the pseudocritical point, named in lots of papers as the large specific heat region, and is hereafter abbreviated as LSHR), the predicated values of many heat transfer correlations, such as Yamagata et al.'s correlation [25], Domin et al. correlation [51] and Swenson et al. correlation [17], are much higher than the corresponding experimental values, implying low prediction capability of these correlations in the LSHR. Careful analysis of the forms of Yamagata et al. correlation [25], Domin's correlation [51], and Swenson et al.'s correlation [17] shows that only one thermophysical correction property factor is employed in the abovementioned three correlations. None of the 34 correlations could give good prediction accuracy in the whole enthalpy region of SCW. It is well known that the thermophysical properties of SCW experience dramatic change in the vicinity of pseudocritical point (i.e., in the LSHR), and with this view in mind, it is suggested that one thermophysical property correction factor might not be sufficient, and, however, proper and enough correction factors should be used in the corrections in order to capture the effect of the dramatic variation in thermophysical properties on the heat transfer of SCW in LSHR.

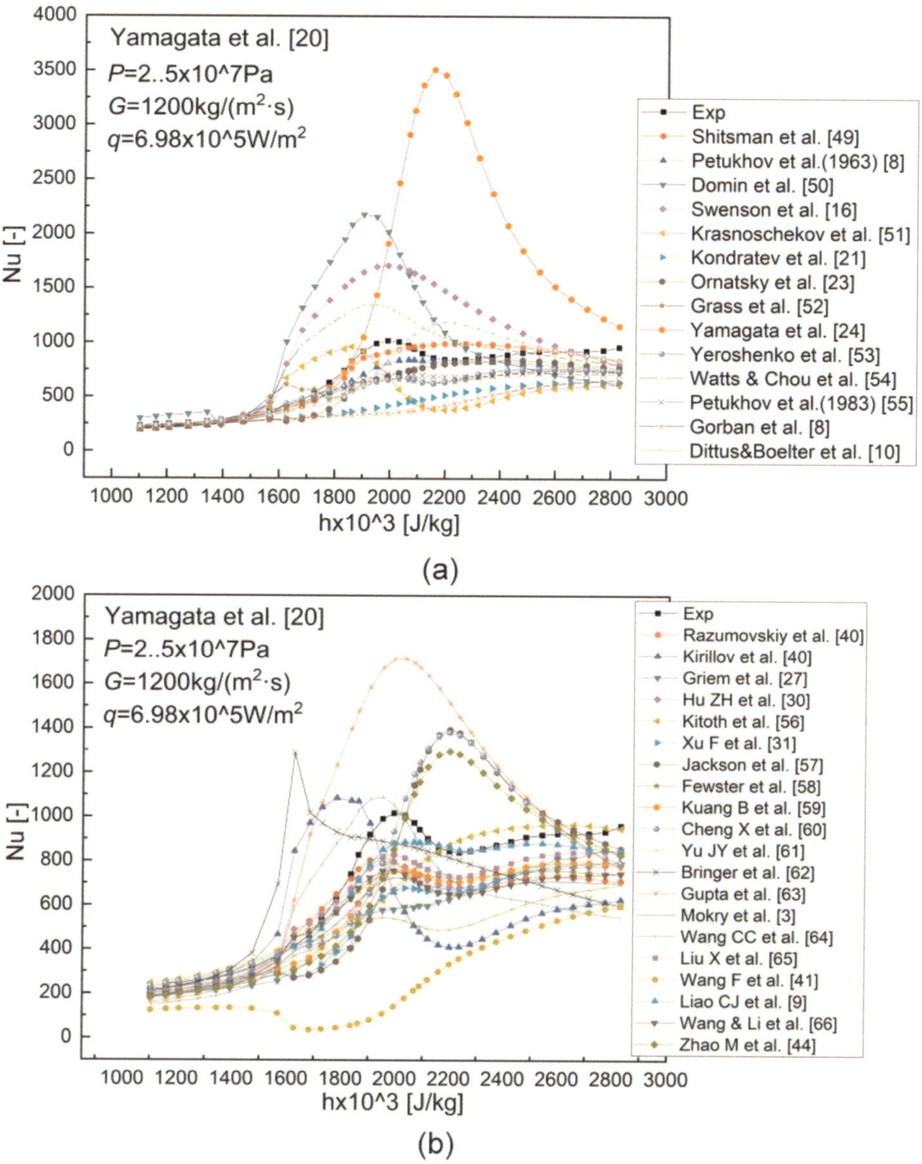

Figure 2.
Comparison of the calculated values to the corresponding experimental data (Yamagata et al. [25], EHT case). (a) Prediction results of the first 14 correlations in ***Table 3*** *(b)Prediction results of the last 20 correlations in* ***Table 3****.*

Figure 3 shows that most of the 34 heat transfer correlations for SCW could not provide high prediction accuracy under DHT conditions, except for Grass et al.'s correlation [53] and Petukhov et al.'s correlation [56]. It is seen from **Figure 3** that the predicted results of Grass et al.'s correlation [53] and Petukhov et al.'s correlation [56] agree pretty well with the corresponding experimental data under DHT conditions through the whole enthalpy region of SCW studied here. Unfortunately, as seen in **Figure 2**, the prediction results of the above two correlations do not agree satisfactorily with the corresponding experimental data under the EHT conditions, especially in the vicinity of pseudocritical point (in the LSHR). The reason for this result might be attributed to that most of the experimental data adopted in the development Petukhov et al.'s correlation [56] were under DHT regimes. Little

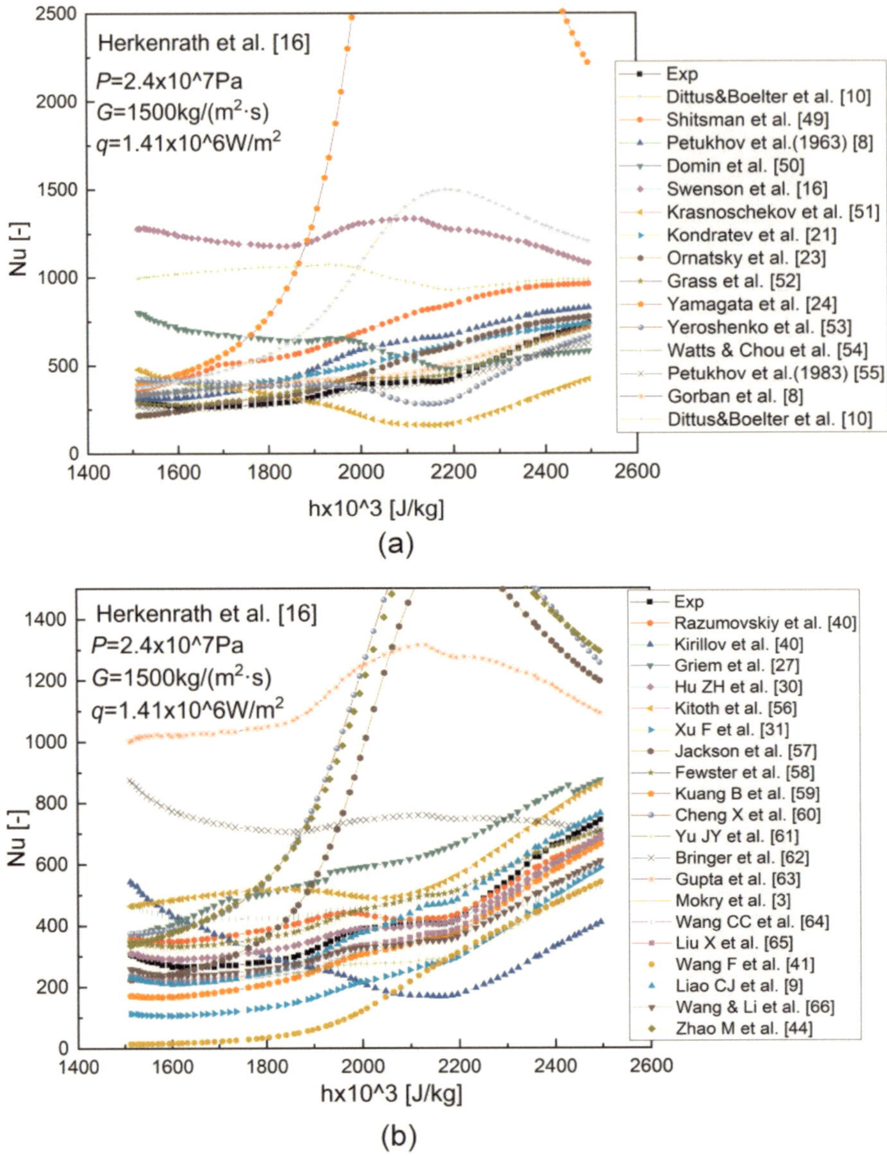

Figure 3.
Comparison of the calculated values to the corresponding experimental data (Herkenrath et al. [21], DHT case). (a) Prediction results of the first 14 correlations in ***Table 3*** *(b) Prediction results of the last 20 correlations in* ***Table 3***

information can be found about the experimental data used for developing Grass et al.'s correlation [53].

It is confusing that under EHT regimes, the prediction results of Miropolskii and Shitsman correlation [50] and Yu et al.'s correlation [62] agree relatively well with the corresponding experimental (see **Figure 2**) data, however, under DHT regimes, the prediction results of these two correlations are in quite different trend from that of the experimental data, indicating that these two correlations could not capture the variation characteristics of the experimental data under the DHT regimes (see **Figure 3**). As reported early by Pioro et al. [8] and Lei et al. [2], experiment data under DHT regime were excluded in the development of many of the heat transfer correlations, and this exclusion of the data under DHT regimes results in the low prediction accuracy under DHT regimes. It is seen again from

Author	σ_1 (%)	σ_2 (%)	σ_3 (%)	ρ_{xy}	% of data within the error bands		
					±10%	±20%	±30%
Dittus-Boelter et al. [10]	63.65	73.22	130.58	0.63	24.09	44.18	56.42
Shitsman et al. [50]	41.26	46.01	62.73	0.87	27.86	47.54	58.34
Petukhov et al. [56]	21.87	30.45	43.89	0.88	33.00	56.49	69.16
Domin et al. [51]	104.07	108.80	103.59	0.66	10.47	20.58	27.29
Swenson et al. [17]	124.64	127.80	122.01	0.67	14.40	24.35	30.17
Krasnoschekov et al. [52]	8.38	38.51	48.32	0.73	15.51	31.10	47.13
Kondratev et al. [22]	3.05	36.60	55.67	0.71	22.39	41.17	57.53
Ornatsky et al. [24]	8.51	25.29	38.31	0.90	35.26	59.41	73.23
Grass et al. [53]	12.90	25.84	35.85	0.87	33.96	58.78	71.51
Yamagata et al. [25]	224.43	226.89	461.56	0.42	15.67	28.81	37.73
Yeroshenko et al. [54]	9.98	21.85	30.41	0.90	34.45	59.81	76.28
Watts-Chou et al. [55]	101.98	106.21	113.46	0.69	14.95	26.94	33.73
Petukhov et al. [56]	2.66	15.03	20.42	0.92	46.26	74.20	87.30
Bringer et al. [63]	−8.17	33.42	45.33	0.70	20.84	39.92	56.12
Gorban et al. [8]	13.86	23.62	29.23	0.89	32.34	55.74	71.14
Razumovskiy et al. [41]	9.37	39.78	51.68	0.70	14.83	29.87	47.67
Kirillov et al. [41]	14.23	37.20	58.65	0.77	26.07	47.14	63.61
Griem et al. [28]	9.44	17.16	22.19	0.91	42.71	70.55	82.98
Hu et al. [31]	29.98	40.05	56.33	0.83	21.21	40.03	57.30
Kitoh et al. [57]	−14.18	20.45	22.89	0.93	39.15	58.84	72.20
Xu et al. [32]	52.14	69.88	133.56	0.65	22.48	41.18	56.38
Jackson [58]	16.39	24.13	29.53	0.90	31.54	55.12	73.00
Fewster et al. [59]	−0.42	16.41	22.62	0.93	45.37	68.69	83.67
Kuang et al. [60]	78.42	85.38	153.65	0.61	21.99	39.65	53.28
Cheng et al. [61]	−11.99	17.94	18.53	0.90	30.75	61.90	84.39
Yu et al. [62]	78.11	87.44	103.32	0.59	12.91	25.36	35.84
Gupta et al. [64]	104.79	111.99	111.23	0.70	13.59	25.09	33.42
Mokry et al. [3]	−5.03	13.26	16.58	0.93	46.04	78.81	93.30
Wang et al. [65]	22.92	39.26	44.58	0.80	14.26	30.07	49.74
Liu et al. [66]	6.10	14.81	19.29	0.93	45.46	75.86	88.79
Liao et al. [9]	−54.71	60.61	37.7	0.69	3.41	7.38	14.33
Wang et al. [42]	6.19	16.14	20.47	0.94	40.22	71.62	86.90
Wang and Li [67]	0.39	13.33	18.45	0.92	50.56	79.01	90.48
Zhao et al. [45]	60.03	76.49	146.38	0.59	23.65	44.20	57.80

Table 4.
Quantitative analysis of the existing heat transfer correlations.

the above results that the applicable scope of each heat transfer correlation is limited by the scope of the experimental database employed. Thus, it is of great importance to develop a new heat transfer correlation with high prediction accuracy over a wide range of experimental parameters covering the NHT regime, the EHT regime, and the DHT regime [4].

In order to gain comprehensive understanding of prediction performance of the existing correlations, quantitative analyses are conducted by employing four parameters, i.e., σ_1, σ_2, σ_3, and ρ_{xy} as defined in the previous section of this chapter, and the results of σ_1, σ_2, σ_3, and ρ_{xy} for each correlation are listed in **Table 4**.

It is seen from **Table 4** that the predicted values of most of the existing correlations falling into the ±10% error band are lower than 50%, and that falling into the ±30% error band are lower than 90%, indicating that no of these heat transfer correlations could give satisfactory predict accuracy under both DHT and EHT regimes. Generally, the prediction performance of Mokry et al.'s correlation [3]and Wang and Li's correlation [67] are compareatively the best among the 34 correlations. More than 90% of the calculated values of these two correlations fall into the ±30% error band. However, it should be noted that only about 50% of the calculated values of these two correlations fall into the ±10% error band, and the prediction accuracy the correlations needs to be improved further.

The Domin's correlation [51] and the Swenson et al.'s correlation [17] exhibit special prediction features, as seen in **Figure 2**. It is seen from **Table 4** that the predicted values of Domin's correlation [51] and Swenson et al.'s correlation [17] falling into ±30% error band are lower than 50%. This result is in accordance with that in **Figure 2**, that is, under EHT conditions, the prediction accuracy of these correlations are quite low in the LSHR of SCW although relatively good prediction performance is observed in the enthalpy region lower than 1600 kJ/kg and higher than 2800 kJ/kg of SCW. As seen in **Figure 3**, the Domin's correlation [51] and Swenson et al.'s correlation [17] could not provide prediction accuracy either under DHT conditions.

4. Development of the new heat transfer correlation

As we discussed earlier, insufficient description of the severe variation of thermophysical properties of SCW in the LSHR is one of the main reasons for the low prediction accuracy of the existing heat transfer correlations. Proper correction terms should be selected carefully to reflect the impact of thermophysical properties of SCW in LSHR on the heat transfer.

Based on the analysis of the existing heat transfer correlations, a general form of heat transfer correlations for SCW is suggested as follows:

$$\mathrm{Nu} = C_0\,\mathrm{Re}_b^{C_1}\mathrm{Pr}_b^{C_2}F_1^{k_1}\cdots F_n^{k_n} \tag{6}$$

where $F_1 \cdots F_n$ are correction terms, defined as one of the parameters such as ρ_w/ρ_b, μ_w/μ_b, λ_w/λ_b, C_p, and $\mathbf{Gr}^*$, and so on, and C_0, C_1, C_2, k_1, $\cdots$ k_2 are constant indices for the corresponding correction terms. As mentioned in the previous section, one thermophysical correction factor might not be sufficient to capture the effect of the severe variation of thermophysical properties of SCW on the heat transfer in LSHR. On the other hand, introducing too many correction terms into the correlation does not always mean a high prediction accuracy.
It was shown that strong linear correlations existed among the correction terms, and such linear correlation might limit, even reduce, the prediction performance of the correlations [68]. Based on the multicollinearity analysis as conducted in

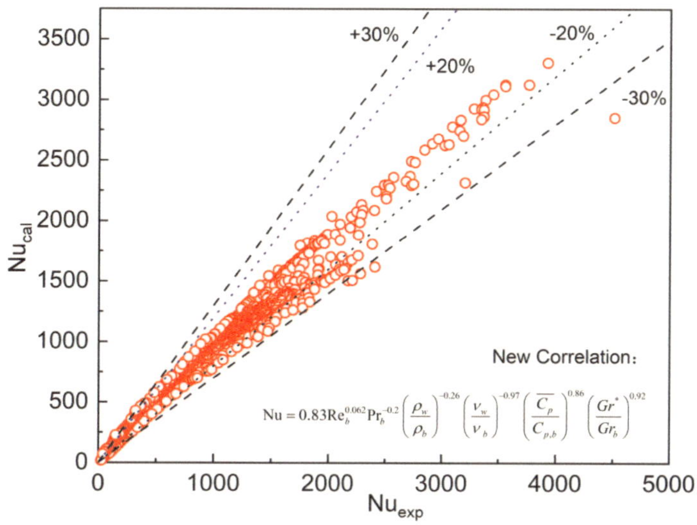

Figure 4.
Comparison between experimental and calculated **Nu** *using new heat transfer correlation.*

Correlation	σ_1(%)	σ_2(%)	σ_3(%)	ρ_{xy}	% of data within the error bands		
					±10%	±20%	±30%
New Correlation	0.77	6.71	9.23	0.991	78.57	95.18	99.39

Table 5.
Quantitative analysis of the new heat transfer correlation.

[68], four correction terms are finally adopted in the new heat transfer correlation, i.e.,

$$Nu = C_0\,Re_b^{C_1}Pr_b^{C_2}\left(\frac{\rho_w}{\rho_b}\right)_1^{k_1}\left(\frac{\nu_w}{\nu_b}\right)^{k_2}\left(\frac{\overline{C_p}}{C_{p,b}}\right)^{k_3}\left(\frac{Gr^*}{Gr_b}\right)^{k_4} \tag{7}$$

Experimental data including both the DHT regimes and the EHT regimes (as listed in **Table 2**) are adopted simultaneously in this study to develop the new heat transfer correlation, in hope to get high prediction accuracy under both the DHT and the EHT regimes. Based on the experimental database (**Table 2**), the constant indices in Eq. (7) are determined via the Levenberg–Marquardt method [69]. The final new heat transfer correlation for SCW is obtained as Eq. (8):

$$Nu = 0.83\,Re_b^{0.062}Pr_b^{-0.2}\left(\frac{\rho_w}{\rho_b}\right)^{-0.26}\left(\frac{\nu_w}{\nu_b}\right)^{-0.97}\left(\frac{\overline{C_p}}{C_{p,b}}\right)^{0.86}\left(\frac{Gr^*}{Gr_b}\right)^{0.92} \tag{8}$$

Figure 4 displays the comparison of the calculated **Nu** by using the new heat transfer correlation to the corresponding experimental results. It is seen from **Figure 4** that the relative errors of most of the calculated values are much lower than ±20%. **Table 5** gives the quantitative evaluation of the prediction performance of the new heat transfer correlation. It is clearly seen from **Table 5** that the new correlation has a mean absolute deviation of 9.23%, and 78.57% of the predicted results fall into the 10% error band, and 95.18% of the predicted results fall into the 20% error band, and 99.39% of the predicted results fall into the 30% error band, indicating that the prediction accuracy of the new correlation is better than that of the existing heat transfer correlations.

5. Conclusions

1. Comprehensive reviews of experimental data of heat transfer of SCW and heat transfer correlations for SCW are conducted in this study. Totally, 250 experimental cases and 12,704 data points about heat transfer of SCW are collected.

2. About 34 heat transfer correlations for SCW are evaluated. It is found that remarkable discrepancies exist among the results predicted by different correlations, and there is no correlation which could give satisfactory prediction accuracy under different heat transfer conditions.

3. A new heat transfer correlation is proposed for heat transfer of SCW flowing in vertically upward tubes under both DHT and EHT regimes, and the prediction performance of the new correlation is better than that of other correlations when both the DHT and EHT regimes are considered at the same time.

Acknowledgements

The authors acknowledge the support of the National Basic Research Program of China (973 Program, Grant No. 2015CB251502).

Conflict of interest

The authors declared that there is on conflict of interest.

Appendices and Nomenclature

Bu	buoyance correction factor, $\mathrm{Bu} = \mathrm{Gr}/\left(\mathrm{Re}^{2.7}\overline{\mathrm{Pr}}^{0.5}\right)$
C	constant indices
C_p	specific heat at constant pressure, J/(kg·K)
$\overline{C_p}$	average specific heat, $(h_w - h_b)/(T_w - T_b)$, J/(kg·K)
d	inner tube diameter, m
f	friction coefficient
g	acceleration due to gravity, m/s^2
Gr	Grashof number
Gr*	acceleration correction factor, $\beta g D^4 q_w/\lambda \nu^2$
$\overline{\mathrm{Gr}}$	average Grashof number, $\overline{\mathrm{Gr}_b} = g d^3 (\rho_b - \bar{\rho})\rho_b/\nu_b^2$
h	enthalpy, J/kg
k	constant indices
K_v	acceleration correction factor

Nu	Nusselt number
P	Pressure, Pa
Pr	Prandtl number
$\overline{\mathrm{Pr}}$	average Prandtl number, $\overline{C_{p,b}}\mu_b/\lambda_b$
q^*	buoyance correction factor, $q_w\beta/GC_{p,b}$
q^+	thermal acceleration parameter, $q_w(\beta/C_p)/\overline{\rho u}$
Re	Reynolds number
T	thermodynamic temperature, K
t	Celsius temperature, °C
σ_1	mean relative deviation, MRD
σ_2	mean absolute deviation, MAD
σ_3	standard deviation, SD
F	correction terms

Greek symbols

α	heat transfer coefficient, W/(m^2·K)
β	thermal expansion coefficient, K^{-1}
λ	thermal conductivity W/(m·K)
μ	dynamic viscosity, Pa·s
ν	kinematic viscosity, m^2/s
ρ	density, kg/m^3
$\overline{\rho}$	average density, $\overline{\rho} = \frac{1}{T_w - T_b}\int_{T_b}^{T_w}$, kg/m^3
ρ_{xy}	correlation coefficient

Subscripts

b	at bulk temperature experimental
cal	calculated value
exp	experimental value
pc	pseudo-critical point
w	at inner wall temperature

Abbreviations

DHT	deteriorated heat transfer
EHT	enhanced heat transfer
LSHR	the large specific heat region
MAD	mean absolute deviation
MRD	mean relative deviation
NHT	normal heat transfer
SD	standard deviation
SCFs	supercritical pressure fluids
SCW	supercritical water
SCWRs	supercritical water-cooled reactors

Author details

Huixiong Li*, Xiangfei Kong, Xianliang Lei, Qian Zhang, Changjiang Liao, Lingtong Gou and Dongfeng Sun
State Key Laboratory of Multiphase Flow in Power Engineering, Xi'an Jiaotong University, Xi'an, Shaanxi, China

*Address all correspondence to: huixiong@mail.xjtu.edu.cn

References

[1] Kelly JE. Generation IV international forum: A decade of progress through international cooperation. Progress in Nuclear Energy. 2014;**77**:240-246. DOI: 10.1016/j.pnucene.2014.02.010

[2] Lei X, Guo Y, Zhang W, Li H, Li L. Development of heat transfer correlation for supercritical water in vertical upward tubes. Heat Transfer Engineering. 2018;**40**:652-666. DOI: 10.1080/01457632.2018.1436656

[3] Mokry S, Pioro I, Kirillov P, Gospodinov Y. Supercritical-water heat transfer in a vertical bare tube. Nuclear Engineering and Design. 2010;**240**: 568-576. DOI: 10.1016/j.nucengdes.2009.09.003

[4] Pioro IL, Duffey RB. Experimental heat transfer in supercritical water flowing inside channels (survey). Nuclear Engineering and Design. 2005; **235**:2407-2430. DOI: 10.1016/j.nucengdes.2005.05.034

[5] Zhang D, Guo L, Zhao J, Jin H, Cao W, Wang R, et al. Kinetics study for sodium transformation in supercritical water gasification of Zhundong coal. International Journal of Hydrogen Energy. 2018;**43**:13869-13878. DOI: 10.1016/j.ijhydene.2017.12.011

[6] Loewenberg MF, Laurien E, Class AG, Schulenberg T. Supercritical water heat transfer in vertical tubes: A look-up table. Progress in Nuclear Energy. 2008;**50**(2):532-538. DOI: 10.1016/j.pnucene.2007.11.037

[7] Zahlan H, Tavoularis S, Groeneveld DC. A look-up table for trans-critical heat transfer in water-cooled tubes. Nuclear Engineering and Design. 2015;**285**:109-125. DOI: 10.1016/j.nucengdes.2014.12.027

[8] Pioro IL, Khartabil HF, Duffey RB. Heat transfer to supercritical fluids flowing in channels—Empirical correlations (survey). Nuclear Engineering and Design. 2004;**230**: 69-91. DOI: 10.1016/j.nucengdes.2003.10.010

[9] Liao C. Investigation on heat transfer correlations and the criterion for the onset of heat transfer deterioration of supercritical pressure water in vertical upward pipes [master]. Xi'an: Xi'an Jiaotong University; 2014

[10] Dittus FW, Boelter LMK. Heat Transfer in Automobile Radiator of the Tubular Type. Vol. 2. Berkeley, Calif: University of California Press; 1930. pp. 3-22. DOI: doi.org/10.1016/0735-1933(85)90003-X

[11] Jackson JD. Fluid flow and convective heat transfer to fluids at supercritical pressure. Nuclear Engineering and Design. 2013;**264**: 24-40. DOI: 10.1016/j.nucengdes.2012.09.040

[12] Pioro IL, Duffey RB. Heat Transfer and Hydraulic Resistance at Supercritical Pressures in Power-Engineering Applications. New York: ASME Press; 2007. DOI: 10.1115/1.802523

[13] Kurganov VA, Zeigarnik YA, Maslakova IV. Heat transfer and hydraulic resistance of supercritical pressure coolants: Part III: Generalized description of SCP fluids normal heat transfer, empirical calculating correlations, integral method of theoretical calculations. International Journal of Heat and Mass Transfer. 2013;**67**:535-547. DOI: 10.1016/j.ijheatmasstransfer.2013.07.056

[14] Pioro IL, Duffey RB, Dumouchel TJ. Hydraulic resistance of fluids flowing in channels at supercritical pressures (survey). Nuclear Engineering and

Design. 2004;**231**:187-197. DOI: 10.1016/j.nucengdes.2004.03.001

[15] Shitsman ME. Impairment of the heat transmission at supercritical pressures(heat transfer process examined during forced motion of water at supercritical pressures). High Temperature Science. 1963;**1**:237-244

[16] Bishop A, Sandberg R. Forced-convection heat transfer to water at near-critical temperatures and supercritical pressures. In: L, Westinghouse Electric Corp. Tong, Pittsburgh, Pa: Atomic Power Div; 1964

[17] Swenson HS, Carver JR, Kakarala Cd. Heat transfer to supercritical water in smooth-bore tubes. Journal of Heat Transfer. 1965;**87**:477-483. DOI: 10.1115/1.3689139

[18] Vikrev YV, Lokshin VJT. An experimental study of temperature conditions in horizontal steam generating tubes at supercritical pressures. Teploenergetika. 1964;**11**: 79-82

[19] Vikhrev YV, Barulin YD, Kon'Kov AS. A study of heat transfer in vertical tubes at supercritical pressures. Thermal Engineering. 1967;**14**:116-119

[20] Styrikovich MA, Margulova TK, Miropol'Skii ZL. Problems in the development of designs of supercritical boilers. Teploenergetika. 1967;**14**(6):4-7

[21] Herkenrath H, Mörk-Mörkenstein P, Jung U, Weckermann F. Wärmeübergang an Wasser bei Erzwungener Strömung im Druckbereich von 140 bis 250 bar, EURATOM, Berlin (1967). EUR. 1967; **3658**:132

[22] Kondratev N. Heat transfer and hydraulic resistance with supercritical water flowing in tubes. Thermal Engineering. 1969;**16**:73

[23] Ackerman JW. Pseudoboiling heat transfer to supercritical pressure water in smooth and ribbed tubes. Journal of Heat Transfer. 1970;**92**:69. DOI: 10.1115/1.3449698

[24] Ornatsky AP, Glushchenko LP, Siomin ET, et al. The research of temperature conditions of small diameter parallel tubes cooled by water under supercritical pressures. Proceedings of the 4th International Heat Transfer Conference. Vol. VI. Paris-Versailles, France: Elsevier Publishing Company; 1970. Paper No. B 8.11

[25] Yamagata K, Nishikawa K, Hasegawa S, Fujii T, Yoshida S. Forced convective heat transfer to supercritical water flowing in tubes. International Journal of Heat and Mass Transfer. 1972; **15**:2575-2593. DOI: 10.1016/0017-9310(72)90148-2

[26] Lee RA, Haller KH. Supercritical water heat transfer developments and applications. Proceedings of the 5th International Heat Transfer Conference. Vol. IV. Tokyo, Japan; 3–7 September 1974. pp. 335-339. Paper No. B7.7

[27] A. Polyakov: Heat transfer under supercritical pressures, in: Advances in Heat Transfer, Vol. 21, Elsevier, 1991. pp. 1–53. DOI: 10.1016/S0065-2717(08)70333-2

[28] Griem H. A new procedure for the prediction of forced convection heat transfer at near- and supercritical pressure. Heat and Mass Transfer. 1996; **31**:301-305. DOI: 10.1007/BF02184042

[29] Koshizuka S, Oka Y. Computational analysis of deterioration phenomena and thermal-hydraulic design of SCR. In: Proc. Proceedings of the First International Symposium on Supercritical Water-Cooled Reactors, Design And Technology

[30] Yoshida S. Heat transfer to supercritical pressure fluids flaming in tubes. Proc. Proc. SCR-2000. pp. 72–78

[31] Hu Z. Heat transfer characteristics of vertical upflow and inclined tubes in the supercritical pressure and near-critical region [doctor]. Xi'an: Xi'an Jiaotong University; 2001

[32] Xu F. Study of water flow and heat transfer characteristics through pipes under supercritical pressure [doctor]. Xi'an: Xi'an Jiaotong University; 2004

[33] Kirillov P, Pometko R, Smirnov A, Grabezhnaia V, Pioro I, Duffey R. Experimental study on heat transfer to supercritical water flowing in 1-and 4-m-long vertical tubes, Proc. Proceedings of GLOBAL 2005, Oct 9–13, 2005, Tsukuba, Japan; 2005. pp. 9-13

[34] Zhu X, Bi Q, Yang D, Chen T. An investigation on heat transfer characteristics of different pressure steam-water in vertical upward tube. Nuclear Engineering and Design. 2009; **239**:381-388. DOI: 10.1016/j. nucengdes.2008.10.026

[35] Wang J. Investigation on heat transfer and frictional pressure drop at low mass flux in the water-wall tubes of supercritical boiler [doctor]. Xi'an: Xi'an Jiaotong University; 2010

[36] Wang J, Li H, Yu S, Chen T. Investigation on the characteristics and mechanisms of unusual heat transfer of supercritical pressure water in vertically-upward tubes. International Journal of Heat and Mass Transfer. 2011; **54**:1950-1958. DOI: 10.1016/j. ijheatmasstransfer.2011.01.008

[37] Mokry S, Pioro I, Farah A, King K, Gupta S, Peiman W, et al. Development of supercritical water heat-transfer correlation for vertical bare tubes. Nuclear Engineering and Design. 2011; **241**:1126-1136. DOI: 10.1016/j. nucengdes.2010.06.012

[38] Pan J, Yang D, Dong Z, Zhu T, Bi Q. Experimental investigation on heat transfer characteristics of water in vertical upward tube under supercritical pressure. Nuclear Power Engineering. 2011;**32**:75-80

[39] Li H, Yang Y, Gu H, Zhao M, Lu D. Experimental study of supercritical water flow and heat transfer in vertical tube. Proc. Progress Report on China Nuclear Science and Technology. pp. 874-880

[40] Li H, Zhao M, Hu Z, Zhang Y, Wang F. Experimental study of supercritical water heat transfer deteriorations in different channels. Annals of Nuclear Energy. 2018;**119**: 240-256. DOI: 10.1016/j. anucene.2018.05.009

[41] Wang L. Research on heat transfer characteristics of supercritical water in the vertical tube [doctor]. Shanghai: Shanghai Jiaotong University; 2012

[42] Wang F, Yang J, Gu H, Zhao M, Li H. D. Lu: Experimental research on heat transfer performance of supercritical water in vertical tube, atomic energy. Science and Technology. 2013;**47**:933-939

[43] Huang Z, Li Y, Zeng X, Yan X, Xiao Z. Experimental and numerical simulation of supercritical water heat transfer in vertical upward circular tube. Atomic Energy Science Technology. 2012;**46**:799-803

[44] Li Y, Zeng X, Huang Z, Yan X, Huang Y, Xiao Z. Experimental study on heat transfer of supercritical water in simple channels. Nuclear Power Engineering. 2013;**34**:101-107

[45] Zhao M, Gu H, Cheng X. Experimental study on heat transfer of

supercritical water flowing downward in circular tubes. Annals of Nuclear Energy. 2014;**63**:339-349. DOI: 10.1016/j.anucene.2013.07.003

[46] Gu H, Zhao M, Cheng X. Experimental studies on heat transfer to supercritical water in circular tubes at high heat fluxes. Experimental Thermal and Fluid Science. 2015;**65**:22-32. DOI: 10.1016/j.expthermflusci.2015.03.001

[47] Shen Z, Yang D, Xie H, Nie X, Liu W, Wang S. Flow and heat transfer characteristics of high-pressure water flowing in a vertical upward smooth tube at low mass flux conditions. Applied Thermal Engineering. 2016;**102**: 391-401. DOI: 10.1016/j.applthermaleng.2016.03.150

[48] Shen Z, Yang D, Chen G, Xiao F. Heat transfer characteristics of high temperature and pressure water in vertical downward tube. CIESC Jorunal. 2013;**64**:2386-2393

[49] Qu M, Liang Z, Zhao Y, Wan L, Yang D. Experimental and numerical investigation on heat transfer characteristics of ultra-supercritical water in smooth tube with low mass flux. Journal of Xi'an Jiaotong University. 2018;**52**:52-59

[50] Miropolskii L, Shitsman M. Heat transfer to water and steam at variable specific heat (in near-critical region). Soviet Physics-Technical Physics. 1957; **2**:2196-2208

[51] Domin G. Wärmeübergang in Kritischen Und Berkritischen Bereichen Von Wasser in Rohren. Brennst-Wäirme-Kraft. 1963;**15**:527-532

[52] Krasyakova LY, Raykin YM, Belyakov II. Investigation of temperature regime of heated tubes at supercritical pressure. Soviet Energy Technology. 1967;(1):1-4

[53] Grass G, Herkenrath H, Hufschmidt W. Anwendung des Prandtlschen Grenzschichtmodells auf den Wärmeübergang an Flüssigkeiten mit stark temperaturabhängigen Stoffeigenschaften bei erzwungener Strömung. Wärme-und Stoffübertragung. 1971;**4**:113-119

[54] Yeroshenko V, Yaskin LJC. Applicability of various correlations for the prediction of turbulent heat transfer of supercritical helium, 21 1981. pp. 94-96. DOI: 10.1016/0011-2275(81)90055-2

[55] Watts MJ, Chou CT. Mixed convection heat transfer to supercritical pressure water. Proceedings of the 7th International Heat Transfer Conference. Vol. 3. Munchen, Germany: 1982. pp. 495-500. Paper No. 6–10

[56] Petukhov BS, Kurganov VA, Ankudinov VB. Heat transfer and flow resistance in the turbulent pipe flow of a fluid with near-critical state parameters. High Temperature Science. 1983;**21**: 92-100

[57] Kitoh K, Koshizuka S, Oka Y. Refinement of transient criteria and safety analysis for a high-temperature reactor cooled by supercritical water. Nuclear Technology. 2001;**135**:252-264. DOI: 10.13182/NT01-A3220

[58] Jackson JD. Consideration of the heat transfer properties of supercritical pressure water in connection with the cooling of advanced nuclear reactors. In: Proceedings of the 13th Pacific Basin Nuclear Conference. Shenzhen City, China; 21-25 October 2002. (Abstracts)

[59] Fewster J, Jackson JD. Experiments on supercritical pressure convective heat transfer having relevance to SPWR Proceedings of ICAPP'04. Pittsburgh, PA, USA; 13–17 June 2004. (Paper 4342)

[60] Kuang B, Zhang Y, Cheng X. A new, wide-ranged heat transfer correlation of water at supercritical pressures in vertical upward ducts[C]//Seventh

International Topical Meeting on Nuclear Reactor Thermal Hydraulics, Operation and Safety (NUTHOS-7). Seoul, South Korea; Oct. 2008. pp. 5-9

[61] Cheng X, Yang Y, Huang SF. A simplified method for heat transfer prediction of supercritical fluids in circular tubes. Annals of Nuclear Energy. 2009;**36**(8):1120-1128. DOI: 10.1016/j.anucene.2009.04.016

[62] Yu J, Jia B, Wu D, Wang D. Optimization of heat transfer coefficient correlation at supercritical pressure using genetic algorithms. Heat and Mass Transfer. 2009;**45**:757-766. DOI: 10.1007/S00231-008-0475-4

[63] Bringer RP,Smith JM. Heat transfer in the critical region. Vol. 3. 1957. pp. 49-55. DOI: 10.1002/aic.690030110

[64] Gupta S, Farah A, King K, Mokry S, Pioro I. Developing new heat-transfer correlation for supercritical-water flow in vertical bare tubes, Proc. 18th International Conference on Nuclear EngineeringAmerican Society of Mechanical Engineers. 2010. pp. 809-817. DOI: 10.1115/ICONE18-30024

[65] Wang C. Heat transfer analysis and the evaluation of the correlations for supercritical pressure water in vertical pipes [master]. Xi'an: Xi'an Jiaotong University; 2012

[66] Liu X. Heat transfer of water under supercritical pressures in vertical upward tubes [doctor]. Shanghai: Shanghai Jiaotong University; 2012

[67] Wang C, Li H. Evaluation of the heat transfer correlations for supercritical pressure water in vertical tubes. Heat Transfer Engineering. 2014; **35**:685-692. DOI: 10.1080/01457632.2013.837756

[68] Kong X, Li H, Liao C, Lei X, Zhang Q. A new correlation for heat transfer coefficient prediction of supercritical pressure water flowing in vertical upward tubes. Proc. ASME 2016 Heat Transfer Summer Conference, HT 2016, collocated with the ASME 2016 Fluids Engineering Division Summer Meeting and the ASME 2016 14th International Conference on Nanochannels, Microchannels, and Minichannels, July 10, 2016 - July 14, 2016American Society of Mechanical Engineers, p. Heat Transfer Division. DOI: 10.1115/HT2016-7304

[69] Pujol J. The solution of nonlinear inverse problems and the Levenberg-Marquardt method. Geophysics. 2007; **72**:W1-W16. DOI: 10.1190/1.2732552

Chapter 3

Heat Transfer Correlations of Supercritical Fluids

Fangbo Li, Binbin Pei and Bofeng Bai

Abstract

The drastic changes of thermophysical properties in the pseudo-critical region of supercritical fluids bring very big challenges to the traditional Dittus-Boelter-type heat transfer correlations. In this chapter, we will talk about the principles and applications of two kinds of heat transfer correlations of supercritical fluids: the empirical type and the semiempirical type. For the empirical correlations, the modification methods taking into account the variable properties and body force effects will be introduced. We will focus on the proposal of nondimensional parameters describing the buoyancy effect and flow acceleration; while for the semiempirical ones, we mainly talk about the new kind of correlation which is based on the momentum and energy conservations in the mixed convective flow.

Keywords: heat transfer correlation, empirical type, semiempirical type, buoyancy effect, flow acceleration

1. Introduction

Establishing an accurate heat transfer correlation of supercritical fluid is the prerequisite of the design of the apparatus and devices, such as the supercritical boilers, generation IV reactors, and heat exchangers [1–4]. However, the drastic change of the thermodynamic properties in the pseudo-critical region and the complex mechanism of turbulent heat transfer lead to difficulties in predicting the heat transfer performance accurately. There are three distinguished heat transfer regimes for supercritical fluid: normal heat transfer (normal HT), improved heat transfer (improved HT), and deteriorated heat transfer (deteriorated HT). The normal HT regime is characterized by the heat transfer coefficients similar with those predicted by the Dittus-Boelter (DB) correlation at subcritical condition; the improved HT regime can be characterized by the higher heat transfer coefficients and lower wall temperatures compared with the normal HT; and the deteriorated HT regime is accompanied with lower heat transfer coefficients than the normal HT; hence, the wall temperature peak can be observed. **Figure 1** illustrates the heat transfer performances of supercritical CO_2 under different mass fluxes and heat fluxes [5] of the upward flow in a circular tube with an inner diameter of 6.32 mm and heated length of 2.65 m. At relatively lower heat flux/mass flux ratio, the heat transfer will be obviously improved due to the large specific heat in the

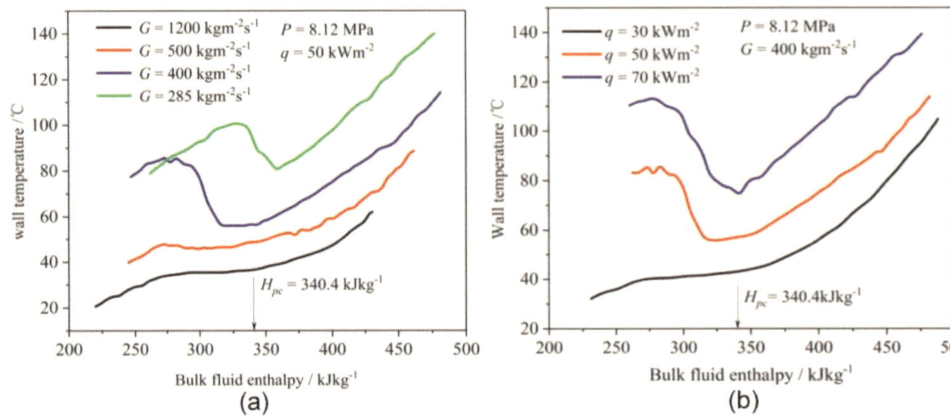

Figure 1.
Heat transfer performances of supercritical CO_2 of the upward flow under different (a) mass fluxes and (b) heat fluxes in a circular tube with an inner diameter of 6.32 mm and heated length of 2.65 m. Symbols: G, mass fluxes (kg m^{-2} s^{-1}); P, pressure (MPa); q, wall heat flux (kW m^{-2}); H, enthalpy (kJ kg^{-1}). Data in this figure come from Ref. [1].

pseudo-critical region. The heat transfer coefficients will be higher than the ones calculated by the Dittus-Boelter correlation. As the heat flux/mass flux ratio increases, the mixing convective heat transfer occurs, and the wall temperature peak caused by the heat transfer deterioration can be observed. The experiment conducted by Kurganov and Kaptil'ny [6] offered an insight into the mechanism of heat transfer deterioration, the M-shaped velocity profile was observed, and it was considered to be closely associated with the flattened velocity gradient, reduced turbulent shear stress, and turbulent kinetic energy due to the body force (also known as the buoyancy effect). Bae et al. [7, 8] obtained the budgets of the turbulent kinetic energy in the deteriorated heat transfer by using direct numerical simulation, and the mechanism of buoyancy and flow acceleration were analyzed.

The heat transfer correlations are established on the basis of heat transfer mechanism, and a vast majority of correlations are intended for use only for normal HT and improved HT regimes. They can be categorized into two types: the empirical type and the semiempirical type. In the empirical correlations, the correction terms composed of different thermophysical properties (such as the density, thermal conductivity, specific heat, and viscosity) are introduced to the heat transfer correlation of the constant-property fluid (i.e., DB correlation) [9]. However, the performance of the empirical correlations deteriorates with the increasing of the heat flux/mass flux ratio; thus, the empirical correlations considering the nondimensional numbers which reflect the buoyancy and thermally induced flow acceleration effects have also been proposed [10]. It is shown that the predictions on the mixing convection cases can be improved by this method. In order to further improve the performances of the empirical correlations under the strong buoyancy effect, some researchers tried to establish the semiempirical correlations based on the theoretical analysis approaches [11, 12]. The existing semiempirical correlations mainly aim to derive the qualitative relationship between the heat transfer impairment and turbulent shear stress reduction; the coefficients appeared in the correlation are obtained by fitting the experimental data.

In this chapter we mainly talk about the heat transfer correlations of supercritical fluid. Both the empirical type and the semiempirical type will be introduced here. We mainly focus on the principles, applications, and comparisons of these correlations.

2. Empirical heat transfer correlations

For the improved HT and normal HT regimes, abundant empirical correlations have been proposed, and most of them can give satisfying predictions [13–15]. Morky et al. [9] and Gupta et al. [16] proposed the correlations based on the Buckingham Π theorem, and the comparison with the experimental results confirmed their capability on the calculating the heat transfer improvement regime. However, as the increasing of the heat flux, the effect of the mixing convection increases. The fluid near the heated wall is accelerated under the effect of the buoyancy force and the flow acceleration, leading to the attenuation of heat transfer [17]. On the basis of this, researchers tried to modify the correlations by considering the nondimensional numbers reflecting the buoyancy force and the flow acceleration. Bae and Kim [10] and Komita et al. [18] analyzed the momentum balance under the mixed convection case and proposed a nondimensional parameter Bu to describe the effect of the buoyancy on the shear stress; then, they proposed a new function to describe the relationship between the Nusselt numbers and Bu based on the experimental data. This heat transfer correlation takes the form of the piecewise function divided by the value of the Bu number. Deev et al. [19] discussed the heat transfer of supercritical water in the channel. Two nondimensional criteria considering the effects of the viscous force and inertial force on heat transfer were proposed, and the weight constant treating the superposition between the forced and natural convection was introduced. Such practices were also conducted by Cheng et al. [20], Yu et al. [21], and Kuang et al. [22]. In their work, the dimensionless numbers such as the buoyancy number and the acceleration number were proposed to correct the deviation of the real heat transfer from that of the normal HT regime. **Figure 2** shows the approach establishing the empirical correlations. The dimensionless groups are firstly chosen, and the coefficients in the correlation can be obtained by the linear regression analysis. Through variable transformations the dimensionless groups are reduced, and the final form can be obtained. Some typical empirical correlations are listed in **Table 1**.

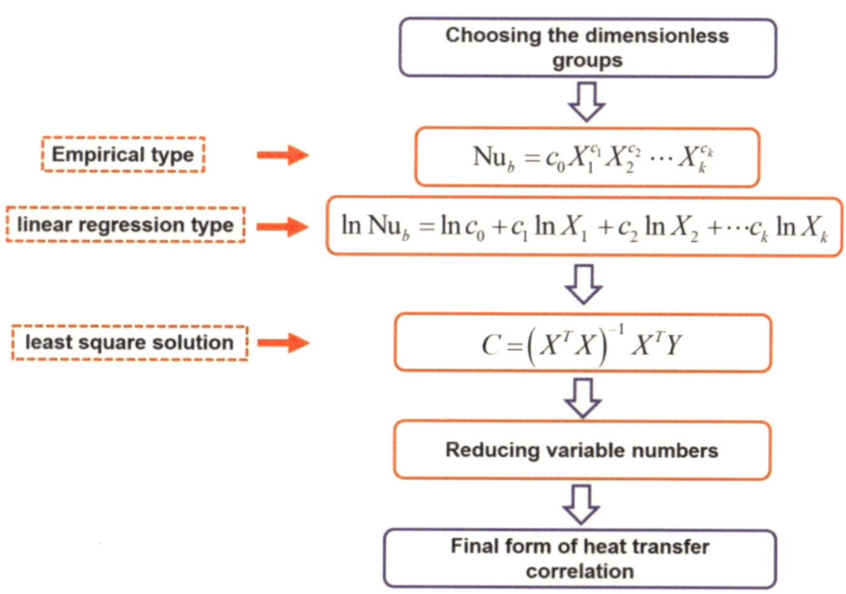

Figure 2.
The approach deriving the empirical correlations.

Ref.	Fluid	Parameters range	Correlation
Mokry et al. [9]	H_2O	P = 22.8–29.4 MPa G = 200–1500 kg m^{-2} s^{-1} q = 70–1250 kW m^{-2} d = 3–38 mm	$\mathrm{Nu}_b = 0.0061\,\mathrm{Re}_b^{0.914}\overline{\mathrm{Pr}}_b^{0.654}(\rho_w/\rho_b)^{0.518}$ $\overline{\mathrm{Pr}}_b = \frac{\overline{c_p}\mu_b}{\lambda_b}, \overline{c_p} = \frac{H_w-H_b}{T_w-T_b}$
Gupta et al. [16]	H_2O	P = 24 MPa G = 200–1500 kg m^{-2} s^{-1} q = 70–1250 kW m^{-2}	$\mathrm{Nu}_w = 0.004\,\mathrm{Re}_w^{\,0.923}\overline{\mathrm{Pr}}_w^{0.773}$ $(\mu_w/\mu_b)^{0.366}(\rho_w/\rho_b)^{0.186}$
Jackson [23]	H_2O	After review of the existing literatures and data	$\mathrm{Nu}_b = 0.0183\,\mathrm{Re}_b^{\,0.82}\mathrm{Pr}_b^{0.5}(\rho_w/\rho_b)^{0.3}(\overline{c_p}/c_{pb})^n$ where the exponent n is taken as i. $n = 0.4 + 0.2(T_w/T_{pc} - 1)(1 - 5(T_b/T_{pc} - 1))$ for $T_{pc} < T_b < 1.2T_{pc}$; ii. $n = 0.4 + 0.2(T_w/T_{pc} - 1)$ for $T_b < T_{pc} < T_w$; iii. n = 0.4 for $T_b < T_w < T_{pc}$ and $1.2T_{pc} < T_b < T_w$.
Bae and Kim [10]	CO_2	P = 7.75–8.55 MPa G < 1200 kg m^{-2} s^{-1} q_w < 150 kW m^{-2}	$\mathrm{Nu}_b = \mathrm{Nu}_{\mathrm{varp}}f(\mathrm{Bu})\mathrm{Nu}_{\mathrm{varp}} = 0.021\,\mathrm{Re}_b^{0.82}\mathrm{Pr}_b^{0.5}(\rho_w/\rho_b)^{0.3}(\overline{c_p}/c_{pb})^n$ i. For $5 \times 10^{-8} < \mathrm{Bu} < 7 \times 10^{-7}, f(\mathrm{Bu}) = (1 + 10^8\mathrm{Bu})^{0.032}$ ii. For $7 \times 10^{-7} < \mathrm{Bu} < 1 \times 10^{-6}, f(\mathrm{Bu}) = 0.0185\mathrm{Bu}^{-0.435}$ iii. $1 \times 10^{-6} < \mathrm{Bu} < 1 \times 10^{-5}, f(\mathrm{Bu}) = 0.75$ iv. For $1 \times 10^{-5} < \mathrm{Bu} < 3 \times 10^{-5} f(\mathrm{Bu}) = 0.0119\mathrm{Bu}^{-0.36}$ v. For $3 \times 10^{-5} < \mathrm{Bu} < 1 \times 10^{-4} f(\mathrm{Bu}) = 32.4\mathrm{Bu}^{0.4}$
Cheng et al. [20]	H_2O	d = 10, 20 mm P = 22.5–25 MPa G < 3500 kg m^{-2} s^{-1} q_w < 2000 kW m^{-2} T_b = 300–450°C	$\mathrm{Nu_b} = 0.023\,\mathrm{Re}_b^{0.8}\mathrm{Pr}_b^{1/3}F$ $F = \min(F_1, F_2)$ $F_1 = 0.85 + 0.776(10^3\pi_A)^{2.4}$ $F_2 = \frac{0.48}{(10^3\pi_{A,pc})^{1.55}} + 1.21\left(1 - \frac{\pi_A}{\pi_{A,pc}}\right)$ $\pi_A = \frac{\beta}{c_p}\frac{q}{G}$
Deev et al. [19]	H_2O	d = 2.67–12.0 mm P = 22.6–26 MPa G < 2000 kg m^{-2} s^{-1} q_w = 200–2250 kW m^{-2}	$\mathrm{Nu}_b = 0.023\,\mathrm{Re}_b^{0.8}\mathrm{Pr}_b^{0.4}(\rho_w/\rho_b)^{0.25}(\overline{c_p}/c_{pb})^n YY = (1-\xi)Y_1 + \xi Y_2$ $Y_1 = 1 + a_1\exp(b_1K_h^{\,2} + c_1K_h)$ $Y_2 = 1 + a_2\exp(b_2K_h^{\,2} + c_2K_h)$ $\zeta = \exp(-0.5(K_{Am})^2)$ $K_h = K_{Am}(H_b - H_{pc})/H_{pc}$

Ref.	Fluid	Parameters range	Correlation
			$a_1 = 1.5, b_1 = -30, c_1 = 3$ $a_2 = -0.15, b_2 = -125, c_2 = -25$ $K_{Am} = (10^{-5}\mathrm{Re}_{pc})^{0.5}/(10^3 K_{qm})$ $K_{qm} = q_w\beta_{pc}/(Gc_{p,pc})$
Yu et al. [21]	H_2O	After review of the existing literatures and data	$\mathrm{Nu}_b = 0.01378\,\mathrm{Re}_b^{0.9078}\overline{\mathrm{Pr}}_b^{0.6171}$ $(\rho_w/\rho_b)^{0.4356}(\mathrm{Gr}^*)^{-0.012}(q^+)^{0.0605}$ $\overline{\mathrm{Pr}}_b = \frac{\overline{c_p}\mu_b}{\lambda_b}, \overline{c_p} = \frac{H_w - H_b}{T_w - T_b},$ $\mathrm{Gr}^* = \frac{g\beta d^4 q_w}{\lambda_b \nu_b^2}, q^+ = \frac{\beta q_w}{G\overline{c_p}}$
Kuang et al. [22]	H_2O	After review of the existing literatures and data	$\mathrm{Nu}_b = 0.0239\,\mathrm{Re}_b^{0.759}\overline{\mathrm{Pr}}_b^{0.833}(\rho_w/\rho_b)^{0.31}(\lambda_w/\lambda_b)^{0.0863}$ $(\mu_w/\mu_b)^{0.832}(\mathrm{Gr}^*)^{0.014}(q^+)^{-0.021}$ $\overline{\mathrm{Pr}}_b = \frac{\overline{c_p}\mu_b}{\lambda_b}, \overline{c_p} = \frac{H_w - H_b}{T_w - T_b},$ $\mathrm{Gr}^* = \frac{g\beta d^4 q_w}{\lambda_b \nu_b^2}, q^+ = \frac{\beta q_w}{G\overline{c_p}}$

Table 1.
Typical empirical heat transfer correlations of supercritical fluids.

3. Semiempirical heat transfer correlations

As is mentioned before, as the heat flux increases, the heat transfer deterioration phenomenon will happen. Heat transfer deterioration is characterized with lower values of the heat transfer coefficients and very high wall temperatures. A widely used quantitative expression to define the heat transfer deterioration of supercritical fluid [24–26] is shown by Eq. (1):

$$h < c \cdot h_{DB} = c \cdot 0.023\frac{\lambda_b}{d}\,\mathrm{Re}_b^{0.8}\mathrm{Nu}_b^{0.4} \tag{1}$$

where h refers to the heat transfer coefficient of the heat transfer deterioration and h_{DB} is the heat transfer coefficient given by the Dittus-Boelter correlation (i.e., normal HT regime). The coefficient c is taken as 0.3 by Schatte et al. [24]. Unfortunately, under the deteriorated HT regime, most of the empirical correlations will lose their accuracy. When deteriorated HT regime occurs, the dominated mechanism of the turbulent heat transfer is the interaction among the buoyancy, flow acceleration, and variable thermophysical properties. Unfortunately the empirical correlations can hardly reflect the heat transfer mechanism. Thus, their performances are unsatisfying, as is shown in **Figure 3**.

3.1 Correlation by Jackson

Jackson [12] investigated the semiempirical correlation considering the effect of the mixed convective heat transfer in a vertical heated tube. The author proposed an equivalent "buoyancy-free forced convective flow" which has the same heat transfer effect with the real supercritical fluid flow. He tried to establish the relationship between the heat transfer coefficients attenuation and the shear stress reduction based on the equivalent Nusselt and Reynolds numbers of the equivalent flow with the aid of the empirical equation under forced convective condition. Integrating the equations of momentum across the entire flow and the near-wall layer of thickness δ_t, respectively, then reorganizing the two equations leads to

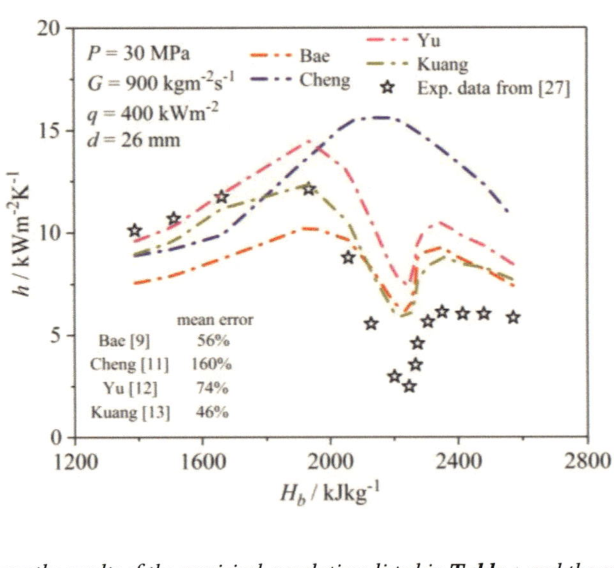

Figure 3.
*The comparison among the results of the empirical correlations listed in **Table 1** and the experimental results from Hu [27]. This case describes the upward flow of supercritical water in a circular tube with an inner diameter of 26 mm and heated length of 2.0 m. Data in this figure come from Ref. [11].*

$$\tau_w - \tau_{\delta_t} = \pm\delta_t(\rho_b - \rho_{ave})g \tag{2}$$

where τ_w is the wall shear stress, τ_{δ_t} is the shears stress at $y = \delta_t$, and the integrated average value of the density ρ_{ave} across the near-wall layer is defined by $\rho_{ave} = \int_0^{\delta_t} \rho dy/\delta_t$. In Eq. (2) the positive sign indicates the upward flow, and the negative one is applied for the downward flow. Then the author assumed that the effect of the buoyancy in modifying the distribution of shear stress and turbulence production has led to such an "equivalent" flow which is not influenced by the buoyancy but has a bulk velocity which is either higher or lower than the real flow. The wall shear stress τ'_w for this "equivalent" buoyancy-free flow can be replaced by the modified shear stress τ_{δ_t} at the location $y = \delta_t$ in a turbulent shear flow. Then the author proposed a model which reflects the shear stress distributions in the supercritical fluid flow:

$$\tau_{\delta_t}/\tau_w = \left(\mathrm{Re}'_b/\mathrm{Re}_b\right)^{2-m_1}\left(F'_{VP1}/F_{VP1}\right) \tag{3}$$

where the F'_{VP1} and F_{VP1} are the modifications on the variable properties for the "equivalent" buoyancy-free flow and the real flow, respectively. The Nusselt number of the "equivalent" forced convection can be expressed in terms of the Reynolds number, Re'_b, using the empirical equation:

$$\mathrm{Nu}_B = K_2\left(\mathrm{Re}'_b\right)^{m_2}\mathrm{Pr}_b^{n_2}F'_{VP2} \tag{4}$$

where K is the constant and then

$$\mathrm{Nu}_F = K_2\,\mathrm{Re}_b^{m_2}\mathrm{Pr}_b^{n_2}F_{VP2} \tag{5}$$

Combining Eqs. (4) and (5) follows that

$$\mathrm{Nu}_B/\mathrm{Nu}_F = \left(\mathrm{Re}'_b/\mathrm{Re}_b\right)^{m_2}\left(F'_{VP2}/F_{VP2}\right) \tag{6}$$

Combining Eqs. (2) and (5) the author obtained

$$\mathrm{Nu}_B/\mathrm{Nu}_F = \left[\left(F'_{VP2}/F'^{m_4}_{VP1}\right)/\left(F_{VP2}/F^{m_4}_{VP1}\right)\right]\left(\tau_{\delta_t}/\tau_w\right)^{m_4} \tag{7}$$

Or equivalently

$$\mathrm{Nu}_B = K_2\,\mathrm{Re}_b^{m_2}\mathrm{Pr}_b^{n_2}F_{VP2}\left(F'_{VP3}/F_{VP3}\right)\left(\tau_{\delta_t}/\tau_w\right)^{m_4} \tag{8}$$

The above equation shows the relationship between the heat transfer and the reduced shear stress across the near-wall layer. Based on this, the author considered three cases of mixing convective heat transfer in a vertical heated tube to fluid flowing either in the upward direction (i.e., the buoyancy-aided case) or the downward direction (i.e., the buoyancy-opposed case). Firstly, a model of the buoyancy-influenced heat transfer was proposed:

$$\left(\frac{\mathrm{Nu}_B}{\mathrm{Nu}_F}\right)\left(\frac{F_{VP3}}{F'_{VP3}}\right) = \left[\left|1 \mp \left(C_B \mathrm{Bo}_b^* F_{VPB}\right)\left[\left(\frac{\mathrm{Nu}_B}{\mathrm{Nu}_F}\right)\left(\frac{F_{VP3}}{F'_{VP3}}\right)\right]^{-2.1}\right|\right]^{0.45} \tag{9}$$

where C_B is a constant, Bo_b is the buoyancy number, and F_{VPB} is the modification on the variable properties for the buoyancy-influenced flow. As can be seen from Eq. (9), $(\mathrm{Nu}_B/\mathrm{Nu}_F)\left(F_{VP3}/F'_{VP3}\right)$ is a unique function of a single parameter

$(C_B \mathrm{Bo}_b^* F_{VPB})$, and the author referred to this function as $\psi(C_B \mathrm{Bo}_b^* F_{VPB})$. It can be evaluated by assigning a series of values to the product $(C_B \mathrm{Bo}_b^* F_{VPB})$ in Eq. (9) and calculating the corresponding values of $(\mathrm{Nu}_B/\mathrm{Nu}_F)(F_{VP3}/F'_{VP3})$. **Figure 4(a)** shows the effect of the buoyancy on heat transfer predicted by Eq. (9) by the three curves (the red line for the downward flow and the black line for the upward flow). Note that by **Figure 4(a)**, we can see if $(C_B \mathrm{Bo}_b^* F_{VPB})$ is less than 0.04, the function $\psi(C_B \mathrm{Bo}_b^* F_{VPB})$ will be within about 2% of unity; thus, the heat transfer mode will be the variable-property forced convective heat transfer.

The second model proposed by the author was about the heat transfer due to the thermally induced flow acceleration. It shows how the acceleration of a heated flow causes the reduced turbulence production, the deteriorated HT, and the turbulence laminarization. The final form of this model is given as

$$\left(\frac{\mathrm{Nu}_A}{\mathrm{Nu}_F}\right)\left(\frac{F_{VP3}}{F'_{VP3}}\right) = \left[\left|1 - (C_A \mathrm{Ac}_b F_{VPA})\left[\left(\frac{\mathrm{Nu}_B}{\mathrm{Nu}_F}\right)\left(\frac{F_{VP3}}{F'_{VP3}}\right)\right]^{-1.1}\right|\right]^{0.45} \tag{10}$$

where C_A is a constant, Ac_b is the flow acceleration number, and F_{VPA} is the modification on the variable properties for the acceleration-influenced flow. As is shown in Eq. (10), the product $(\mathrm{Nu}_A/\mathrm{Nu}_F)(F_{VP3}/F'_{VP3})$ is a unique function of the parameter $(C_A \mathrm{Ac}_b F_{VPA})$, and we will describe this function as $\psi(C_A \mathrm{Ac}_b F_{VPA})$. The variation of $\psi_A(C_A \mathrm{Ac}_b F_{VPA})$ with $(C_A \mathrm{Ac}_b F_{VPA})$ can be readily determined using a similar approach to that described in the buoyancy-influenced heat transfer case. The result is shown in **Figure 4(b)**. The function $\psi(C_A \mathrm{Ac}_b F_{VPA})$ is predicted to fall from unity as $(C_A \mathrm{Ac}_b F_{VPA})$ increases from zero. This indicates that the heat transfer is attenuated as the flow acceleration increases, and note that this phenomenon is independent on the flow direction. The author argued that with the increase of the flow acceleration, the excess pressure acting on the flow is able to balance the shear stress; thus, the core layer will not experience any shear stress and soon leads to the turbulence laminarization.

The third model was presented which describes the heat transfer under the condition where the influence of the buoyancy and the thermally induced bulk flow acceleration are combined together. Using the approach developed earlier to relate the Nusselt number ratio and the shear stress ratio, the author proposed

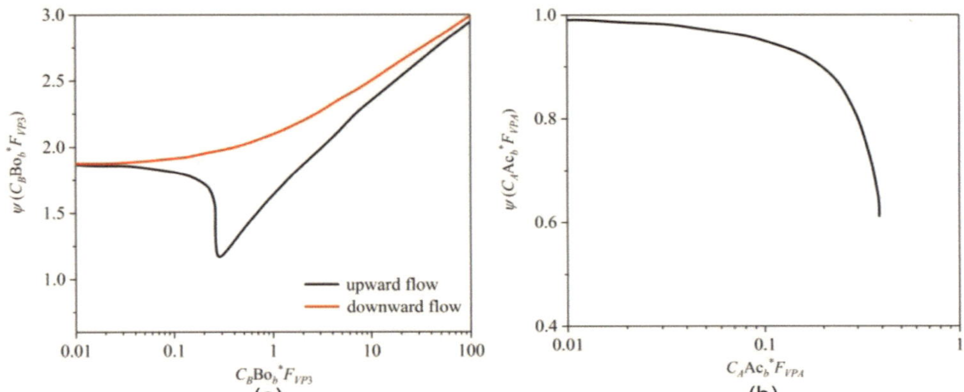

Figure 4.
Effect of the theoretical buoyancy and flow acceleration factors on heat transfer [12], (a) the effect of buoyancy, and (b) the effect of flow acceleration. Data in this figure come from Ref. [12].

$$\left(\frac{\mathrm{Nu}_{BA}}{\mathrm{Nu}_F}\right)\left(\frac{F_{VP3}}{F'_{VP3}}\right)^{-1.1} = \left[1\mp\left(C_B\mathrm{Bo}_b^* F_{VPB}\right)\left(\frac{\mathrm{Nu}_{BA}}{\mathrm{Nu}_F}\right)^{-2.1}\left(\frac{F_{VP3}}{F'_{VP3}}\right)^{-2.1} - r\left(C_B\mathrm{Bo}_b^* F_{VPB}\right)\left(\frac{\mathrm{Nu}_{BA}}{\mathrm{Nu}_F}\right)^{-2.1}\left(\frac{F_{VP3}}{F'_{VP3}}\right)^{-2.1}\right]^{0.45} \quad (11)$$

Here the parameter r indicates the ratio of $(\mathrm{Nu}_{BA}/\mathrm{Nu}_F)^{-1.1}(C_A\mathrm{Ac}_b F_{VPA})$ to $(\mathrm{Nu}_{BA}/\mathrm{Nu}_F)^{-2.1}(C_B\mathrm{Bo}_b^* F_{VPB})$ (i.e., the ratio of the strengths of the buoyancy effect to the flow acceleration effect). In **Figure 5**, the author evaluated the parameter $(\mathrm{Nu}_{BA}/\mathrm{Nu}_F)(F_{VP3}/F'_{VP3})$ for the specified values of $(C_B\mathrm{Bo}_b^* F_{VPB})$ and r and produced the curves of $\phi_{BA}(C_B\mathrm{Bo}_b^* F_{VPB}) = (\mathrm{Nu}_{BA}/\mathrm{Nu}_F)(F_{VP3}/F'_{VP3})^{-1.1}$ against $(C_B\mathrm{Bo}_b^* F_{VPB})$ and r. Thus, for the upward flow, the acceleration influence promotes the buoyancy effect, whereas for the downward flow, the acceleration opposes the buoyancy effect. These trends are illustrated in **Figure 5** for three values of r = 0, 0.5, and 0.8.

3.2 Correlation by Li and Bai

Li and Bai [11] proposed a semiempirical correlation based on the momentum and energy integration equations in the thermal boundary layer. The authors observed the direct numerical simulation (DNS) data under deteriorated HT regime in Bae et al. [7], Hu [27], Peeters et al. [28], and Zhang et al. [29], and they found that in the turbulent core, the temperature gradient is negligible; therefore, the fluid temperature in the core is approximately the bulk temperature, especially when $T_b < T_{pc} < T_w$. While in the thermal boundary layer δ, the fluid temperature increases from T_b to T_w (where δ is the thickness of the thermal boundary layer). This can be explained as follows: the fluid within the pseudo-critical region absorbs considerable energy due to the high c_p, leading to the small temperature gradient in the turbulent core [7]. Based on these characteristics, Li and Bai built a “two-layer model” for deteriorated HT (as is shown in **Figure 6**) [11, 30]. The turbulence in the tube is divided into two layers: the thermal boundary layer and the core layer. In the thermal boundary layer, the buoyancy is strong and leads to the shear stress

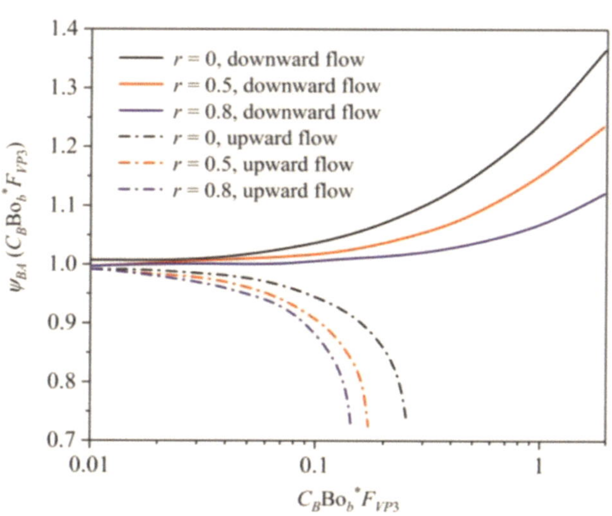

Figure 5.
Heat transfer for mixed flow with combined influence of flow acceleration and buoyancy effect [12]. Data in this figure come from Ref. [12].

redistribution. However, in the core layer, the temperature and density gradients are small; hence, the buoyancy effect can be neglected.

Figure 6 shows the straight tube the authors analyzed, and the flow direction is vertically upward. As is shown, the element *abcd* with a thickness of δ and a length of dx in the thermal boundary layer is chosen to build up the momentum equation:

$$-\frac{dp}{dx}-\frac{\tau_w}{\delta}+\frac{\tau_\delta}{\delta}-\overline{\rho}g=\frac{1}{\delta}\frac{d}{dx}\left(\int_0^\delta \rho u^2 dy\right) \tag{12}$$

In the above equation, τ_w is the shear stress at the wall and $\overline{\rho}$ is the average density across the thermal boundary layer. The wall shear stress and friction factor here are calculated based on Blasius equation, as is shown in Eqs. (13) and (14):

$$\tau_w = 0.5 f_b \rho_b u_b^2 \tag{13}$$

$$f_b = 0.079\,\mathrm{Re}_b^{-0.25} \tag{14}$$

Assume that the pressure gradient on the cross section keeps constant. In a similar way, for the bulk flow, we could get

$$-\frac{dp}{dx}-\frac{2\tau_w}{R}-\rho_b g=(\rho u)_b\frac{du_b}{dx}=G\frac{du_b}{dx} \tag{15}$$

Subtracting Eq. (15) and Eq. (12), the pressure gradient could be eliminated and the following result could be obtained:

$$\frac{2\tau_w}{R}-\frac{\tau_w-\tau_\delta}{\delta}+(\rho_b-\overline{\rho})g=\frac{1}{\delta}\frac{d}{dx}\left(\int_0^\delta \rho u^2 dy\right)-G\frac{du_b}{dx} \tag{16}$$

It is obvious that for the vertically upward flow, the radial part of the velocity (u_r) is small enough to be neglected. Thus, the term (ρu) in the integration sign equals to the total mass flux G. Then Eq. (16) can be simplified as

$$\frac{2\tau_w}{R}-\frac{\tau_w-\tau_\delta}{\delta}+(\rho_b-\overline{\rho})g=\frac{G}{\delta}\frac{d}{dx}\left(\int_0^\delta u dy\right)-G\frac{du_b}{dx} \tag{17}$$

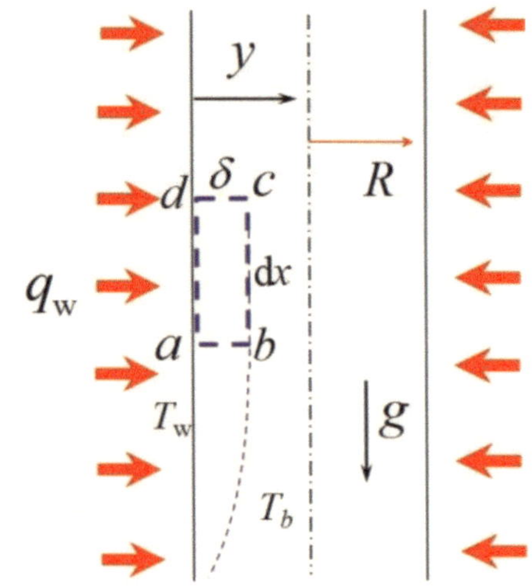

Figure 6.
Physical model of the straight tube.

Note that the thickness of the thermal boundary layer δ is negligible compared with the radius of the tube R. Concerning with this fact

$$\frac{\tau_w - \tau_\delta}{\delta} = (\rho_b - \overline{\rho})g + \left[G\frac{\mathrm{d}u_b}{\mathrm{d}x} - \frac{G}{\delta}\frac{\mathrm{d}}{\mathrm{d}x}\left(\int_0^\delta u\mathrm{d}y\right)\right] \tag{18}$$

The effects of the buoyancy and flow acceleration on the shear stress redistribution are reflected by the two terms at the right side of Eq. (18), respectively. According to Negoescu et al. [31], in the upward flow, the influence of the flow acceleration is much weaker compared with the buoyancy. In order to simplify Eq. (18), the authors adopted the assumption that the axial velocity u_b linearly varies within the thermal boundary layer thickness δ. This leads to

$$\frac{\tau_w - \tau_\delta}{\delta} = (\rho_b - \overline{\rho})g + \left[G\frac{\mathrm{d}u_b}{\mathrm{d}x} - \frac{G}{2\delta}\left(u_b\frac{\mathrm{d}\delta}{\mathrm{d}x} + \delta\frac{\mathrm{d}u_b}{\mathrm{d}x}\right)\right] \tag{19}$$

The gradient of the bulk velocity can be obtained by the analysis on the energy balance of the bulk fluid:

$$\frac{\mathrm{d}u_b}{\mathrm{d}x} = G\frac{\mathrm{d}(1/\rho_b)}{\mathrm{d}T_b}\frac{\mathrm{d}T_b}{\mathrm{d}x} = \frac{G}{\rho_b}\frac{\beta_b}{c_{pb}}\frac{\mathrm{d}H_b}{\mathrm{d}x} = \frac{4q_w\beta_b}{\rho_b c_{pb} d} \tag{20}$$

The authors made further assumption that $\delta(\mathrm{d}u_b/\mathrm{d}x)$ in Eq. (19) is much smaller than $u_b(\mathrm{d}\delta/\mathrm{d}x)$. This can be interpreted as follows: the axial thickness variation of the thermal boundary layer (δ) is negligible compared with the axial bulk velocity (u_b) variation. Then Eq. (19) could be simplified as

$$\frac{\tau_w - \tau_\delta}{\tau_w} = \frac{2(\rho_b - \overline{\rho})g\delta}{f_b\rho_b u_b^2} + \frac{\delta}{f_b u_b}\frac{\mathrm{d}u_b}{\mathrm{d}x} \tag{21}$$

Thus, the drop of the shear stress across the thermal boundary layer can be rewritten as the following pattern:

$$\frac{\tau_w - \tau_\delta}{\tau_w} = \left[\frac{2\overline{\mathrm{Gr}_b}}{\mathrm{Re}_b^2} + \frac{4k_T}{\mathrm{Re}_b\mathrm{Pr}_b}\right]\frac{\delta}{f_b d} \tag{22}$$

where $\overline{\mathrm{Gr}_b}$ is the average Grashof number and k_T is a nondimensional parameter reflecting the expansion of the fluid:

$$\overline{\mathrm{Gr}_b} = \frac{\rho_b(\rho_b - \overline{\rho})gd^3}{\mu_b^2},$$
$$\overline{\rho} = \begin{cases} \frac{1}{2}(\rho_w + \rho_b) \text{ for } T_w < T_{pc} \text{ or } T_b > T_{pc} \\ \frac{1}{T_w - T_b}\left[\rho_b(T_{pc} - T_b) + \rho_w(T_w - T_{pc})\right] \text{ for } T_b \le T_{pc} \le T_w \end{cases} \tag{23}$$

$$k_T = \frac{\beta_b q_w d}{\lambda_b} \tag{24}$$

Therefore, Eq. (22) can also be rewritten as

$$\frac{\tau_w - \tau_\delta}{\tau_w} = \left[\frac{2\overline{\mathrm{Gr}_b}}{\mathrm{Re}_b^2} + \frac{4\beta_b d q_w}{\mathrm{Re}_b\mu_b c_{pb}}\right]\frac{\delta}{f_b d} \tag{25}$$

In order to build the connection between the shear stress and heat transfer, the authors analyzed the energy balance of the element *abcd* shown in **Figure 6**:

$$q_w + \rho_b u_b \frac{\mathrm{d}}{\mathrm{d}x}\left(\int_0^\delta c_p T \mathrm{d}y\right) + c_{pb} T_b \frac{\mathrm{d}}{\mathrm{d}x}\left(\int_0^\delta \rho u \mathrm{d}y\right) = -\lambda_b \left(\frac{\mathrm{d}T}{\mathrm{d}y}\right)_{y=\delta} - \frac{\mu_t}{\mathrm{Pr}_t}\left(\frac{\partial H}{\partial y}\right)_{y=\delta} \tag{26}$$

The physical interpretation of the budgets in the above equation can be given as follows: the first term is the wall heat flux (i.e., energy input from the boundary *ad*); the second term is the net energy input by the convection through the boundaries *ab* and *cd*; the third term is the fluid energy from the core layer through the boundary *bc*. The axial heat conduction is ignored here. The fourth and fifth terms are the energy output by the heat conduction and turbulence diffusion through the boundary *bc*, respectively. As is indicated by Kurganov and Kaptil'ny [6], the turbulent viscosity in the thermal boundary layer drastically attenuates under the deteriorated HT regime. A factor $C_{\mu 1}$ is introduced here to reflect the ratio of the turbulent viscosities under the deteriorated HT regime and normal HT regime. Moreover, based on the mixing length assumption, the turbulent viscosity at the boundary *bc* is

$$\mu_t = C_\mu \rho_b l^2 \left|\frac{\partial u}{\partial y}\right|_{y=\delta} = C_\mu \rho_b l^2 \frac{|\tau_\delta|}{\mu_t}, \mu_t = C_{\mu 1}\left(\rho_b l^2 |\tau_\delta|\right)^{1/2} \tag{27}$$

In order to give the qualitative description of μ_t, the authors applied the Bo* number and ρ_w/ρ_b into Eq. (27) to reflect the influence of the buoyancy force and the variable thermophysical properties:

$$C_{\mu 1} = \varphi \mathrm{Bo}_b^{*m}\left(\frac{\rho_w}{\rho_b}\right)^n, \mathrm{Bo}_b^* = \overline{\mathrm{Gr}}_b / \mathrm{Re}_b^{2.7} \tag{28}$$

where *m* and *n* are the exponents. Then the Nikuras-Van Driest mixing length model was chosen here [32]:

$$l_m/R = \left(0.14 - 0.08\eta^2 - 0.06\eta^4\right)\left[1 - \exp\left(-y^+/26\right)\right], \eta = (R - y)/R \tag{29}$$

The wall distance y^+ was taken as the dimensionless thickness of the thermal boundary layer, and it is defined as

$$y^+ = \frac{\rho_l u_\tau}{\mu_l} y = \frac{\sqrt{\rho_l \tau_w}}{\mu_l} y \tag{30}$$

For the variable-property turbulence, the local value of the thermophysical properties is more advantageous [33, 34]. Therefore, in Eq. (30) the local definitions of the density and viscosity were adopted. Considering the thermal boundary layer thickness (δ) minuteness, the authors omitted the terms higher than the second order in the Taylor series of the two terms at the right side of Eq. (26):

$$-\lambda_b\left(\frac{\mathrm{d}T}{\mathrm{d}y}\right)_{y=\delta} \approx \lambda_b \frac{T_w - T_b}{\delta}, -\frac{\mu_t}{\mathrm{Pr}_t}\left(\frac{\mathrm{d}H}{\mathrm{d}y}\right)_{y=\delta} \approx \frac{\mu_t}{\mathrm{Pr}_t}\frac{H_w - H_b}{\delta} \tag{31}$$

The terms on the left side of Eq. (26) can be dealt as follows:

$$\rho_b u_b \frac{\mathrm{d}}{\mathrm{d}x}\left(\int_0^\delta c_p T dy\right) \approx \rho_b u_b \frac{\mathrm{d}}{\mathrm{d}x}(H_{ave}\delta) = G\left(\delta\frac{\mathrm{d}H_{ave}}{\mathrm{d}x} + H_{ave}\frac{\mathrm{d}\delta}{\mathrm{d}x}\right) \tag{32}$$

$$c_{pb} T_b \frac{\mathrm{d}}{\mathrm{d}x}\left(\int_0^\delta \rho u \mathrm{d}y\right) = G H_b \frac{\mathrm{d}\delta}{\mathrm{d}x} \tag{33}$$

where H_{ave} is the average specific enthalpy across the thermal boundary layer. Then the heat transfer coefficient can be obtained based on Eq. (26):

$$h = \frac{\lambda_b}{\delta} + \frac{\mu_t \overline{c_p}}{\mathrm{Pr}_t \delta} - \frac{G}{T_w - T_b}\left(\delta\frac{\mathrm{d}H_{ave}}{\mathrm{d}x} + H_{ave}\frac{\mathrm{d}\delta}{\mathrm{d}x} + H_b\frac{\mathrm{d}\delta}{\mathrm{d}x}\right) \tag{34}$$

The μ_t in Eq. (26) is determined by Eq. (25) and Eqs. (20)–(22). The fitting curve among $C_{\mu 1}$, Bo* number, and the density ratio (ρ_w/ρ_b) is obtained by the dataset of the vertically upward water flow in the circular tube under the deteriorated HT reported by Mokry et al. [9], Hu [27], Wen [35], and Wang et al. [36]. The total number of the data points is 700, with 495 data points as the fitting set and 205 data points as the assessment set. The parameters range of the dataset is given in **Table 2**.

In Eq. (26), the thermal boundary layer thickness δ is expressed as

$$\delta = \frac{\mu_b}{\sqrt{\rho_b \tau_w}}\delta^+ \tag{35}$$

According to the analysis on the temperature profiles based on the data presented in the Ref. [29] which has similar working conditions, the dimensionless excess temperature exceeds 0.9 at $\delta^+ = 300$ in the tube turbulence (for details, see Ref. [11]). For simplicity, the authors assumed an approximate value of 300 for δ^+, and then the expression of δ can be expressed as

$$\delta = \delta^+ \mu_b/\sqrt{\rho_b \tau_w} = 300\mu_b/\sqrt{\rho_b \tau_w} \tag{36}$$

According to Eqs. (36), (13), and (14), it can be found that the thickness of the thermal boundary layer $\delta = 300\mu_b/\sqrt{\rho_b \tau_w} \sim 10^{-4}$ m is $R \approx 100\delta$. Thus, the assumption that the thermal boundary layer thickness is negligible compared with the tube radius is reasonable. The order of magnitude of the terms in Eq. (19) is listed as follows based on the dataset:

$$\begin{aligned} &u_b \sim 1\ \mathrm{m\ s^{-1}}, G \sim 10^3\ \mathrm{kg\ m^{-2}\cdot s^{-1}}, \delta = 300\mu_b/\sqrt{\rho_b\tau_w} \sim 10^{-4}\ \mathrm{m}, \\ &\frac{\mathrm{d}\delta}{\mathrm{d}x} = 1321 G^{-0.875} d^{0.125} \mu^{-0.125} \frac{\mathrm{d}\mu}{\mathrm{d}T}\frac{\mathrm{d}T_b}{\mathrm{d}x} \sim 10^{-6}, u_b\frac{\mathrm{d}\delta}{\mathrm{d}x} \sim 10^{-6}\ \mathrm{m\ s^{-1}} \end{aligned} \tag{37}$$

Refs.	P (MPa)	G (kg m^{-2} s^{-1})	q (kW m^{-2})	d (mm)	H_b (kJ kg^{-1})	Heated length (m)
Mokry et al. [9]	24.1	129–334	200–500	10	1450–2600	4.0
Hu [27]	23–30	300–500	600–900	26	1300–2900	2.0
Wen [35]	23–26	332–700	446–600	7.6	1300–2900	2.64
Wang et al. [36]	23	450–536	450–458	10	1300–2400	2.5

Table 2.
The experimental data sources used in the paper.

$$\begin{aligned}&q_w \sim 5\times10^5\ \mathrm{W\,m^{-2}}, \rho_b \sim 5\times10^2\ \mathrm{kg\,m^{-3}}, d\sim10^{-2}\ \mathrm{m},\\ &\beta\sim10^{-1}\ \mathrm{K^{-1}}, c_p\sim10^4\ \mathrm{J\,kg^{-1}\,K^{-1}}, \frac{\mathrm{d}u_b}{\mathrm{d}x}\sim1\ \mathrm{s^{-1}}, \delta\frac{\mathrm{d}u_b}{\mathrm{d}x}\sim10^{-4}\end{aligned} \tag{38}$$

The authors got that the $u_b(\mathrm{d}\delta/\mathrm{d}x)$ could be ignored compared with the term $\delta(\mathrm{d}u_b/\mathrm{d}x)$. Therefore, the assumption simplifying Eq. (19) is reasonable. Similarly, the order of magnitude of the parameters in Eqs. (32) and (33) could be analyzed as follows:

$$\begin{aligned}&G\sim10^3\ \mathrm{kg\,m^{-2}\,s^{-1}}, q_w\sim5\times10^5\ \mathrm{W\,m^{-2}},\\ &H\sim10^6\ \mathrm{J\,kg^{-1}}, d\sim10^{-2}\ \mathrm{m}, \frac{\mathrm{d}H_{ave}}{\mathrm{d}x}\approx\frac{\mathrm{d}H_b}{\mathrm{d}x}=\frac{4q_w}{Gd}\sim10^5\ \mathrm{J\,kg^{-1}}\end{aligned} \tag{39}$$

$$\delta=300\mu_b/\sqrt{\rho_b\tau_w}\sim10^{-4}\ \mathrm{m}, \frac{\mathrm{d}\delta}{\mathrm{d}x}=1321G^{-0.875}d^{0.125}\mu^{-0.125}\frac{\mathrm{d}\mu}{\mathrm{d}T}\frac{\mathrm{d}T_b}{\mathrm{d}x}\sim10^{-6} \tag{40}$$

Therefore,

$$\rho_b u_b\frac{\mathrm{d}}{\mathrm{d}x}\left(\int_0^\delta c_pT\mathrm{d}y\right)\approx G\left(\delta\frac{\mathrm{d}H_{ave}}{\mathrm{d}x}+H_{ave}\frac{\mathrm{d}\delta}{\mathrm{d}x}\right)\sim10^4\ \mathrm{W\,m^{-2}} \tag{41}$$

$$c_{pb}T_b\frac{\mathrm{d}}{\mathrm{d}x}\left(\int_0^\delta\rho u\mathrm{d}y\right)=GH_b\frac{\mathrm{d}\delta}{\mathrm{d}x}\sim10^3\ \mathrm{W\,m^{-2}} \tag{42}$$

The second and third terms at the left side of Eq. (26) are far less than the q_w in the left side; thus, they could also be ignored. Eq. (34) could be simplified as

$$h=\frac{\lambda_b}{\delta}+\frac{\mu_t\overline{c_p}}{\mathrm{Pr}_t\delta} \tag{43}$$

Plugging Eqs. (25), (27)–(30), and (36) into Eq. (43), the final form of the model could be obtained after some arrangements:

$$\mathrm{Nu}_b=C_A\,\mathrm{Re}_b^{0.875}\left\{1+C_{\mu1}C_B\,\mathrm{Re}_b^{0.875}\frac{\overline{\mathrm{Pr}_b}}{\mathrm{Pr}_t}\left(\frac{l_m}{d}\right)\left|1-C_d\left(\frac{2\overline{\mathrm{Gr}_b}}{\mathrm{Re}_b^{2.625}}+\frac{4k_T}{\mathrm{Re}_b^{1.625}\mathrm{Pr}_b}\right)\right|^{0.5}\right\} \tag{44}$$

where $C_A=6.67\times10^{-4}, C_B=0.2, C_d=1.9\times10^4, \mathrm{Pr}_t=0.9$. The fitting equation of $C_{\mu1}$ was obtained as follows:

For $3\times10^{-6}<\mathrm{Bo}^*_{\max}<8\times10^{-5}$,

$$C_{\mu1}=1.115\mathrm{Bo}_b^{*\,0.147}(\rho_w/\rho_b)^{1.365} \tag{45}$$

For $8\times10^{-5}<\mathrm{Bo}^*_{\max}<3\times10^{-4}$,

$$C_{\mu1}=0.0016\mathrm{Bo}_b^{*\,-0.416}(\rho_w/\rho_b)^{1.325} \tag{46}$$

Here the parameter $\mathrm{Bo}^*_{\max}$ is defined as the maximum of the buoyancy number (Bo^*) under a specified case (i.e., specified P, G, q_w, d). The range of the parameters in the new correlation is P = 23–30 MPa, G = 200–900 kg m^{-2} s^{-1}, q = 129–700 kW m^{-2}, d = 7.6–26 mm, $3\times10^{-6}<\mathrm{Bo}^*_{\max}<3\times10^{-4}$.

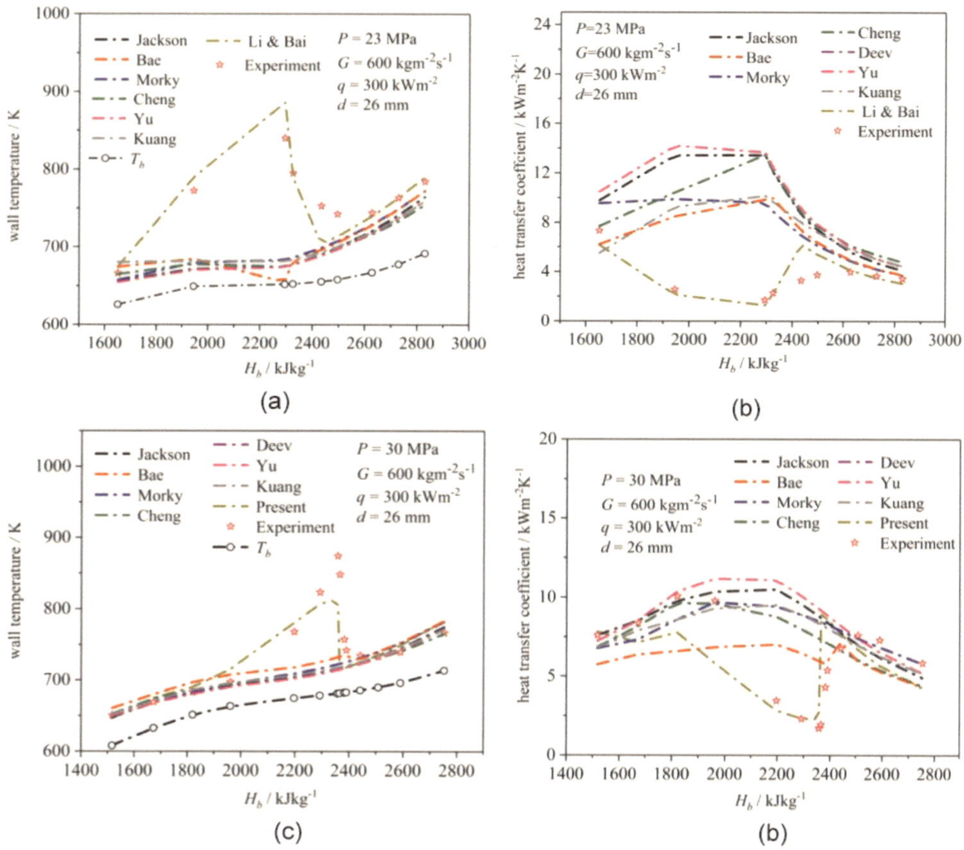

Figure 7.
*Comparisons of the predicted results and experimental results. Experimental data come from Hu [27], as is shown in **Table 2**; (a) and (c) are the wall temperature results; (b) and (d) are the heat transfer coefficients results. Data in this figure come from Ref. [11].*

Figure 7 depicts the results of the model evaluation based on the above assessing procedure. It is clear that all the existing correlations could not obtain the tendencies of the wall temperature under the heat transfer deterioration regime. The semiempirical model proposed by Li and Bai shows comparatively higher accuracy compared with the existing correlations, especially at the peak region of the wall temperature. Most of the existing correlations significantly overestimate the heat transfer coefficients under the heat transfer deterioration regime.

In **Figure 8**, the error graphs of different heat transfer correlations (here the fitting set and the assessment set are both included) are shown. It is shown that the semiempirical correlations introduced here are correlated well with the heat transfer coefficients dataset. Seventy-two percent and 63% of the data points fall into the error bars of ±30% and ±25%, respectively. In order to further investigate the predicting capacity under the deteriorated HT regime, the authors introduced another two statistical parameters: the mean relative error (*MRE*) and the root mean square error (*RMSE*) to evaluate the different correlations performance. Their definitions are given in Eqs. (47)–(49), and the comparison results are shown in **Table 3**. As can be seen, the Li and Bai correlation can give the smallest prediction errors, thus proving its suitability and superiority.

$$RE_i = \frac{|EXP_i - PRE_i|}{EXP_i} \tag{47}$$

$$MRE = \frac{1}{n}\sum_{i=1}^{n}|RE_i| \tag{48}$$

$$RMSE = \sqrt{\frac{1}{n}\sum_{i=1}^{n}\left(RE_i^2\right)} \tag{49}$$

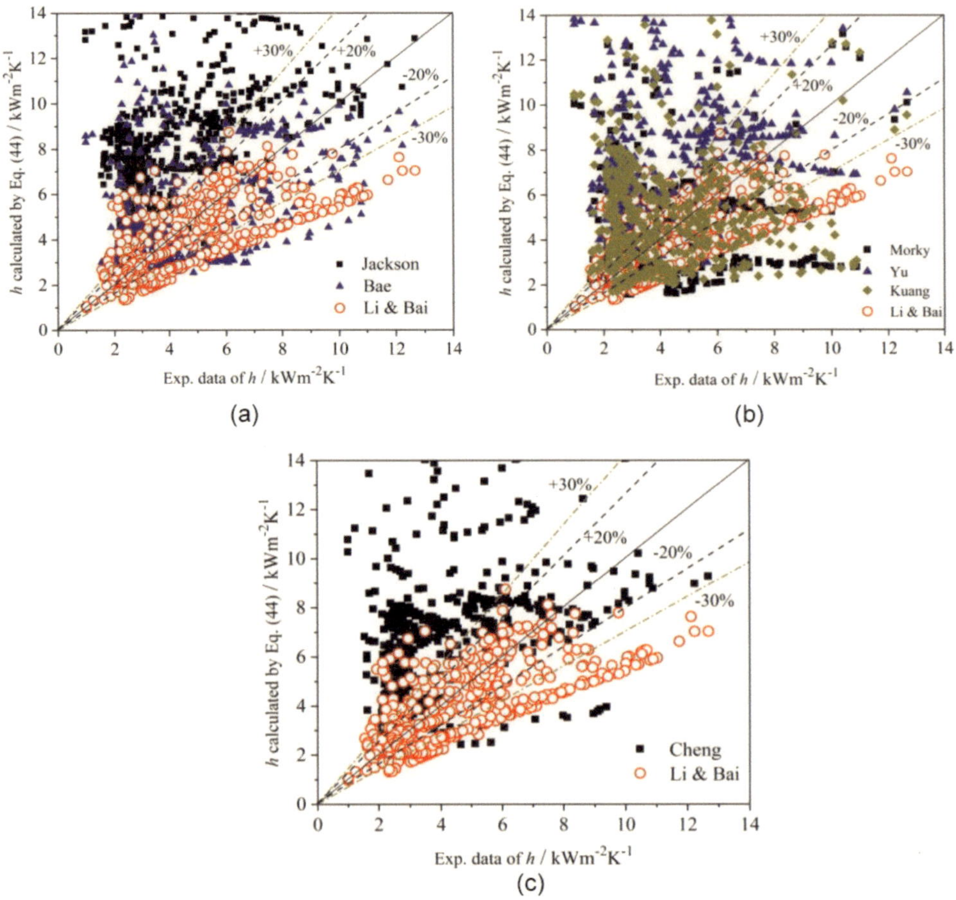

Figure 8.
The errors of the predicted results of different heat transfer correlations. Data in this figure come from Ref. [11]. (a) comparison among Jackson correlation, Bae correlation and Li & Bai correlation; (b) comparison among Morky correlation, Yu correlation, Kuang correlation and Li & Bai correlation; (c) comparison between Cheng correlation and Li & Bai correlation.

Correlation	*MRE*/%	*RMSE*/%
Mokry et al. [9]	44	96
Bae et al. [10]	63	279
Jackson [23]	130	180
Cheng et al. [20]	94	148
Yu et al. [21]	10	163
Kuang et al. [22]	65	113
Li and Bai [11]	27	32

Table 3.
The statistics of the present and existing correlations.

4. Conclusions

The approach to establishing the heat transfer correlations of supercritical fluids is a critical issue since the correlations play very important role in the design and optimization of the systems and devices. In this chapter, we have discussed the principles and applications of the heat transfer correlations of supercritical fluids. The modeling approaches of the correlations of supercritical fluid heat transfer are reviewed, including the nondimensional parameters applied on the modification of the empirical correlations and the "equivalent buoyancy-free flow method" used for the semiempirical correlations.

Then we introduce a new physically based semiempirical correlation which is based on the momentum and energy conservations in the mixing convective flow. Considering the mechanism of heat transfer deterioration, a physical model characterizing the redistribution of the shear stress under the combined effect of buoyancy and flow acceleration was obtained. Then the model about the heat transfer coefficients under the influence of the reduced shear stress was derived by the energy integration equation within the thermal boundary layer. Based on this, a semiempirical heat transfer correlation was proposed and then verified with a wide range of experimental data. Compared with the existing correlations, the prediction accuracy of this newly developed correlation is significantly improved under the heat transfer deterioration regime. The investigation on the different statistical parameters shows that this semiempirical correlation is superior to the empirical ones.

Acknowledgements

This study is supported by the National Key Research and Development Program of China (no. 2016YFB0600100) and the China National Funds for Distinguished Young Scientists (no. 51425603).

Conflict of interest

No conflict.

Abbreviations

DB	Dittus-Bolter
DNS	direct numerical simulation
EXP	experimental results
HT	heat transfer
MRE	mean relative error
PRE	results given by correlations
RE	relative error
RSME	root mean square error

Appendices and nomenclature

Ac	flow acceleration parameter
Bo^*	Jackson buoyancy parameter

$\mathrm{Bo}_m{}^*$	the maximum Jackson buoyancy number under a specified working condition
Bu	Bae and Kim buoyancy parameter
c_1, c_2, c_k	exponents in the empirical correlations (in **Figure 2**)
c_p	specific heat (J kg^{-1} K^{-1})
$\overline{c_p}$	average specific heat in the thermal boundary layer $(=(H_w - H_b)/(T_w - T_b))$ (J kg^{-1} K^{-1})
$C_\mu, C_{\mu 1}$	the correction coefficients under heat transfer deterioration in the Li and Bai correlation
$C_A, C_B, C_{\mu 1}, C_d, m, n$	constants in the Li and Bai correlation
d	inside diameter of the circular tube (m)
f	Fanning resistance coefficient
F, F_1, F_2	constants in the Cheng correlation
$F'_{VP1}, F'_{VP2}, F'_{VP3}$	the modifications on the variable properties for the "equivalent" buoyancy-free flow
$F_{VP1}, F_{VP2}, F_{VP3}$	the modifications on the variable properties for the real flow
F_{VPA}	the modification on the variable properties for the acceleration-influenced flow
F_{VPB}	the modification on the variable properties for the buoyancy-influenced flow
g	gravitational acceleration (m s^{-2})
G	mass flux (kg m^{-2} s^{-1})
Gr	Grashof number
h	local heat transfer coefficient (kW m^{-2} K^{-1})
H	enthalpy (kJ kg^{-1})
k_T	the dimensionless number reflecting the expansion of the fluid
l_m	the mixing length (m)
m_1, m_2, m_4, n_2	the exponents in the Jackson's semiempirical model
Nu	Nusselt number
P	pressure (MPa)
Pr	Prandtl number
$\overline{\mathrm{Pr}_b}$	average Prandtl number across the thermal boundary layer
Pr_t	turbulent Prandtl number
q	wall heat flux (kW m^{-2})
r	radial coordinate (m)
R	tube radius (m)
Re	Reynolds number
T	temperature (K)
X, Y	nondimensional parameters in the empirical correlation (in **Figure 2**)
$Y, Y_1, Y_2, K_h, K_{Am}, K_{qm}$	constants in the Deev correlation
y	distance from the wall (m)
y^+	nondimensional distance from the wall

Greek symbols

λ	thermal conductivity (W m^{-1} K^{-1})
β	thermal expansion rate (K^{-1})
ρ	density (kg m^{-3})
μ	dynamic viscosity (Pa s)

δ	thermal boundary layer thickness (m)
δ^+	nondimensional thermal boundary layer thickness in wall units
θ	nondimensional excess temperature
φ	the constant in the Li and Bai correlation
ψ	the functional expression for the acceleration-influenced or buoyancy-influenced heat transfer
τ	shear stress (Pa)

Supercripts

T	transpose matrix

Subscripts

A	flow acceleration effect
ave	average value over the thermal boundary layer
B	buoyancy effect
BA	combined effect of the buoyancy and flow acceleration
b	bulk value
max	maximum
F, f	forced convection
l	definition based on the local values
t	turbulence
pc	pseudo-critical point
w	values at the wall

Author details

Fangbo Li, Binbin Pei and Bofeng Bai*
State Key Laboratory of Multiphase Flow in Power Engineering, Xi'an Jiaotong University, Xi'an, China

*Address all correspondence to: bfbai@mail.xjtu.edu.cn

References

[1] Zhang Q, Li HX, Zhang WQ, Li LX, Lei XL. Experimental study on heat transfer to the supercritical water upward flow in a vertical tube with internal helical ribs. International Journal of Heat and Mass Transfer. 2015; **89**:1044-1053. DOI: 10.1016/j.ijheatmasstransfer.2015.05.109

[2] Guo LJ, Jin H, Lu YJ. Supercritical water gasification research and development in China. Journal of Supercritical Fluids. 2015;**96**:144-150. DOI: 10.1016/j.supflu.2014.09.023

[3] Zhao Z, Su S, Si N, Hu S, Wang Y, Xu J, et al. Exergy analysis of the turbine system in a 1000 MW double reheat ultra-supercritical power plant. Energy. 2017;**119**:540-548. DOI: 10.1016/j.energy.2016.12.072

[4] Li ZH, Wu YX, Tang GL. Comparison between heat transfer to supercritical water in a smooth tube and in an internally ribbed tube. International Journal of Heat and Mass Transfer. 2015;**84**:529-541. DOI: 10.1016/j.ijheatmasstransfer.2015.01.047

[5] Bae YY, Kim HY, Kang DJ. Forced and mixed convection heat transfer to supercritical CO2 vertically flowing in a uniformly-heated circular tube. Experimental Thermal and Fluid Science. 2010;**34**:1295-1308. DOI: 10.1016/j. expthermflusci. 2010. 06. 001

[6] Kurganov VA, Kaptil'ny AG. Velocity and enthalpy fields and eddy diffusivities in a heated supercritical fluid flow. Experimental Thermal and Fluid Science. 1992;**5**:465-478. DOI: 10.1016/0894-1777(92)90033-2

[7] Bae JH, Yoo JY, McEligot DM. Direct numerical simulation of heated CO2 flows at supercritical pressure in a vertical annulus at Re = 8900. Physics of Fluids. 2008;**20**:055108. DOI: 10.1063/1.2927488

[8] Bae JH, Yoo JY, Choi H. Direct numerical simulation of turbulent supercritical flows with heat transfer. Physics of Fluids. 2005;**20**:105104. DOI: 10.1063/1.2047588

[9] Mokry S, Pioro IL, Farah A, King K, Gupta S, Peiman W, et al. Development of supercritical water heat-transfer correlation for vertical bare tubes. Nuclear Engineering and Design. 2011; **241**:1126-1136. DOI: 10.1016/j.nucengdes.2010.06.012

[10] Bae YY, Kim HY. Convective heat transfer to CO2 at a supercritical pressure flowing vertically upward in tubes and an annular channel. Experimental Thermal and Fluid Science. 2009;**33**:329-339. DOI: 10.1016/j.expthermflusci.2008.10.002

[11] Li FB, Bai BF. A model of heat transfer coefficient for supercritical water considering the effect of heat transfer deterioration. International Journal of Heat and Mass Transfer. 2019;**133**:316-329. DOI: 10.1016/j.ijheatmasstransfer.2018.12.121

[12] Jackson JD. Models of heat transfer to fluids at supercritical pressure with influences of buoyancy and acceleration. Applied Thermal Engineering. 2017;**124**:1481-1491. DOI: 10.1016/j.applthermaleng.2017.03.146

[13] Yamagata K, Nishikawa K, Hasegawa S, Fujii T, Yoshida S. Forced convective heat transfer to supercritical water flowing in tubes. International Journal of Heat and Mass Transfer. 1972; **15**:2575-2593. DOI: 10.1016/0017-9310(72)90148-2

[14] Vikhrev YV, Barulin YD, Kon'Kov AS. A study of heat transfer in

vertical tubes at supercritical pressures. Thermal Engineering. 1967;**14**(9): 116-119

[15] Pioro IL, Khartabil HF, Duffey RB. Heat transfer to supercritical fluids flowing in channels-empirical correlations (survey). Nuclear Engineering and Design. 2004;**230**: 69-91. DOI: 10.1016/j.nucengdes.2003.10.010

[16] Gupta S, Farah A, King K, Mokry S, Pioro I. Developing new heat-transfer correlation for supercritical-water flow in vertical bare tubes. In: Proceedings of ICONE-18; 17–21 May 2010; Xi'an, China.

[17] Yoo JY. The turbulent flows of supercritical fluids with heat transfer. Annual Review of Fluid Mechanics. 2013;**45**:495-525. DOI: 10.1146/annurev-fluid-120710-101234

[18] Komita H, Morooka S, Yoshida S, Mori H. Study on the heat transfer to the supercritical pressure fluid for supercritical water cooled power reactor development. In: Proceedings of the NURETH-10, 2005; Seoul, Korea

[19] Deev VI, Kharitonov VS, Baisov AM, Churkin AN. Universal dependencies for the description of heat transfer regimes in turbulent flow of supercritical fluids in channels of various geometries. Journal of Supercritical Fluids. 2018;**135**:160-167. DOI: 10.1016/j.supflu.2018.01.019

[20] Cheng X, Yang YH, Huang SF. A simplified method for heat transfer prediction of supercritical fluids in circular tubes. Annals of Nuclear Energy. 2009;**36**:1120-1128. DOI: 10.1016/j.anucene.2009.04.016

[21] Yu J, Jia B, Wu D, Wang D. Optimization of heat transfer coefficient correlation at supercritical pressure using genetic algorithms. Heat and Mass Transfer. 2009;**45**:757-766. DOI: 10.1007/s00231-008-0475-4

[22] Kuang B, Zhang Y, Cheng X. A wide-ranged heat transfer correlation of water at supercritical pressures in vertical upward tubes. In: Proceedings of the China-Canada Joint Workshop on SCWR, 2008; Shanghai

[23] Jackson JD. Fluid flow and convective heat transfer to fluids at supercritical pressure. Nuclear Engineering and Design. 2013;**264**: 24-40. DOI: 10.1016/j.nucengdes.2012.09.040

[24] Schatte GA, Kohlhepp A, Wieland C, Spliethoff H. Development of a new empirical correlation for the prediction of the onset of the deterioration of heat transfer to supercritical water in vertical tubes. International Journal of Heat and Mass Transfer. 2016;**102**:133-141. DOI: 10.1016/j.ijheatmasstransfer.2016.06.007

[25] Koshizuka S, Takano N, Oka Y. Numerical analysis of deterioration phenomena in heat transfer to supercritical water. Transactions of the JSME. 1995;**38**:3077-3084. DOI: 10.1016/0017-9310(95)00008-W

[26] Cheng X, Kuang B, Yang YH. Numerical analysis of heat transfer in supercritical water cooled flow channels. Nuclear Engineering and Design. 2007;**237**:240-252. DOI: 10.1016/j.nucengdes.2006.06.011

[27] Hu ZH. Heat transfer characteristics of vertical upflow and inclined tube in the supercritical pressure and near-critical pressure region [thesis]. Xi'an: Xi'an Jiaotong University; 2001

[28] Peeters JW, Pecnik R, Rohde M, et al. Turbulence attenuation in simultaneously heated and cooled annular flows at supercritical pressure.

Journal of Fluid Mechanics. 2016;**799**: 505-540. DOI: 10.1017/jfm.2016.383

[29] Zhang G, Zhang H, Gu H, Yang Y, Cheng X. Experimental and numerical investigation of turbulent convective heat transfer deterioration of supercritical water in vertical tube. Nuclear Engineering and Design. 2012; **248**:226-237. DOI: 10.1016/j.nucengdes.2012.03.026

[30] Kim DE, Kim MH. Two layer heat transfer model for supercritical fluid flow in a vertical tube. Journal of Supercritical Fluids. 2011;**58**:15-25. DOI: 10.1016/j.supflu.2011.014.

[31] Negoescu CC, Li Y, Al-Duri B, Ding Y. Heat transfer behavior of supercritical nitrogen in the large specific heat region flowing in a vertical tube. Energy. 2017;**134**:1096-1106. DOI: 10.1016/j.energy.2017.04.047

[32] Patankar SV, Ivanovic M, Sparrow EM. Analysis of turbulent flow and heat transfer in internally finned tubes and annuli. Journal of Heat Transfer-Transactions of the ASME. 1979;**101**:29-42. DOI: 10.1115/1.3450925

[33] Chu X, Laurien E, McEligot DM. Direct numerical simulation of strongly heated air flow in a vertical pipe. International Journal of Heat and Mass Transfer. 2016;**101**:1163-1176. DOI: 10.1016/j.ijheatmasstransfer.2016.05.038

[34] Huang PG, Coleman GN, Bradshaw P. Compressible turbulent channel flows: DNS results and modelling. Journal of Fluid Mechanics. 1995;**305**:185-218. DOI: 10.1017/S0022112095004599

[35] Wen QL. Experimental study of flow and heat transfer characteristics at SCWR typical conditions [thesis]. Shanghai, Shanghai Jiao Tong University; 2015

[36] Wang F, Yang J, Gu H, Zhao M, Li H, Lu D. Experimental research on heat transfer performance of supercritical water in vertical tube. Atomic Energy Science and Technology. 2013;**47**:933-939

Chapter 4

Resource Upgrading in Advanced Supercritical Fluid (Supercritical Fluid with Catalyst and Cosolvent): Liquid Fuels from Biomass in Sub and Supercritical Water and Carbohydrate Up-Conversion in Ionic Liquid and Supercritical Fluids Mixtures

Masaru Watanabe, Masayoshi Wagatsuma, Keisuke Suzuki, Takuma Kato, Yasuto Goto, Yukihiro Kanaguri and Yuya Hiraga

Abstract

Liquid fuels from biomass and up-conversion of biomass in advanced supercritical fluid are reviewed in this chapter. Lignin can be converted into heavy hydrocarbons in subcritical water extraction. Lipid, which is triglyceride, is catalytically converted into straight-chain hydrocarbons of free fatty acid (decarboxylation) formed by hydrolysis. Carbohydrate is also hydrothermally converted into furan ring compound and fatty acids. Protein is converted into amino acids in hydrothermal water and depolymerization of protein is favored with rapid heating and denaturation agency such as alkaline earth metals. Free amino acids are further decomposed into carboxylic acid through deamination and into amine through decarboxylation. To inhibit Maillard reactions, which result in polymerization, the deamination of amino acid at low temperature was favored and a solid catalyst was quite active for deamination of free amino acids at quite low temperature hydrothermal water. Cellulose was dissolved in some ionic liquids with high mass percentages (5–20 wt%) and converted into monomers and useful components such as furan ring compounds and supercritical fluid cosolvent such as hydrothermal water in ionic liquids supported improvement of reaction efficiency. For hydrogenation of biomass, it was confirmed that hydrogen solubility was enhanced with supercritical carbon dioxide and it must be helpful for hydrogen reaction with biomass molecule.

Keywords: liquid fuels, biomass, iso-conversion, ionic liquid, hydrogen solubility

1. Introduction

Renewable carbon resource is only biomass on the earth and its utilization must be enhanced to replace fossil carbon resources into biomass. In particular, among a various energy demands, transport occupies 30% of total carbon footprint in a personal daily life of an American [1]. For short-distance travel at near a house and a town, electrical vehicle such as EV (electric vehicle), HV (hybrid vehicle), PHEV (plug-in hybrid electric vehicle), and so on (so-called xEV) should be spread, but still a long-distance travel with an airplane and a large passenger ship requires liquid fuels such as diesel and heavy oils even though secondary batteries would be further compacted and lightened. For managing of the demands, liquid fuels must be obtained from renewable and sustainable carbon resources, and the only answer is biomass. For transformation of biomass into liquid fuels, one of the useful techniques should be sub and supercritical fluids technique with catalyst and cosolvent, which is called advanced supercritical fluid technology here.

In this chapter, we review our previous studies concerning liquid fuels from biomass and up-conversion of biomass (mainly carbohydrates) in advanced supercritical fluid technology (supercritical fluid with catalyst such as solid aid-base catalyst or cosolvent such as ionic liquid) are introduced.

In a previous study, lignin, which is a major component of woody biomass, can be converted into heavy hydrocarbons in subcritical water (or hydrothermal) extraction [2]. The dimensionless severity number ($k_r \cdot R_o$: k_r [time^{-1}] is kinetic constant at reference condition and R_o [time] is severity factor) can be used to predict conversion of lignin into liquid fuel. The mean molecular weight of the recovered lignin (so-called "hydrothermal lignin" because "lignin" is usually named with the name of solvent, which is used for recovery, such as "alkali-lignin" and "organosolv-lignin") was about 1 kDa, which is 6-mer of syringyl structure. The hydrothermal lignin should be used as heavy oil or source of aromatic structure after post-treatment such as a conventional catalytic cracking and hydrogenation. In this review, the lignin recovery and dissolution into hydrothermal water (obtained lignin is "hydrothermal lignin") is briefly introduced in Section 2, and the importance of dimensionless severity number is emphasized.

Protein-rich biomass can be converted into liquid hydrocarbons through catalytic hydrothermal cracking (CHTC) and conventional hydrogenation/distillation process (total process is so-called "iso-conversion") [3–5]. That is, the first step of "iso-conversion" is CHTC, and the main reaction of CHTC should be catalytic decarboxylation because hydrolyzed lipid is mainly to form free fatty acid, and carboxyl group in the free fatty acid must be detached to form fuel component, such as hydrocarbons. In our previous study, it was found that lipid, which is triglyceride, is catalytically converted into straight-chain hydrocarbons of free fatty acid (decarboxylation) formed by hydrolysis and one of the active materials of catalytic decarboxylation is zirconia [6]. It was also found that zirconia is also active for decarboxylation of small fatty acid such as acetic acid in supercritical water [7]. In "iso-conversion," the remaining components after extraction of lipid from protein-rich biomass, which are carbohydrate and protein, are proposed to be fermented to form small organic acids or olefinic molecules to combine together with unsaturated hydrocarbon (olefin) produced from lipid decarboxylation. Anaerobic fermentation is suggested to get small molecules in "iso-conversion" process, but it requires large space and long time [3]. Then, additional hydrothermal process for carbohydrate and protein conversion into small molecules such as furan ring compounds and fatty acids are proposed by our research group, and a small review of the proposed process is described in Section 3 of this chapter.

Carbohydrate is also hydrothermally converted into furan ring compound [8–13], which is used as liquid fuel, and fatty acids, which are the similar compounds produced from lipid. The furan ring compound such as hydroxymethyl furfural (HMF) was obtained with high yield in hydrothermal water in the presence of appropriate catalyst such as TiO_2 under microwave irradiation [12]. Microwave irradiation affected HMF formation from fructose, and the selectivity of HMF was enhanced under microwave heating in hydrothermal conversion [13].

Protein is converted into amino acids in hydrothermal water, and depolymerization of protein is favored with rapid heating and denaturation agency such as alkaline earth metals [14]. Free amino acids are further decomposed into carboxylic acid through deamination and amine through decarboxylation. However, in high temperature reactions, carbohydrates and protein (and these monomers) are easily combined together to form melanoidins, which inhibit fragmentations of both of the components (such as carbohydrates and proteins), through so-called Maillard reaction, which is a group of amino-carbonyl reactions. The deamination of amino acid at low temperature is favorable because Maillard reaction is not developed. However, deamination requires high temperature in subcritical water. For example, alanine deamination is meaningfully developed over 513 K [15]. To know the favorable acid or basic condition for enhancing amino acid deamination, pH dependence on alanine reaction was investigated. As a result, as introduced in Section 3, lower temperature than 513 K, the amino acid degradation is remarkably slow. By screening of several types of solid acid additives, it was surprisingly found that a type of solid catalyst was quite active for deamination of free amino acids not only alanine but also glycine, leucine, and serine at quite low temperature hydrothermal water.

Further decomposition of small free fatty acid into alkene and CH_4 is favored for iso-conversion because these can be acted as alkane source or hydrogen source for long chain hydrocarbon formation. By integration of each unit reactions/operation (hydrolysis, deamination, decarboxylation, hydrogenation, and distillation), liquid fuels such as straight-chain and aromatic-ring compounds are obtained from woody/protein-rich biomass. One of the images for process integration (process intensification) shows in **Figure 1**.

In Section 3 of this chapter, we briefly introduce protein conversion into small fragments with catalyst in hydrothermal water. Then the effect of acid and alkali on subcritical water conversion of alanine, which is the smallest chiral amino acid, is also roughly explained as the basic knowledge of alanine conversion in advanced supercritical fluid technologies. With a solid acid catalyst, alanine conversion is enhanced and the discussion why the solid acid catalyst can promote alanine conversion in low temperature hydrothermal water is provided.

The other advanced supercritical technology is combined utilization of supercritical or hydrothermal water with cosolvent such as ionic liquid in the absence and presence of additives (homogeneous or heterogeneous). To improve reaction efficiency (increase of rate, yield, and selectivity of a target component), catalyst is also used in the ionic liquid-supercritical (hydrothermal) fluid mixture. Cellulose, which is the most abundant biomass on the earth, is dissolved in some ionic liquids with high mass percentages (5–20 wt%) and converted into monomers and further conversion into useful components such as furan ring compounds. In our research group, the combinational utilization of ionic liquid with hydrothermal water (or other solvents with and without catalyst) for cellulose (and its monomer, glucose, and the derivative) conversion into small fragments and useful chemicals were reported [16–25]. There are several merits for use of supercritical fluid technology (not only supercritical region but also subcritical and hydrothermal water) with

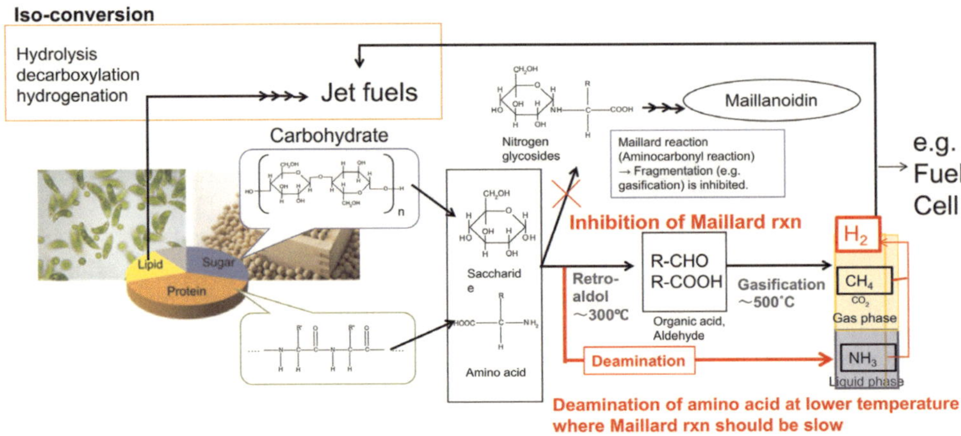

Figure 1.
Proposed process for liquid fuel production from protein-rich biomass by hydrothermal conversion together with "iso-conversion."

ionic liquid as cosolvent: reduction of viscosity of ionic liquid solution [26], control of acidity and basicity of ionic liquid, solubility control of material in ionic liquid, separation of target molecule from ionic liquid into supercritical fluids, and so on. Our research group also progress the study of the separation of some components (biomass molecules or biomass refinery are targeted) from ionic liquid solution with supercritical carbon dioxide [27–30]. Here, we only briefly introduced some research results concerning biomass conversion into useful molecule in ionic liquid-hydrothermal water mixture with additives (Section 4).

In the conclusion of this chapter, we will conclude this chapter and mention outlook in this field.

2. Hydrothermal lignin: aromatic hydrocarbons from woody biomass

2.1 Reducing molecular weight of lignin

Lignin is one of main components of lignocellulosic biomass and a natural phenolic polymer with propyl-phenol groups. One big drawbacks of natural lignin in originally existing woody biomass is large molecular weight (because high molecular weight lignin has high viscosity or it is solid at room temperature and it has a complicated structure), and molecular weight reduction of lignin will provide many advantages in handling and processing as a chemical feedstock. In the paper [2], we proposed an environmental-friendly process for recovery of lignin from woody biomass, and it should have high efficiency with producing simple structure lignin, which is highly desired as a biorefinery feedstock. Here, the brief introduction of the process and the index of the operational parameters (how to know the optimum conditions) is summarized.

2.2 Experimental for hydrothermal lignin preparation

In the paper [2], low-grade wood from Japanese cedar production was provided from a domestic wood manufacturer and had a lignin content of 44.6 wt% (the sample should be mainly bark of the Japanese cedar). The bark was crushed to have a coarse particle size of less than 5 mm. The elemental balance of the Japanese cedar

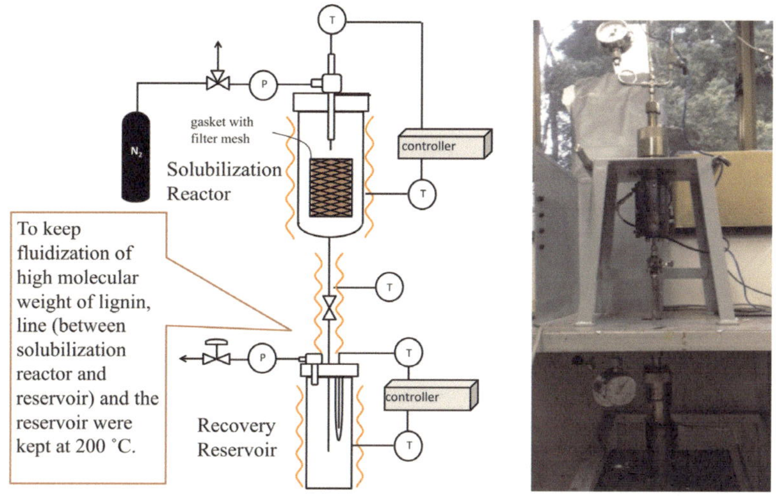

Figure 2.
Coupled system of two batch reactors.

was 49, 5.9, 0.1, 1, and 42.2 wt% for C, H, N, S, and O content, respectively. Ash content was 2 wt%.

A coupled system of two batch reactors was used (the photograph and schematic diagram of the reactor is shown in **Figure 2**, and its detail can be seen elsewhere [2]), and the sample was loaded into the upper reactor. The lower reactor was reservoir and used to recover hydrothermal extracts. The reservoir was maintained at a temperature of 473 K to inhibit the aggregation of lignin because lignin typically is melted into liquid phase at temperatures greater than 423 K in the presence of water [31]. Typically, a 7.5 g sample of bark (covered with stainless mesh) was loaded in the upper reactor along with 150 mL of pure water. After purging air inside the reactor, N_2 gas was loaded. The upper reactor was heated up to 473–573 K. The pressure in the upper reactor (sample-loaded reactor) was close to the saturated pressure of water at the reaction temperature. After the desired treatment time, the valve between the two reactors was opened and the fractionated components in the reactor were recovered in the reservoir. The reservoir was then rapidly cooled with a water bath (ca. 10 min) and was then detached from the sample-loaded reactor. The residue in the upper (sample-loaded) reactor with the stainless mesh was weighed after drying at 333 K overnight. Connection lines between the two reactors were rinsed. The water-insoluble fraction among the recovered solutions was hydrothermal-soluble lignin (namely "hydrothermal lignin"). The soluble Klason lignin (hydrothermal lignin) was quantified, and the molecular characteristics (molecular weight with GPC and molecular structure with NMR) were evaluated.

As a result of the molecular characteristics, the hydrothermal lignin obtained from bark of Japanese cedar by hydrothermal extraction at 523 K under the saturated pressure of water, 3.98 MPa, about was mainly 6-mer (about 1 kDa of mean molecular weight) with syringyl-rich structure [2].

2.3 Dimensionless severity number

The kinetic study for lignin recovery (how much percentage of Klason lignin can be recovered based on the Klason lignin in the raw material, the bark of Japanese cedar) was performed, and severity factor was used for consideration of the effect of reaction temperature and reaction time. The severity factor, R_o [min], is defined as below:

$$R_o = exp\left(\frac{T - T_r}{\omega}\right)t \quad (1)$$

where T [K] is reaction temperature, T_r [K] is the reference temperature (here is 373 K), t [min] is reaction time, and ω is the function of activation energy, E_a [kJ mol^{-1}], gas constant, R = 8.314 J mol^{-1} K^{-1}, and T_r, as shown in the following equation.

$$\omega = \frac{R \bullet T_r^2}{E_a} \quad (2)$$

According to the derivation of the severity model, lignin recovery (LR) can be calculated by the next equation.

$$Lignin\ recovery\ (LR)[wt\%] = \{1 - exp(-k_r \bullet R_o)\} \times 100 \quad (3)$$

where k_r is the rate constants in min^{-1} at reference temperature T_r (373 K), as described in the following equation:

$$k_r = k_o exp\left(-\frac{E_a}{RT_r}\right) \quad (4)$$

where k_o [min^{-1}] is pre-exponential factor of the reference rate constant.

By fitting the equations to the experimental data (as shown in **Figure 3**), all the kinetic parameters were decided as to be 135 kJ mol^{-1} for E_a and 6.8×10^{-5} min^{-1} for k_r. The experimental data at the most severe condition ($k_r \cdot R_o$ value is over 1, and these data were obtained at 573 K for 10 min) were far from the calculated value, and the reason why the data are not fitted with the calculated value is because the cellulose in the bark was carbonized and the remained Klason lignin in the residue was estimated to be high.

As shown in **Figure 3**, lignin solubilization (lignin recovery) in hydrothermal water drastically changed between 0.1 < $k_r \cdot R_o$ value <10. The similar trend was seen in the previous reports [2, 32, 33]. From Eqs. (1), (2), and (4), the dimensionless severity number ($k_r \cdot R_o$) is written by the below equation:

$$k_r \bullet R_o = k_o \bullet t \bullet exp\left(\frac{E_a(T - 2T_r)}{R \bullet T_r^{\ 2}}\right) \quad (5)$$

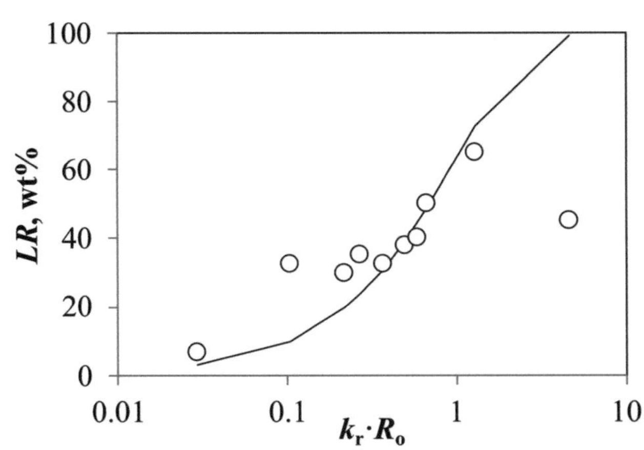

Figure 3.
Dimensionless severity number for lignin recovery from bark of Japanese cedar (473–573 K for 3–40 min).

Eq. (5) shows that the dimensionless severity number ($k_r{\cdot}R_o$) is also the function of t and T as well as severity factor (R_o) as shown in Eq. (1). While R_o varies a wide range of values and the optimum R_o for the target reaction is not easy to estimate, meaningful transitional change should occur at the range of 0.1 to 10 for the dimensionless severity number $k_r{\cdot}R_o$, as reported here and in the literatures [2, 32, 33]. Therefore, the dimensionless severity number ($k_r{\cdot}R_o$) is quite useful, and the severity analysis using $k_r{\cdot}R_o$ can be adopted to various type of the reactions. Our research group is going to study on utilization of the dimensionless severity number ($k_r{\cdot}R_o$) for various kind of reactions and its application and usefulness with a variety of the examples will be reported in near future.

3. Protein hydrolysis and amino acid degradation: low temperature catalytic hydrothermal cracking (LT-CHTC) for liquid fuel production

3.1 How to convert protein-rich biomass into liquid fuels

Based on the concept of "sustainable development goals (SDGs)" by UN and "circular economy" by EU, waste materials (by-product of main activity) must be used in effective and useful application and ideally must be back to primary resource production. In our daily life, huge amount of food waste generates and its utilization/valorization is still insufficient. This situation is not only for food waste (including house waste, municipal waste, sewage sludge, which are by-product of our daily life) but also agricultural by-product. One of the big problems of these wastes is to contain high amount of water, which is hard to exclude from the wastes. Water inside the waste (namely wet biomass) assists to cultivate fungi, and some of the wet biomass wastes are easy to rot with a bad odor and toxic component formation. Even with huge amount of energy loss, the large part of the wet wastes is forcedly burned to eliminate without effective utilization as energy or thermal purposes.

Microalgae, which has fast grow rate with CO_2 stabilization through photosynthesis and attracts much attention as renewable resource, is also wet biomass as the same as daily-life wastes with containing a large amount of water mentioned above. The total difference of microalgae from the waste wet biomass is that cultivation and harvesting of microalgae can be controlled to level not only for yield but also composition. Thus, microalgae is a primary and industrial renewable resource. However, problem related to water is to be solved as well as wet biomass waste.

Sub and supercritical water technologies enable water (also in the wet biomass) to eliminate from the structure of the biomass and to use for hydrolysis of the natural polymer in the biomass as reactant. As mentioned above, supercritical water plays the important role in the CHTC for producing olefinic hydrocarbon from lipid and free fatty acid, as solvent and reactant. But, for feeding lipid into the process, extraction of the lipid from microalgae must be done. Typical lipid extraction processes from microalgae are solvent extraction and mechanical press, both of which require dry to improve lipid yield before the extraction. Also, in iso-conversion concept, the residue of the extraction of lipid (protein and carbohydrate are main components) will be treated by anaerobic fermentation to form small molecules such as organic acids. Advantage of the biochemical process is high selectivity, but disadvantage is low reaction rate. For the purpose of liquid fuels production, broad product distribution in the range of small molecular weight is acceptable. Or rather, rapid transformation process has to be necessary for the residue conversion of the residual biomass containing protein and carbohydrate.

Thus, the whole biomass including lipid, protein, and carbohydrate should be treated by sub and supercritical water technologies in the presence of an appropriate catalyst (so it should be advanced supercritical technology). Here, we focus on effective conversion of carbohydrates and protein into small fragments, and low temperature catalytic hydrothermal cracking (LT-CHTC) must be designed with the concept of inhibition of Maillard reaction, which is undesired reaction because complex nitrogen-containing components are produced and these will be repolymerized into non-decomposable materials like char or coke.

3.2 Carbohydrate conversion into precursors of liquid fuels

Our research group has studied protein and carbohydrate conversion in hydrothermal water (subcritical water) condition with some catalysts.

It has been revealed that hydrolysis of cellulose, which is main component of carbohydrate in biomass, is rapid in sub and supercritical water without additive [34, 35]. The hydrolysis product of cellulose, glucose, is converted into furan ring compounds [8–13] and further small molecules (including hydrogen gas) in hydrothermal and supercritical water in the presence and absence of catalyst/additive [34, 36–43].

3.3 Protein conversion into small fragments

Protein-rich biomass, defatted soybean meal, was converted into oligopeptides with the advanced supercritical technology (catalytic hydrothermal conversion) [14]. Protein is easily denatured to aggregate into brownish compound, which is hard to be hydrolyzed, with heat. To alter the reactivity of peptide under the heat, rapid heating technique is helpful [14, 44]. In addition, denaturation agent is also useful to improve and accelerate hydrolysis rate of protein into small fragments [14]. The effectiveness of the denaturation agent is known as Hofmeister series, which tells that alkaline earth metal ion such as Ca^{2+} and Mg^{2+} is a strong denaturation agent [45, 46]. Further, pH of the protein solution is important factor for keeping solubility in water, and alkali condition is favor for soy protein solubilization in water [47]. In our previous study, the multiple techniques (microwave technique for rapid heating, Mg^{2+} addition for Hofmeister series and alkali condition) for keeping reactivity and solubility of soy protein were employed for improving the yield of oligomer (over 70% of oligopeptide yield based on the soy protein in the defatted soy bean meal) from defatted soy bean meal at 190°C under saturated pressure of water for 1 h [14]. The oligopeptide obtained from the advanced supercritical technology for protein conversion (rapid heating with high Hofmeister additive in subcritical water) shows high foaming ability, antioxidant activity, and high inhibitory activity on angiotensin I converting enzyme (ACE) [14]. We also confirmed that the oligopeptide can be converted into free amino acid with over 80% yield at an advanced supercritical fluid technology (patent application and data are not shown), and it will be reported in near future elsewhere.

3.4 Alanine degradation in subcritical water

In the concept of the total protein-rich biomass conversion into liquid fuel with advanced supercritical fluid technology (new "iso-conversion" for total protein-rich biomass), free amino acid produced through hydrolysis of protein and oligopeptides is further degraded into small molecular weight components such as organic acids and aldehydes (**Figure 1**). To know the optimum condition of amino acid degradation in subcritical water at a wide range of reaction temperature, alanine is selected as a standard amino acid. In a literature [48], it was found that the rate of thermal decomposition of amino acids were correlated with Taft rule, which allows the prediction of the rate of different amino acid based on a reaction rate of standard amino acid with substituent parameter, σ^*, and reaction parameter. It is expected that the same relation of degradation rate of amino acid in subcritical water can be found (it will be revealed in near future by our research group) and alanine is appropriate as the standard.

3.4.1 pH dependence of rate constant of alanine degradation

Here, pH dependence of the reaction rate of alanine was investigated because an appropriate additive for improvement of alanine conversion will be looked for. For the objective, an overall rate constant for 1st order alanine degradation at a wide range of pH condition [49] was considered:

$$k_{\mathrm{Ala}} = \frac{k_{\mathrm{Ala}^+}[\mathrm{H}^+]^2 + k_{\mathrm{Ala}^\pm}Ka_1[\mathrm{H}^+] + k_{\mathrm{Ala}^-} \bullet Ka_1Ka_2}{[\mathrm{H}^+]^2 + Ka_1[\mathrm{H}^+] + Ka_1Ka_2} \tag{6}$$

where k_{ala} is the rate constant of 1st order overall alanine degradation, $k_{\mathrm{ala}+}$ is the rate constant of 1st order overall cationic alanine (Ala^+) degradation, $k_{\mathrm{ala}\pm}$ is the rate constant of 1st order overall zwitterion alanine ($\mathrm{Ala}^\pm$) degradation, and $k_{\mathrm{ala}-}$ is the rate constant of 1st order overall anionic alanine (Ala^-) degradation. $[\mathrm{H}^+]$ is the concentration of proton in the system. Ka_1 is the dissociation constant between cationic (Ala^+) and zwitterion alanine ($\mathrm{Ala}^\pm$), and Ka_2 is the dissociation constant between zwitterion ($\mathrm{Ala}^\pm$) and anionic alanine (Ala^-), as shown in the below equation:

$$\mathrm{Ala}^+ \overset{Ka_1}{\leftrightarrow} \mathrm{Ala}^\pm \overset{Ka_2}{\leftrightarrow} \mathrm{Ala}^- \tag{7}$$

In the literature [49], the rate of decarboxylation (which is main reaction at sub and supercritical water over 573 K [15]) was measured with monitoring CO_2 formation by in situ FTIR. Here, in Eq. (6), it is considered that pH dependence of the 1st order rate constant of alanine evaluated from the rate of decarboxylation was almost the same as the rate of overall alanine degradation (k_{ala}). The dissociation constant of Ka_1 and Ka_2 in subcritical condition can be calculated by revised HKF model with the reported parameters [50]. We performed the alanine degradation in subcritical water (260~360°C at 20 MPa) not only neutral condition but also acid (5 mmol dm^{-3} of H_2SO_4 solution) and alkali (20 mmol dm^{-3} of NaOH solution) conditions to know the intrinsic first-order rate constant of degradation for all kinds of alanine species ($k_{\mathrm{ala}+}$ for cationic, $k_{\mathrm{ala}\pm}$ for zwitterion and $k_{\mathrm{ala}-}$ for anion). **Figure 4** shows the alanine species distribution in the reaction conditions.

Here, the dissociation constant of H_2SO_4 was correlated by the equation proposed by Oscarson et al. [51]. For NaOH, the dissociation constant was calculated by the revised HKF (Helgeson-Kirkham-Flowers) model [52].

In the alkali condition (**Figure 4C**), only the anionic alanine (Ala^-) is present. At the neutral condition (**Figure 4B**), the composition of zwitterion alanine ($\mathrm{Ala}^\pm$) is major. With increasing temperature, the composition of anionic alanine (Ala^-) is drastically increased and the composition of anionic alanine (Ala^-) is higher than that of zwitterion alanine ($\mathrm{Ala}^\pm$). In the acidic condition (**Figure 4A**), the composition of cationic alanine (Ala^+) is major species and its composition becomes higher with increasing temperatures.

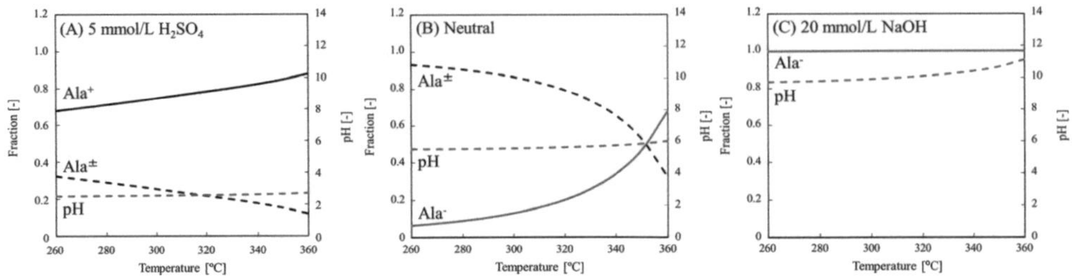

Figure 4.
Alanine species distribution in subcritical water at (A) acid, (B) neutral, and (C) alkali conditions at 20 MPa.

In this part, pH dependence of the overall rate constant of alanine degradation at a wide range of reaction temperature and pH condition is explained. To correlate the rate constant, a correlation was performed:

i. At alkali condition, k_{ala-} was measured and it was the same as the overall rate constant, k_{ala}.

ii. The apparent rate constant was measured at neutral condition as k_{ala}, which includes both contribution of anionic and zwitterion alanine as below.

$$-\frac{d[\mathrm{Ala}]}{dt} = k_{\mathrm{Ala}}[\mathrm{Ala}]_0 = k_{\mathrm{Ala}^-}[\mathrm{Ala}^-] + k_{\mathrm{Ala}^\pm}[\mathrm{Ala}^\pm] \tag{8}$$

By using k_{ala-} measured in alkali condition and the concentration of each species ($Ala^{\pm}$ and Ala^{-}) calculated as shown in **Figure 4B**, $k_{Ala\pm}$ was obtained by the below equation.

$$k_{\mathrm{Ala}^\pm} = \frac{k_{\mathrm{Ala}}[\mathrm{Ala}]_0 - k_{\mathrm{Ala}^-}[\mathrm{Ala}^-]}{[\mathrm{Ala}^\pm]} \tag{9}$$

where $[Ala]_0$ was the initial concentration of alanine. All brackets, [X], are correspondent to concentration of species X in mole m^{-3}.

iii. For k_{Ala+} was obtained as the same manner of $k_{Ala\pm}$ mentioned above.

3.4.2 Experimental for alanine conversion in subcritical water

Figures 5 and **6** show a photograph and schematic diagram of a flow apparatus that used for subcritical water decomposition of alanine, respectively. The reactor consisted of two pumps, preheater, reactor, cooler, and back-pressure regulator. The preheater was made of SUS316 stainless steel (length 40 m., id: 1 mm, od: 1/16 inch) and placed in a GC oven (HP6890, Agilent Technologies). The reactor was also made of SUS316 stainless steel (length 100 m, id: 1 mm, od: 1/16 inch, volume of reactor: 81.2 cm^3) placed in another GC oven (6890N, Agilent Technologies). The shell tube heat exchanger for cooling made of SUS316 stainless steel (length 30 m., id: 1 mm, od: 1/16 inch) was located in a chiller (LE-600, Advantec). The pressure was controlled at 20 MPa by the back-pressure regulator (TESCOM). The sample solution and water were supplied by pumps (PU-2100S, Jasco and NP-KX-500, Nihon Seimitsu Co., Ltd.). For rapid heating, water was fed to mixing part with flow ratio of 1:9 or 1:3 (sample solution:water) and preheated. The preheater temperature was decided by consideration of enthalpy balance to reach the mixing temperature to the targeted temperature, which was in the range of 260–360°C.

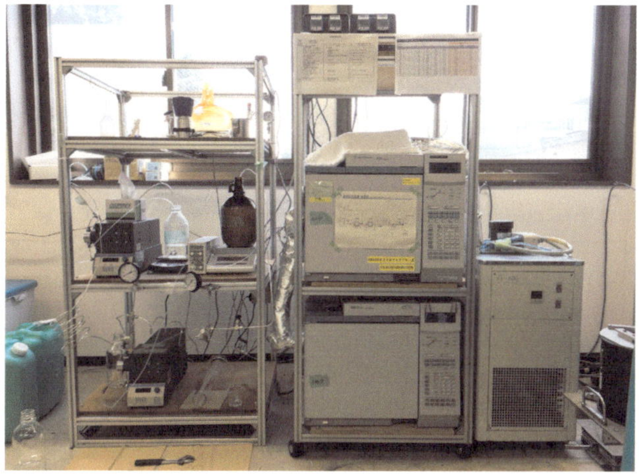

Figure 5.
Photograph of continuous flow reactor.

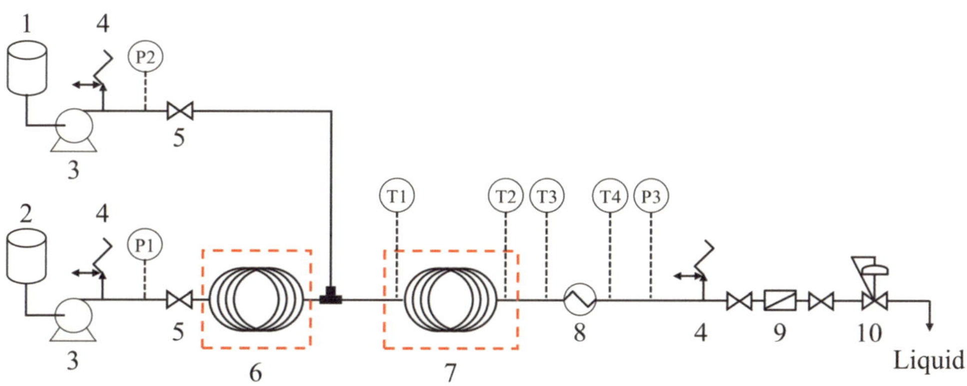

Figure 6.
Schematic diagram of continuous-flow system (1: alanine solution, 2: water, NaOH, or H_2SO_4, 3: pump, 4: relief valve, 5: valve, 6: preheating unit, 7: reaction unit, 8: cooling unit, 9: filter, 10: back-pressure regulator).

The effluent was cooled down in the cooler and collected. Reaction time was varied in the range of 60–300 s by changing flow rates of the reaction fluids and calculated based on the water density at reaction temperature (260–360°C) and pressure (20 MPa).

Alanine in the outlet solution was measured by HPLC with an ODS column (Shim-pack VP-ODS, Shimazu) and PDA detector (L-7455, Hitachi) at wavelength of 200 nm. The column oven was 60°C. The carrier were sodium 50 mM of sodium phosphoric acid with 7.2 mM of hexane sulphonate set to pH 2.5 (solution A) and HPLC grade acetonitrile (solution B) with flow ratio of 96:4 (A:B) and the flow rate was 1.0 mL/min.

Conversion of alanine was calculated from the concentration of alanine before and after the reaction and the rate constants (k_{Ala-}, $k_{Ala\pm}$ and k_{Ala+}) by already mentioned protocol: (i) k_{Ala-} was measured at alkali condition, (ii) $k_{Ala\pm}$ was correlated from k_{Ala} measured in neutral condition, and (iii) k_{Ala+} was obtained at acidic condition by correlation with the measured k_{Ala} measured and $k_{Ala\pm}$.

3.4.3 Rate constant of alanine at wide temperature and pH conditions

Figure 7 shows the Arrhenius plots of the intrinsic rate constant of each alanine species degradation in subcritical water at 20 MPa. The data for k_{Ala+} and k_{Ala-} were largely scattered, and the correctness of the data must further be investigated. Here, we analyzed the trend of the rate constant for each species as the order and the temperature dependence (namely activation energy) should be correct. The activation energy of the rate constant of zwitterion, $k_{Ala\pm}$, was the highest among all the rate constants shown in **Figure 7**. Thus the reactivity of zwitterion alanine becomes higher than the others. **Figure 8** shows the simulation of pH dependence of the overall rate constant of alanine degradation in subcritical water at 20 MPa and various temperatures.

At the range of reaction temperatures from 340 (**Figure 8A**) to 280°C (**Figure 8D**), the simulation of pH dependence of the overall rate constant of alanine degradation, k_{Ala}, (drawn with the bold line) well explained the experimental data (plots in the figures). As shown in **Figure 8**, over 300°C, zwitterion alanine is the most active species and surprisingly an acid or alkali catalyst is not useful for enhancing degradation rate. Below 280°C, alanine degradation should be enhanced by an acid or alkali catalyst. But lower temperature, the reaction rate is so low and decomposition of alanine requires long time. For the pretreatment of protein-rich biomass to produce the precursor of hydrocarbons such as organic acids, the enhancement of the reaction rate must be necessary. To seek an appropriate additive, some types of solid acid additives were screened.

3.5 Amino acid deamination with solid acid additives in hydrothermal water

3.5.1 Previous studies on solid catalytic reaction of amino acid in hydrothermal (subcritical) water

Since amino acid is monomer unit of protein and key compound in life, amino acid reaction in hydrothermal system is highly motivated to know the origin of life because protein is basic molecule for life and polypeptide formation, namely protein synthesis, should be the first step. Hydrothermal vent under deep sea is seemed to be one of the spaces where is the original place of life. To form protein, survivability of amino acid at such severe condition (hydrothermal water is partially higher than 200°C) is crucial for the origin of life, and the stability of amino acid at hydrothermal water condition is important. In the origin of life point of view, amino acid touches on the surface of inorganic materials at the deep sea near the hydrothermal vent and the effect of the inorganics on the stability of amino acid has

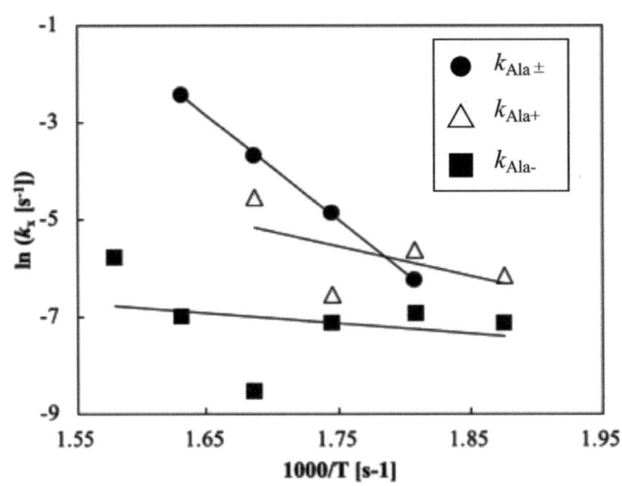

Figure 7.
Intrinsic rate constant of each alanine species at 20 MPa.

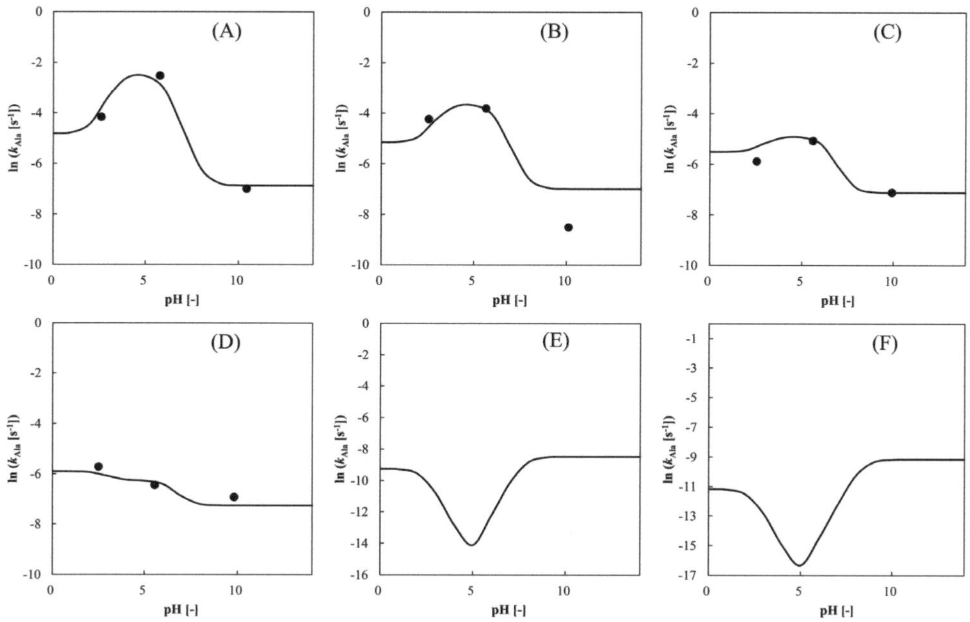

Figure 8
pH dependence of the overall rate constant of alanine degradation, k_{Ala}, at 20 MPa: (A) 340°C, (B) 320°C, (C) 300°C, (D) 280°C, (E) 150°C, and (F) 100°C.

been studied. It was reported that iron compounds enhance dimerization to form alanyl-alanine and diketopiperazine at hydrothermal condition [53]. Glycine and alanine are reacted at 100°C for 35 days with metal ferrites such as $NiFeO_4$ to obtain 30% of yield of dimerization products [53].

As amino acids to be key compounds in biomass refinery, there have also been some reports concerning conversion of amino acids into chemicals with solid additives. It was reported that Pt/TiO_2 promotes the reductive deamination of glutamic acid derivatives to form organic acids at 225°C for 24 h [54]. By the same research group of [54], decarbonylation of amino acids was attempted by Ru-based catalyst [55]. Reductive (hydrogenation) decarbonylation of valine to isobutylamine was developed in the presence of Ru-catalyst with isobutylamine yield of 87% in hydrothermal water at 150°C for 2 h.

3.5.2 Sulfonyl solid acid catalytic reactions of alanine in hydrothermal (subcritical) water

As described above, there have been several papers concerning some solid catalytic reactions of amino acids in hydrothermal water; however, solid acid or base reactions on hydrothermal (subcritical water) reaction of amino acids was not investigated so far.

Based on the analysis of the pH dependence of degradation of alanine, solid acid or base catalyst must be effective. The batch experiments were performed with microwave hydrothermal small reactor (**Figure 9**). The detail of the apparatus and procedures are found elsewhere [13]. Here, conversion of alanine was investigated in hydrothermal water with and without sulfonyl solid catalyst at 150°C for 1–60 min. Basically, 5.0 g of 110 mmol dm^{-3} of amino acid solution was loaded in a glass reactor with 0.5 g of solid sulfonyl additive such as sulfonyl carbon (detail is not shown because of the patent application) and cationic (having sulfonyl group) ionic exchange (Amberlyst 45) resin. As shown in **Figure 10**, alanine is decomposed

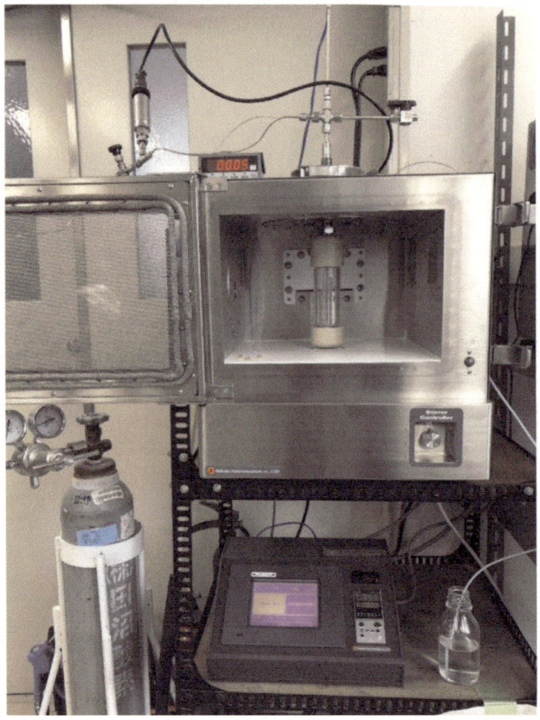

Figure 9.
Microwave hydrothermal small reactor.

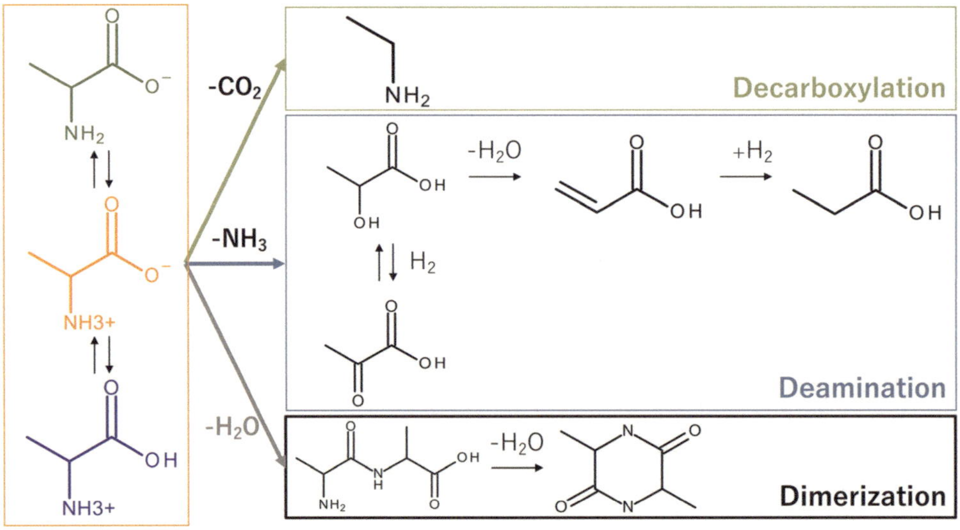

Figure 10.
Main pathways of alanine degradation in hydrothermal water.

through decarboxylation, deamination, and dimerization. To know which pathway is developed, decarboxylation product, CO_2, was detected by CO_2 meter and fingerprint of deamination, NH_3, was analyzed by an ion chromatography.

Figure 11 shows the results. Amberlyst 45 was inert for deamination of amino acids at hydrothermal water at 150°C. In the presence of the sulfonyl carbon, deamination of alanine was progressed. Not shown here, the other pathways of alanine degradation were not found for all the cases in these experiments.

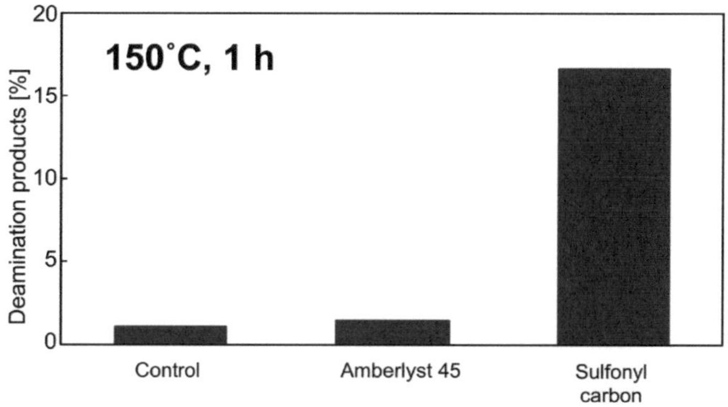

Figure 11.
Yield of deamination product, NH_3, of alanine in hydrothermal water with and without sulfonyl solid additive at 150°C for 1 h (under saturated pressure of water, 0.48 MPa).

On Amberlyst 45, 2.95 eq kg^{-1} of sulfonyl group is bearded, while 1.90 eq kg^{-1} of sulfonyl group is attached on the surface of sulfonyl carbon. The catalytic activity for deamination of alanine is probably related to oxygenated function group (OFGs) such as hydroxyl and carboxyl because the sulfonyl carbon has 2.40 mmol g^{-1} of OFGs (Amberlyst 45 has no OFG except for sulfonyl group). More detail study for amino acid deamination on the sulfonyl carbon catalyst is now on-going and it will be reported in near future.

3.5.3 Sulfonyl carbon reactions of okara in hydrothermal (subcritical) water

Through the experiments, it was revealed that sulfonyl carbon was active for deamination of amino acids. It must be resulted in the activation of alanine at strong acid condition (not only by sulfonyl group but also by OFGs) as predicted (**Figure 8**). The LT-CHTC, mainly for deamination, is favored as liquid fuel pretreatment as proposed in **Figure 1**. Here, to know the usefulness of the reaction for a real biomass, "okara," which is a residue of soy bean cake ("Tofu") production, was deaminated in the presence of the sulfonyl carbon. The elemental composition of okara used in this study is C:H:N:S:O (wt%) = 47.2:7.29:5.85:0.43:39.2 (the weight percentage of oxygen atom was subtraction of total weight % of C, H, N. and S from 100 and it should include ash, which will be analyzed by TGA). As a result, a large portion of amino group in okara was detached as ammonia, NH_3, (detail is not shown at this moment because of the reason for patent application). Thus, the sulfonyl carbon catalytically assisted NH_2-group abstraction as NH_3. It means that Maillard reaction during okara conversion was less happened and much organic acids were formed. For this study, now the detail is being studied and it will be also reported in near future.

4. Up-conversion of carbohydrate in ionic liquid and supercritical fluids mixtures

Cellulose and chitin are abundant biomass and these up-conversion (valorization) into useful compounds are aggressively being investigated. Our research group has studied cellulose and chitin conversion with advanced supercritical technologies. As one of advanced supercritical technologies, ionic liquid utilization together with supercritical fluid technologies (including subcritical and hydrothermal water) has been studied. Here, two biomass conversion process are focused on: furan ring compound formation from cellulose and hydrogenation.

4.1 HMF formation from cellulose

HMF is a platform compound that can be derived from cellulose in coming biomass society [23, 24]. It is known that the formation of HMF from cellulose is three-step reaction: (1) hydrolysis of cellulose into glucose, (2) glucose isomerization to fructose, and (3) fructose dehydration to HMF, as shown in **Figure 12**. Step 1 and step 3 are acid-catalyzed reactions, and step 2 is a base-catalyzed reaction. To increase the yield of HMF, optimization of each reaction step is necessary.

For step 1, selective glucose production by cellulose hydrolysis with sequential water addition in the presence of solid acid catalysts in 1-butyl-3-methyl imidazolium chloride ([bmIm][Cl]) under microwave irradiation was investigated at 120°C, where water is hydrothermal condition [23]. It revealed that the large amount of water added in the initial stage was inhibited to form homogeneous phase among three components (cellulose-water-[bmIm][Cl]) and some amounts of cellulose was precipitated. The precipitation of cellulose was resulted in the low yield of glucose because hydrolysis of cellulose was stopped. To keep the cellulose dissolution and improve hydrolysis of cellulose into glucose, sequential water addition where water is added in steps as the reaction proceeds was investigated. Glucose yields in the presence of Amberlyst-15 in [bmIm][Cl] were 75.0 mol% by three step of water addition [23].

Step 2 (glucose isomerization, which requires basic catalyst) was also optimized in the presence of hydrothermal water with [bmIm][Cl] at 120°C, and the effect of water on step 3 (glucose into HMF) was investigated [24]. For isomerization of glucose in the presence of 35 wt% of hydrothermal water in [bmIm][Cl], $MgCO_3$ was the most effective base additives (23.1 mol% of fructose yield with 85.3 mol% selectivity was obtained at 120°C for 30 min). For dehydration of fructose into HMF with Amberlyst 15 (Step 3), the rate of dehydration of fructose was slightly reduced by adding water but not so large effect was confirmed. Also in the paper [24], continuous HMF production process via three-step reaction is proposed with separation of HMF from [bmIm][Cl] and water mixture. It was found that dimethyl ether was a good separation solvent and almost 100% of HMF was recovered. Finally, virtual integration of the three-step process tells us that the yield of HMF from cellulose is estimated to be 32 mol%.

4.2 Hydrogen solubility in ionic liquid with supercritical CO_2

The other important reaction routes from biomass into useful material are hydrogenolysis and hydrogenation of biomass [56]. Ionic liquid media has advantage of biomass conversion because solubilization of cellulose and chitin in some types of ionic liquid is high. One of drawbacks of the ionic liquid process is high viscosity. Supercritical CO_2 assists reduction of viscosity of ionic liquid-cellulose mixture [26]. For the process of hydrogen addition (such as hydrogenolysis and

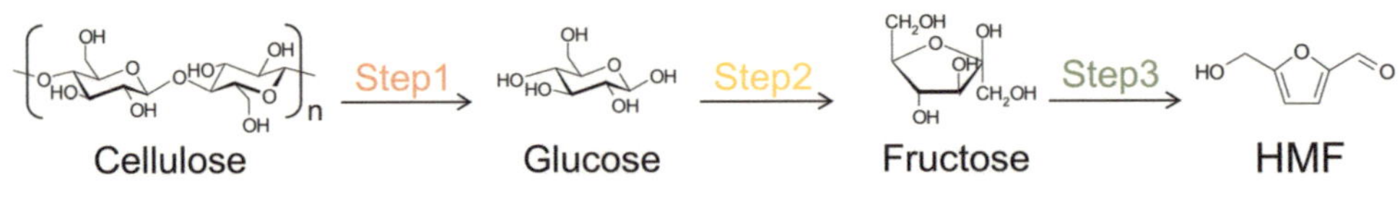

Figure 12.
HMF formation from cellulose.

hydrogenation), the synergetic effect of hydrogen and supercritical CO_2 coexistence must be revealed.

Hydrogen solubility was experimental measured and a simple correlation for hydrogen solubility in ionic liquids in the presence of CO_2 was developed from available binary/ternary data. The correlation could provide reliable estimation of hydrogen solubility enhancement by CO_2 for six ionic liquids at 313–453 K [56]. In the study [56], a definition for enhancement ratio (ER) based on molality and applied it to available hydrogen-CO_2-ionic liquid systems were proposed. It was found that the ER was convenient for examining the trends of solubility change of hydrogen in an ionic liquid with CO_2 concentration or temperature. Also in the study [56], hydrogen solubility in the presence of CO_2 was estimated for biomass soluble ionic liquids.

5. Conclusions

In this chapter, the concept of liquid fuels from biomass in advanced supercritical fluid is firstly explained. To know the optimum condition for various kind of biomass, dimensionless severity number should be useful and the application for lignin recovery in hydrothermal and subcritical water was shown. Then, as the application of one of the advanced supercritical fluid technologies, hydrothermal water with catalyst process for carbohydrate and protein performed by our research group was briefly reviewed. For total biomass utilization as liquid fuel production, protein fragmentation is a key process and some of experimental research results were shown. Particularly, alanine conversion with and without additive was deeply considered. To inhibit Maillard reactions, the deamination of amino acid at low temperature was quite important, and it was found that a sulfonyl carbon was quite active for deamination of alanine at 150°C in hydrothermal water. Cellulose conversion in ionic liquid with supercritical fluid (hydrothermal water) to furan ring compound (HMF) was introduced. For hydrogenation of biomass, hydrogen solubility was controlled in the presence of supercritical carbon dioxide and it shows that ionic liquid with supercritical CO_2 is favored hydrogenolysis and hydrogenation reaction field for up-conversion of biomass.

Conflict of interest

The authors declare no conflict of interest.

Nomenclature

[Ala]	concentration of alanine, mol m^{-3}
$[Ala]_0$	initial concentration of alanine, mol m^{-3}
$[Ala^-]$	concentration of anionic alanine, mol m^{-3}
$[Ala^\pm]$	concentration of zwitterion alanine, mol m^{-3}
E_a	Activation energy, kJ mol^{-1}
$[H^+]$	concentration of proton, mol m^{-3}
K_{a1}	dissociation constant between cationic and zwitterion alanine
K_{a2}	dissociation constant between zwitterion and anionic alanine
k	rate constant, min^{-1}

k_o	pre-exponential factor, min^{-1}
k_r	rate constant, min^{-1}
LR	lignin recovery, wt%
R	gas constant, 8.314 J mol^{-1} K^{-1}
R_o	severity factor, min
T	temperature, K
t	reaction time, min
T_r	reference temperature, 373 K

Greek letters

σ^*	substituent parameter for Taft rule

Subscripts or superscripts

Ala	alanine
ala+	cationic alanine
ala_-	anionic alanine
ala±	zwitterion alanine

Author details

Masaru Watanabe*, Masayoshi Wagatsuma, Keisuke Suzuki, Takuma Kato, Yasuto Goto, Yukihiro Kanaguri and Yuya Hiraga
Research Center of Supercritical Fluid Technology, Tohoku University, Sendai, Japan

*Address all correspondence to: meijin@scf.che.tohoku.ac.jp

References

[1] The American Carbon Footprint: 19t CO_2e. 2010. Available from: http://shrinkthatfootprint.com/american-carbon-footprint [Accessed: 20 August, 2019]

[2] Watanabe M, Kanaguri Y, Smith RL. Hydrothermal separation of lignin from bark of Japanese cedar. Journal of Supercritical Fluids. 2018;**133**:696-703. DOI: 10.1016/j.supflu.2017.09.009

[3] Li L, Coppola E, Rine J, Miller JL, Walker D. Catalytic hydrothermal conversion of triglycerides to non-ester biofuels. Energy & Fuels. 2010;**24**: 1305-1315. DOI: 10.1021/ef901163a

[4] US2008/0071125A1

[5] Biofuels ISOCONVERSION. Available from: https://www.ara.com/products/biofuels-isoconversion [Accessed: 21 August, 2019]

[6] Watanabe M, Iida T, Inomata H. Decomposition of a long chain saturated fatty acid with some additives in hot compressed water. Energy Conversion and Management. 2006;**7**:3344-3350. DOI: 10.1016/j.enconman.2006.01.009

[7] Watanabe M, Inomata H, Smith RL, Arai K. Catalytic decarboxylation of acetic acid with zirconia catalyst in supercritical water. Applied Catalysis A: General. 2001;**219**:149-156. DOI: 10.1016/S0926-860X(01)00677-9

[8] Watanabe M, Iida T, Aizawa Y, Ura H, Inomata H, Arai K. Conversions of some small organic compounds with metal oxides in supercritical water at 673 K. Green Chemistry. 2003;**5**: 539-544. DOI: 10.1039/B304686A

[9] Watanabe M, Aizawa Y, Iida T, Aida TM, Levy C, Sue K, et al. Glucose reactions with acid and base catalysts in hot compressed water at 473K. Carbohydrate Research. 2005;**340**: 1925-1930. DOI: 10.1016/j.carres.2005.06.017

[10] Watanabe M, Aizawa Y, Iida T, Levy C, Aida TM, Inomata H. Glucose reactions within the heating period and the effect of heating rate on the reactions in hot compressed water. Carbohydrate Research. 2005;**340**:1931-1939. DOI: 10.1016/j.carres.2005.05.019

[11] Watanabe M, Aizawa Y, Iida T, Nishimura R, Inomata H. Catalytic glucose and fructose conversions with TiO_2 and ZrO_2 in water at 473K: Relationship between reactivity and acid–base property determined by TPD measurement. Applied Catalysis A: General. 2005;**295**:150-156. DOI: 10.1016/j.apcata.2005.08.007

[12] Qi X, Watanabe M, Aida TM, Smith RL. Catalytical conversion of fructose and glucose into 5-hydroxymethylfurfural in hot compressed water by microwave heating. Catalysis Communications. 2008;**9**:2244-2249. DOI: 10.1016/j.catcom.2008.04.025

[13] Watanabe M, Qi X, Aida TM, Smith RL. Microwave apparatus for kinetic studies and in-situ observations in hydrothermal or high-pressure ionic liquid system: Setup and some experimental results concerning biomass conversion. In: Cao W, editor. The Development and Application of Microwave Heating. London: InTech; 2012. pp. 163-178. DOI: 10.5772/45625

[14] Wagatsuma M, Miyajima A, Yoshii T, Watanabe M, Aida TM, Smith RL. Preparation of soluble peptide from defatted soybean in the presence of base additives in hydrothermal condition and evaluation of its function. Kagaku Kogaku Ronbunshu (in Japanese). 2018;**44**:78-84. DOI: 10.1252/kakoronbunshu.44.78

[15] Sato N, Quitain AT, Kang K, Daimon H, Fujie K. Reaction kinetics of amino acid decomposition in high-temperature and high-pressure water. Industrial and Engineering Chemistry Research. 2004;**43**:3217-3222. DOI: 10.1021/ie020733n

[16] Qi X, Watanabe M, Aida TM, Smith RL. Efficient process for conversion of fructose to 5-hydroxymethylfurfural with ionic liquids. Green Chemistry. 2009;**11**: 1327-1331. DOI: 10.1039/B905975J

[17] Qi X, Watanabe M, Aida TM, Lee Smith R. Efficient catalytic conversion of fructose into 5-hydroxymethylfurfural in ionic liquids at room temperature. ChemSusChem. 2009;**2**:944-946. DOI: 10.1002/cssc.200900199

[18] Qi X, Watanabe M, Aida TM, Smith RL. Fast transformation of glucose and di-/polysaccharides into 5-hydroxymethylfurfural by microwave heating in an ionic liquid/catalyst system. ChemSusChem. 2010;**3**: 1071-1077. DOI: 10.1002/cssc.201000124

[19] Qi X, Watanabe M, Aida TM, Smith RL. Efficient one-pot production of 5-hydroxymethylfurfural from inulin in ionic liquids. Green Chemistry. 2010; **12**:1855-1860. DOI: 10.1039/C0GC00141D

[20] Qi X, Watanabe M, Aida TM, Smith RL. Catalytic conversion of cellulose into 5-hydroxymethylfurfural in high yields via a two-step process. Cellulose. 2011;**18**:1327-1333. DOI: 10.1007/s10570-011-9568-1

[21] Qi X, Watanabe M, Aida TM, Smith RL. Synergistic conversion of glucose into 5-hydroxymethylfurfural in ionic liquid–water mixtures. Bioresource Technology. 2012;**109**: 224-228. DOI: 10.1016/j.biortech.2012.01.034

[22] Iguchi M, Aida TM, Watanabe M, Smith RL. Dissolution and recovery of cellulose from 1-butyl-3-methylimidazolium chloride in presence of water. Carbohydrate Polymers. 2013; **92**:651-658. DOI: 10.1016/j.carbpol.2012.09.021

[23] Ishida K, Matsuda S, Watanabe M, Kitajima H, Kato A, Iguchi M, et al. Hydrolysis of cellulose to produce glucose with solid acid catalysts in 1-butyl-3-methyl-imidazolium chloride ([bmIm][Cl]) with sequential water addition. Biomass Conversion and Biorefinery. 2014;**4**:323-331. DOI: 10.1007/s13399-014-0116-8

[24] Watanabe M, Matsuda S, Kitajima H, Smith RL. Continuous process for HMF production from cellulose with ionic liquid ([BmIm]Cl)-water mixtures. Journal of the Japan Institute of Energy. 2017;**96**:417-429. DOI: 10.3775/jie.96.417

[25] Azuma D, Yoshii T, Watanabe M, Hiraga Y, Smith RL, Ogata M, et al. Effect of Lewis and Brønsted acids on conversion of chitin monomer N-acetyl-D-glucosamine (GlcNAc) to furan derivatives in [Bmim]Cl ionic liquid. Kagaku Kogaku Ronbunshu (in Japanese). 2019;**45**:141-146. DOI: 10.1252/kakoronbunshu.45.141

[26] Iguchi M, Kasuya K, Sato Y, Aida TM, Watanabe M, Smith RL. Viscosity reduction of cellulose + 1-butyl-3-methylimidazolium acetate in the presence of CO_2. Cellulose. 2013;**20**: 1353-1367. DOI: 10.1007/s10570-013-9884-8

[27] Hiraga Y, Endo W, Machida H, Sato Y, Aida TM, Watanabe M, et al. Infinite dilution partition coefficients of benzene derivative compounds in supercritical carbon dioxide + ionic liquid systems: 1-butyl-3-methylimidazolium chloride [bmim][Cl], 1-butyl-3-methylimidazolium acetate [bmim][Ac] and 1-butyl-3-

methylimidazolium octylsulfate [bmim] [$OcSO_4$]. Journal of Supercritical Fluids. 2012;**66**:49-58. DOI: 10.1016/j.supflu.2011.08.003

[28] Hiraga Y, Hayasaka A, Sato Y, Watanabe M, Smith RL. Partition coefficients of furan derivative compounds in 1-n-butyl-3-methylimidazolium chloride ([bmim][Cl])–supercritical CO_2 biphasic systems and their correlation and prediction with the LSER-δ model. Journal of Supercritical Fluids. 2013;**79**: 32-40. DOI: 10.1016/j.supflu.2012.11.010

[29] Hiraga Y, Kato A, Sato Y, Aida TM, Watanabe M, Smith RL. Separation factors for [amim]Cl–CO_2 biphasic systems from high pressure density and partition coefficient measurements. Separation Purification Technology. 2015;**155**:139-148. DOI: 10.1016/j.seppur.2015.03.008

[30] Shida T, Hiraga Y, Sugiyama T, Sato Y, Watanabe M, Smith RL. Measurement and modeling of infinite dilution activity coefficients of organic compounds in an equimolar ionic liquid mixture of [Bmim]Cl and [Bmim][Tf_2N]. Fluid Phase Equilibria. 2019; **488**:72-78. DOI: 10.1016/j.fluid.2019.01.028

[31] Montane D, Salvado J, Farriol X, Jpliez P, Chornet E. Phenomenological kinetics of woods delignification: Application of a time-dependent rate constant and a generalized severity parameter to pulping and correlation of pulp properties. Wood Science and Technology. 1994;**28**:387-402. DOI: 10.1007/BF00225458

[32] Montane D, Overend RP, Chornet E. Kinetic model for non-homogeneous complex systems with a time-dependent rate constant. The Canadian Journal of Chemical Engineering. 1998;**76**:58-68. DOI: 10.1002/cjce.5450760108

[33] Garrote G, Dominguez H, Prajo JC. Interpretation of deacetylation and hemicellulose hydrolysis during hydrothermal treatments on the basis of the severity factor. Process Biochemistry. 2002;**37**:1067-1073. DOI: 10.1016/S0032-9592(01)00315-6

[34] Watanabe M, Inomata H, Arai K. Catalytic hydrogen generation from biomass (glucose and cellulose) with ZrO_2 in supercritical water. Biomass and Bioenergy. 2002;**22**:405-410. DOI: 10.1016/S0961-9534(02)00017-X

[35] Torii N, Okai A, Shibuki K, Aida TM, Watanabe M, Ishihara M, et al. Production of D-glucose from pseudo paper sludge with hydrothermal treatment. Biomass and Bioenergy. 2010;**34**:844-850. DOI: 10.1016/j.biombioe.2010.01.029

[36] Osada M, Sato T, Watanabe M, Adschiri T, Arai K. Low-temperature catalytic gasification of lignin and cellulose with a ruthenium catalyst in supercritical water. Energy & Fuels. 2004;**18**:327-333. DOI: 10.1021/ef034026y

[37] Watanabe M, Takahashi M, Inomata H. Hydrogen production reaction with a metal oxide catalyst in high pressure high temperature water. Journal of Physics: Conference Series. 2008;**121**:082008. DOI: 10.1088/1742-6596/121/8/082008

[38] Takahashi M, Aizawa Y, Ota M, Hoshina T, Watanabe M, Sato Y, et al. Hydrogen production from glucose by partial oxidation in high temperature high pressure water 1: Study on reaction condition and effect of ZnO. Journal of the Japan Institute of Energy. 2008;**87**: 706-712. DOI: 10.3775/jie.87.706

[39] Takahashi M, Ota M, Hoshina T, Watanabe M, Sato Y, Inomata H. Hydrogen production from glucose by partial oxidation in high temperature high pressure water 2: Reaction

mechanism and rate of HCOOH decomposition. Journal of the Japan Institute of Energy. 2008;**87**:713-718. DOI: 10.3775/jie.87.713

[40] Michael Aida T, Ikarashi A, Saito Y, Watanabe M, Smith RL, Arai K. Dehydration of lactic acid to acrylic acid in high temperature water at high pressures. Journal of Supercritical Fluids. 2009;**50**:257-264. DOI: 10.1016/j.supflu.2009.06.006

[41] Aida TM, Shiraishi N, Kubo M, Watanabe M, Smith RL. Reaction kinetics of D-xylose in sub- and supercritical water. Journal of Supercritical Fluids. 2010;**55**:208-216. DOI: 10.1016/j.supflu.2010.08.013

[42] Watanabe M, Aida TM, Smith RL, Inomata H. Hydrogen formation from biomass model compounds and real biomass by partial oxidation in high temperature high pressure water. Journal of the Japan Petroleum Institute. 2012;**55**:219-228. DOI: 10.1627/jpi.55.219

[43] Kitajima H, Higashino Y, Matsuda S, Zhong H, Watanabe M, Aida TM, et al. Isomerization of glucose at hydrothermal condition with TiO_2, ZrO_2, CaO-doped ZrO_2 or TiO_2-doped ZrO_2. Catalysis Today. 2016;**274**:67-72. DOI: 10.1016/j.cattod.2016.01.049

[44] Aida TM, Oshima M, Smith RL. Controlled conversion of proteins into high-molecular-weight peptides without additives with high-temperature water and fast heating rates. ACS Sustainable Chemistry & Engineering. 2017;**5**: 7709-7715. DOI: 10.1021/acssuschemeng.7b01146

[45] Zhang Y, Cremer PS. Interactions between macromolecules and ions: The Hofmeister series. Current Opinion in Chemical Biology. 2006;**10**:658-663. DOI: 10.1016/j.cbpa.2006.09.020

[46] Okur HI, Hladílkova J, Rembert KB, Cho Y, Heyda J, Dzubiella J, et al. Beyond the Hofmeister series: Ion-specific effects on proteins and their biological functions. The Journal of Physical Chemistry. B. 2017;**121**: 1997-2014. DOI: 10.1021/acs.jpcb.6b10797

[47] Ali F, Mondor M, Ippersiel D, Lamarche F. Production of low-phytate soy protein isolate by membrane technologies: Impact of salt addition to the extract on the purification process. Innovative Food Science & Emerging Technologies. 2011;**12**:171-177. DOI: 10.1016/j.ifset.2011.01.013

[48] Yablokov VA, Smel'tsova IL, Faerman VI. Thermal stability of amino acids. Russian Journal of General Chemistry. 2013;**83**:476-480. DOI: 10.1134/S1070363213030122

[49] Li J, Wang X, Klein MT, Brill TB. Spectroscopy of hydrothermal reactions, 19: pH and salt dependence of decarboxylation of α-alanine at 280-330°C in an FT-IR spectroscopy flow reactor. International Journal of Chemical Kinetics. 2002;**34**:271-277. DOI: 10.1002/kin.10045

[50] Kitadai N. Energetics of amino acid synthesis in alkaline hydrothermal environments. Origins of Life and Evolution of the Biosphere. 2015;**45**: 377-409. DOI: 10.1007/s11084-015-9428-3

[51] Oscarson JL, Izatt RM, Brown PR, Pawlak Z, Gillespie SE, Christensen JJ. Thermodynamic quantities for the interaction of SO_4^{2-} with H^+ and Na^+ in aqueous solution from 150 to 320°C. Journal of Solution Chemistry. 1988;**17**: 841-863. DOI: 10.1007/BF00646553

[52] Shock EL, Sassani DC, Willis M, Sverjensky DA. Inorganic species in geologic fluids: Correlations among standard molal thermodynamic properties of aqueous ions and hydroxide complexes. Geochimica et

Cosmochimica Acta. 1997;**61**:907-950. DOI: 10.1016/S0016-7037(96)00339-0

[53] Iqubal MA, Sharma R, Jheeta S, Kamaluddin. Thermal condensation of glycine and alanine on metal ferrite surface: Primitive peptide bond formation scenario. Life. 2017;7(15): 1-20. DOI: 10.3390/life7020015

[54] Schouwer FD, Cuypers T, Claes L, De Vos DE. Metal-catalyzed reductive deamination of glutamic acid to bio-based dimethyl glutarate and methylamines. Green Chemistry. 2017; **19**:1866-1876. DOI: 10.1039/c6gc03222b

[55] Vandekerkhove A, Claes L, Schouwer FD, Van Goethem C, Vankelecom IFJ, Lagrain B, et al. Rh-catalyzed hydrogenation of amino acids to biobased amino alcohols: Tackling challenging substrates and application to protein hydrolysates. ACS Sustainable Chemistry & Engineering. 2018;**6**:9218-9228. DOI: 10.1021/acssuschemeng.8b01546

[56] Hiraga Y, Sato Y, Smith RL. Development of a simple method for predicting CO_2 enhancement of H_2 gas solubility in ionic liquids. Journal of Supercritical Fluids. 2015;**96**:162-170. DOI: 10.1016/j.supflu.2014.09.010

Chapter 5

Gasification Kinetics in Continuous Supercritical Water Reactors

Brian Pinkard, John Kramlich, Per Reinhall and Igor Novosselov

Abstract

Supercritical water gasification (SCWG) is an emerging technology with synergistic applications in renewable energy and waste processing. Supercritical water (SCW) functions as a green reaction medium during the gasification process, serving to dissolve and decompose complex organic molecules via ionic, radical, hydrolysis, and pyrolysis reaction mechanisms. Researchers investigate the decomposition of model compounds in order to predict product yields and conversion efficiencies during the gasification of heterogeneous biomass waste, food waste, sewage sludge, and other available feedstocks. Continuous, laboratory-scale reactors are often employed to study reaction kinetics, pathways, and mechanisms. This chapter synthesizes previous work investigating model compound gasification in continuous supercritical water reactors (SCWRs). A summary of continuous reactor design strategies is presented for practical benefit, followed by a discussion on reaction chemistry in the supercritical water environment. Reaction pathways and mechanisms have been investigated for several model compounds, lending insight toward the conditions needed for the complete conversion of real-world feedstocks. Several studies assume first-order reaction kinetics and propose Arrhenius parameters for the decomposition reaction. The first-order rate assumption must be carefully evaluated, and the applicable temperature range must be specified. Opportunities for further research are discussed.

Keywords: supercritical water, gasification, chemical kinetics, reaction mechanisms, reaction modeling, model compounds

1. Introduction

Supercritical water (SCW) exhibits unique physiochemical properties beneficial for oxidation or gasification of organic (carbon-containing) compounds ranging from simple molecules to complex heterogeneous waste. Near the critical point (374°C, 22.1 MPa), water exists in a high-temperature, dense fluid phase with high concentrations of H^+ and OH^- ions. These conditions serve to facilitate enhanced ionic chemistry and acid-catalyzed reactions [1]. At temperatures above the critical point, the density, viscosity, and ion product of water drop significantly. In this high-temperature, low-density phase water exhibits superb mass transfer properties, and organic compounds become fully miscible and/or soluble [2]. However, ionic reactions are no longer favored. Instead, pyrolysis, hydrolysis, and free radical reaction mechanisms dominate in this higher-temperature region. These two overlapping reaction regimes explain why SCW is of interest as a reaction medium for applications related to the thermochemical conversion of organic waste into heat and/or gaseous fuel.

An understanding of chemical reaction rates, pathways, and mechanisms involved in decomposing model compounds in SCW sheds important insight into the reaction chemistry of complex organic molecules in SCW. Interest in supercritical water gasification (SCWG) for industrial-scale applications is growing due to increased interest in generating low-cost "green" H_2 from renewable feedstocks. However, there are a number of technical barriers in developing large-scale plants; these include (i) controlling char formation, (ii) limiting salt precipitation which rapidly corrodes reactor components, (iii) identifying the optimal process parameters for high conversion efficiency (CE), (iv) identifying suitable gasification catalysts, and (v) designing an effective heat exchanger for waste heat recovery. Studies of model compounds can aid in addressing some of these challenges.

This chapter summarizes previous studies investigating reaction chemistry in continuous supercritical water reactors (SCWRs). Common reactor designs used to investigate reaction chemistry are discussed, enabling the researchers to replicate or to extend the knowledge of the previous studies. A synthesis of these studies yields important insights into common reaction mechanisms, pathways, and decomposition rates of certain compound classes. Opportunities for further investigations are described, and the practical value of these studies is highlighted.

2. Continuous supercritical water reactors for investigating reaction chemistry at the laboratory-scale

Achieving high temperatures and pressures, mitigating corrosion of reactor components, rapid heating and quenching of the reagent, acquiring accurate experimental data, and strategies for achieving a well-mixed, uniform flow must all be considered in the design of a supercritical water reactor for studies of reaction chemistry. Solutions for mitigating some of these challenges have been reported in the literature, but open questions remain regarding the best methods to mitigate char formation and salt precipitation in reactors designed to process complex feedstocks [3].

2.1 Batch vs. continuous reactors

Studies of reaction chemistry in SCW have been conducted using both batch and continuous reactors at the lab-scale. Batch reactors offer a unique opportunity to study reaction kinetics, mechanisms, and pathways of model compounds in the absence of a catalytic surface. The reactor can be constructed using a number of materials, including quartz capillaries and stainless steel tubing, which are filled with reactants and heated to reaction temperatures in a fluidized bath or electric furnace. Reactions occur at fixed conditions for the desired residence time, after which the reaction is quenched, and products are recovered for ex situ analysis. One limitation of batch reactors is that mass transfer (and, therefore, molecular interaction) is limited by molecular diffusion rates.

Continuous reactors are more complicated and expensive to fabricate. For industrial applications of chemical processes, a continuous setup is preferred over a batch setup. Process throughput is much higher, energy efficiency is significantly improved, heat recovery can be utilized, and opportunities exist for in situ process monitoring and control.

Continuous SCWRs are nearly always manufactured from nickel-base alloys (e.g., Inconel 625, Hastelloy C-276), which offer excellent corrosion resistance and good material strength at high temperatures. However, it is important to mention that nickel-base alloys also provide a catalytic surface for gasification reactions. For this reason, reaction rate parameters determined using batch SCWRs are not applicable for continuous SCWRs. DiLeo and Savage demonstrated this by gasifying methanol with and without a nickel wire in a batch quartz capillary reactor. The nickel wire increased methanol conversion from 20% after 2 h to 90% after 5 min (both at 550°C) [4]. Continuous reactors can enhance the catalytic effect even further, especially in turbulent flow regimes.

It should be noted that the catalytic effects within continuous reactors are dependent on reactor geometry and "aging" of the reactor components. Smaller diameter reactors with high surface-to-volume ratios (S/V) show increased catalytic effects due to increased molecular interaction between the reagents and the reactor wall. Also, reactor aging leads to the decreased catalytic activity over time, due to the formation of carbon layers on the reactor wall, leaching of metals, sintering, catalytic deactivation, and other effects.

For these reasons, this section will primarily focus on common designs of continuous reactors at the lab-scale. A representative schematic of a continuous SCWR is provided in **Figure 1**.

2.2 Heating and pressurization

The most reliable way to achieve independent pressure and mass flow control in a continuous SCWR is to operate a constant flow rate pump(s) in series with a back pressure regulator (BPR). Spring-loaded or dome-loaded BPRs are simple to use and reliable [3, 5]. High-performance liquid chromatography (HPLC) pumps are often used for pumping liquid reagents to high pressures with precise flow rate control in

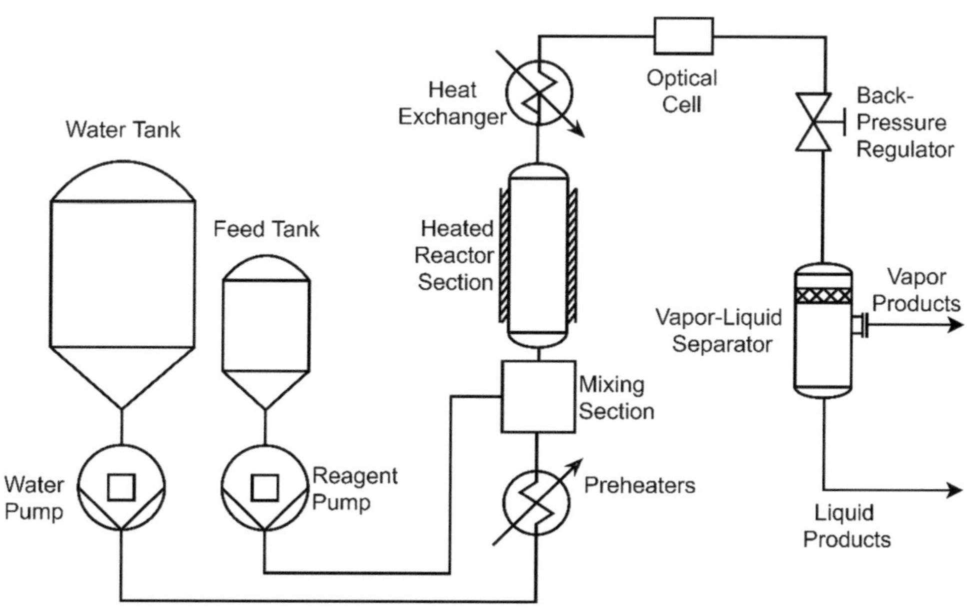

Figure 1.
Representative schematic of a continuous supercritical water gasification reactor with post-critical reagent injection and in situ *monitoring.*

the range of 0.01–30 mL/min [6–9]. Diaphragm, syringe, and piston pumps can be employed when higher flow rates are needed or when pumping a slurry [7, 10].

Reagent heating is important to consider as water can exhibit complex heat transfer characteristics near the critical point. Enhanced or deteriorated heat transfer can occur due to a combination of rapidly changing thermophysical properties and factors such as reactor geometry and the ratio of mass flux to heat flux. Generally, deteriorated heat transfer can be avoided by installing a downward-oriented heating section, which takes advantage of buoyancy effects for more efficient heating. A coiled heating section with a small diameter for high S/V is also desirable for improved heat transfer near the critical point [11, 12].

Resistive heaters, electric furnaces, and immersive fluidized baths have been used to reach the desired reaction temperatures [5, 7, 10, 13–15]. Resistive cartridge heaters are attractive options for preheating, as the tubing can be tightly wound around the cartridge to minimize heat loss. Electric furnaces offer precise control, are well-insulated, and are easy to install. A fluidized bath is a great option for maintaining isothermal conditions in the reactor section but can be expensive and bulky. Some combination of these heating methods is generally sufficient to achieve (a) rapid and efficient heating past the critical point and (b) isothermal reactor conditions.

2.3 Corrosion mitigation

Corollary to its ability to rapidly decompose organic compounds, supercritical water is extremely corrosive to most metals and metal alloys, especially if alkali metals or halogens are present. Thus, corrosion mitigation strategies need to be considered during SCWR design. Many studies have focused on corrosion control methods in SCWRs and SCW heat exchangers [16–18]. Generally, four corrosion mitigation strategies have been proposed and are thoroughly discussed in a review by Marrone et al. [16]. These are (i) preventing corrosive species from interacting with the reactor surface, (ii) forming a corrosion-resistant barrier, (iii) manufacturing the reactor from materials resistant to corrosion, and (iv) tuning operating conditions to minimize severe corrosion conditions. For reactors used to study reaction chemistry of organic compounds that do not contain heteroatoms, it is generally sufficient to rely on the corrosion resistance of the reactor material.

2.4 Mitigating char formation

Char has been reported as a common recalcitrated product formed during the gasification of aromatic compounds or homogeneous biomass components, such as lignin and cellulose [19–23]. Char can rapidly clog reactors, and it should be avoided or suppressed if possible. Broadly, char yields are known to decrease in the presence of certain metal catalysts (such as nickel and ruthenium), which are thought to effectively cleave C–C bonds in the aromatic rings of polycyclic aromatic hydrocarbons (PAHs). Many open questions remain surrounding the exact mechanisms responsible for char formation in SCW. Multiple studies have confirmed that char formation rates are highly dependent on temperature and the initial feedstock concentration. The literature suggests that ionic mechanisms near the critical point are responsible for charring and coking from compounds such as glucose, fructose, and cellulose [20], while free radical mechanisms form char during SCWG of aromatic compounds at higher temperatures, such as phenol, benzene, and lignin [21, 22]. While the industrial implementation of SCWG would require a method for suppressing char formation at high feedstock loadings, researchers can circumvent this issue by performing experiments with low feedstock concentrations [19]. There is a need to study the effect of feedstock type and concentration on char formation rates and char morphology.

2.5 Reagent mixing strategies

For studying reaction chemistry in a lab-scale SCWR, the mixing strategy used to introduce the reagent into the SCW environment should be carefully considered. Mixing can be achieved by (i) premixing water and reagent before heating to supercritical conditions or (ii) injecting reagent directly into supercritical water. If chemical kinetic rates are sought, post-critical injection is a preferred mixing strategy, as it rapidly heats the reagent to reaction temperatures and establishes a definite reaction start time [24].

Premixing is required if the feedstock is solid or viscous and must be pumped as an emulsion or when high reagent loading is considered. Premixed reagents should be rapidly heated, as char and tar formation can be significant when reagents are heated slowly [6, 16].

2.6 Reactor monitoring and data acquisition

The vast majority of SCWG studies rely on ex situ product analysis to quantify yields and determine reaction pathways. Several ex situ techniques exist for analyzing gaseous, liquid, and solid products; for properly characterizing full reaction networks and kinetic rates, all reaction products must be identified and quantified for each experimental condition. Gaseous products (H_2, CO, CO_2, CH_4) are often identified and quantified using gas chromatography (GC) with a thermal conductivity detector (TCD) and flame ionization detector (FID). Liquid products may be identified using HPLC, nuclear magnetic resonance (NMR) spectroscopy, Fourier transform infrared (FTIR) spectroscopy, or Raman spectroscopy. However, some researchers prefer to report the total organic carbon (TOC) concentration in the liquid phase by using a TOC analyzer, which is sufficient for calculating carbon CE. Occasionally solid products are analyzed ex situ, using scanning electron microscopy (SEM), proton-induced X-ray emission (PIXE), Raman spectroscopy, or FTIR spectroscopy [3].

In-line effluent analysis methods are available, such as GC and TOC analysis. These methods may lack the sensitivity and specificity required for the determination of chemical rates, but they can provide real-time input for process control. Alternatively, in situ product analysis greatly speeds the collection of the experimental data. In situ Raman spectroscopy is one of the most promising in situ analysis methods, as it is particularly well-suited for analyzing aqueous mixtures [25]. Water has a strong fluorescence and infrared signal, but a weak Raman signal, allowing product species to be identified and quantified [5]. For example, immersion in situ Raman spectroscopy was used to analyze formic acid decomposition [5], the conversion of ethanol to fuel gas [26], and the oxidation of methanol and isopropyl alcohol in SCW [27, 28]. Note that any spectroscopic methods are susceptible to fouling of the optical access point in the system.

2.7 Performance metrics

From a system-level perspective, effective SCWG is best described as a complete conversion of the mass and energy content of the original feedstock into gaseous products. Three performance metrics are commonly used to quantify this conversion: (i) gasification efficiency (GE), (ii) carbon CE, and (iii) hydrogen efficiency

(HE). GE is defined as the ratio of the total mass of the gaseous product to the initial mass of the feedstock, expressed mathematically as:

$$GE(\%) = \frac{x_{H_2} + x_{CO} + x_{CO_2} + x_{CH_4} + x_{Gas,other}}{x_{feedstock}} * 100 \tag{1}$$

CE is another metric used to quantify completeness of gasification; it is especially relevant if solid or liquid carbonaceous compounds are formed as refractory gasification products. It is defined as the ratio of moles of carbon in the product gas to moles of carbon in the feedstock:

$$CE(\%) = \frac{n_{CO} + n_{CO_2} + n_{CH_4} + xn_{C_xH_y}}{n_{C,feedstock}} * 100 \tag{2}$$

A less frequently used metric is HE, defined as the ratio of moles of hydrogen in the gaseous product to moles of hydrogen in the feedstock:

$$HE(\%) = \frac{2x_{H_2} + 4n_{CH_4} + yn_{C_xH_y}}{n_{H,feedstock}} * 100 \tag{3}$$

HE and GE values from SCWG can be well above 100%, due to a prominent role of the water-gas shift (WGS) reaction during gasification, which can produce H_2 gas via reaction of CO with water.

For determining rates of molecular decomposition in SCW, first-order reaction behavior is commonly assumed. This assumption is typically valid for pyrolysis or hydrolysis reactions or monomolecular decomposition reactions. However, this assumption is not valid for free radical reactions where radical induction and radical pooling behavior are present; more complex reaction modeling is required. The first-order decomposition rate (k) is determined by fitting an exponential decay curve to the reactant concentration varying with residence time, at a given experimental temperature.

Once a range of first-order decomposition rates (k) is determined at various temperatures, Arrhenius parameters can be determined by fitting the $ln(k)$ vs. $1/T$ curve with the following expression:

$$\ln(k) = \ln(A) - \frac{E_A}{RT} \tag{4}$$

This linear curve fit yields the activation energy (E_A) and pre-exponential factor (A) for the first-order decomposition reaction.

3. Gasification kinetics of model compounds in continuous supercritical water reactors

Model compounds serve as useful surrogates for studying the reaction chemistry of biomass constituents in supercritical water. Current interest in converting wet biomass waste into useful fuel has prompted studies of aromatic compounds as model compounds for lignin [22, 29, 30]; glucose and fructose, as cellulose surrogates [7, 14, 31–38]; alcohols, as model compounds for common liquid gasification intermediates and by-products [26, 39–44]; and amino acids—as model compounds for protein [15, 45–50]. Overall, these studies serve as the basis for approximating the operating conditions required to upgrade heterogeneous biomass into high-value fuels, such as "green" hydrogen.

For most studies, reaction mechanisms, pathways, kinetics, and yields are determined by varying the temperature, feedstock concentration, and residence time. Few studies investigate the effect of pressure on reaction chemistry; however, no significant pressure-related trends have been observed. The only time pressure that significantly impacts reaction chemistry is near the critical point, where pressure change can affect the thermophysical properties of SCW, such as density and ion product. For all studies reviewed here, the pressure is taken at 25 MPa unless specified otherwise.

Prevailing reaction mechanisms that deserve mention are the WGS reaction and the methanation reactions. The WGS increases H_2 yields by converting CO to CO_2, expressed as:

$$CO + H_2O \leftrightarrow CO_2 + H_2 \tag{5}$$

Methanation serves to reduce H_2 yields by converting it to methane, via the following two pathways:

$$CO + 3H_2 \leftrightarrow CH_4 + H_2O \tag{6}$$

$$CO_2 + 4H_2 \leftrightarrow CH_4 + 2H_2 \tag{7}$$

Overall, both the WGS and methanation reactions are highly important to the final gaseous product composition.

3.1 Aromatic compounds

One of the most recalcitrant biomass constituents is lignin, a heterogeneous organic polymer with numerous aromatic rings. In order to gain insight into lignin decomposition in SCW, phenol, benzene, and guaiacol have been proposed as lignin surrogates.

Huelsman and Savage [22] gasified phenol in an SCW batch reactor at 500–700°C; the authors identified major reaction products as H_2, CO, CO_2, CH_4, benzene, phenol, PAHs, and char. The presence of benzene and phenol as products indicates two competing reaction mechanisms are at play: aromatic ring growth and ring cleaving. The relative importance of the two mechanisms is highly dependent on the reaction temperature and the concentration of aromatics.

Yong and Matsumura [29] gasified phenol and benzene (separately) in a continuous SCWR at 370–450°C in the residence time range of 0.5–100 s. Observed products from each reagent include benzene, phenol, catechol, naphthalene, char, TOC in the liquid phase, and gaseous products. Catechol and naphthalene are indicative of the aromatic ring growth pathways leading to char formation. Increasing temperature and residence time led to increased yields of gas, TOC, and char. Generally, free radical mechanisms have been thought to be responsible both for decomposition to gaseous products and for ring growth to char. First-order decomposition was assumed, and Arrhenius parameters for general disappearance of phenol and benzene were proposed, as shown in **Table 1**. Yong and Matsumura [30] also gasified guaiacol, another aromatic model compound for lignin, at 300–450°C and residence times of 0.5–40 s. Again, yields of benzene, phenol, catechol, gas, TOC, and char were reported. Char formation was so significant that the initial guaiacol concentration had to be limited to 0.1 wt% to prevent reactor plugging. The higher temperatures and residence times increased yields of gas, TOC, and char. In this temperature range, hydrolysis, pyrolysis, ionic, and free radical

Compound	Temperature range (°C)	Pre-exponential factor "A" (s^{-1})	Activation energy "E_A" (kJ/mol)	Source
Phenol	370–450	7.72×10^1	53.06	[29]
Benzene	370–450	2.78×10^4	91.16	[29]
Guaiacol	300–450	6.52×10^1	32.40	[30]
Glucose	300–460	6.9×10^7	95.54	[34]
Glucose	750–800	1.2×10^3	70	[36]
Methanol	450–650	Not reported	191	[40]
Glycerol	445–600	Not reported	145	[44]
Glycerol	450–650	Not reported	196	[40]
Glycine	200–340	3.51×10^{13}	166	[48]
Glycine	250–450	3.6×10^{11}	160	[15]
Alanine	200–340	2.65×10^{12}	154	[50]
Alanine	250–450	1.4×10^{12}	156	[15]
Serine	200–340	9.85×10^{12}	149	[50]
Aspartic acid	200–340	5.40×10^{13}	148	[50]
Formic acid	320–420	1.58×10^6	85.8	[52]
Formic acid (subcritical)	300–374	4.0×10^1	39.7	[5]
Formic acid (supercritical)	374–430	3.6×10^{12}	175	[5]

Table 1.
First-order Arrhenius parameters of model compound gasification.

reaction mechanisms were all thought to be active. First-order Arrhenius parameters for guaiacol decomposition into intermediate products can be found in **Table 1**.

3.2 Glucose and fructose

Glucose has received considerable attention as a model compound due to its natural prevalence in biomass. Early studies of SCWG by Modell [32] investigated the decomposition of glucose, and early results showed that glucose could be completely converted to gaseous products without significant char formation after 20 s at 600°C [31].

Kabyemela et al. [33] investigated SCWG of glucose at 300–400°C and residence times of 0.02–2 s. Short residence times and low temperatures allowed for the identification of significant intermediate products, such as fructose, saccharinic acids, erythrose, glyceraldehyde, dihydroxyacetone, 1,6-anhydroglucose, pyruvaldehyde, and 5-hydroxymethylfurfural (5-HMF). The presence of numerous molecules containing furan rings, such as 5-HMF, illustrates the potential to form char via ionic mechanisms near the critical point. In a similar study, Aida et al. [14] gasified glucose at 350–400°C and residence times of 0.2–1.7 s with the explicit goal of optimizing furfural and 5-HMF yields. At these relatively low temperatures and short residence times, products were identified as fructose, erythrose, glycolaldehyde, glyceraldehyde, hydroxyacetone, 5-HMF, and furfural. Promdej and Matsumura [34] gasified glucose in the 300–460°C range, proposing Arrhenius parameters for the decomposition reaction (**Table 1**). Reported products include 5-HMF, furfural, and char, with subcritical temperatures favoring char formation.

These studies show good agreement in reported product yields in the region of enhanced ionic chemistry around the critical point.

SCWG of glucose has also been studied at higher temperatures. Goodwin and Rorrer [35] used a microchannel reactor to gasify glucose at 650–750°C; complete conversion was achieved after 2 s at 750°C. Intermediate products were identified as acetic and propanoic acids, 5-HMF, 2,5-hexanedione, phenol, lactic acid, formic acid, and furfural. Small inner reactor diameters were credited with improving heat transfer to the reaction environment and enhancing gasification, likely due to an increased catalytic wall effect. Hendry et al. [36] gasified 10–15 wt% glucose for residence times of 4.0–6.5 s and 750–800°C. Higher temperatures and lower initial concentrations improved conversion rates, which were used to calculate the Arrhenius parameters. There is significant discrepancy between the Arrhenius parameters proposed by Promdej and Matsumura [34] vs. Hendry et al. [36], which is likely due to the different temperature ranges tested. Near the critical point ionic chemistry is dominant, while free radical mechanisms are favored at temperatures well above the critical point.

Both Castello et al. [8] and Weiss-Hortala et al. [51] gasified glucose/phenol mixtures to study the effect of phenol on the gasification of glucose. Broadly, it was found that the presence of phenol inhibited H_2 production while promoting CH_4 production, with an overall decrease in gas yield.

Fructose has also been studied as a model compound for waste fruits and vegetables. Kabyemela et al. [37] subjected fructose to SCW at 300–400°C for 0.02–2 s, to determine decomposition pathways and kinetic rates. Major liquid products were identified as dihydroxyacetone, glyceraldehyde, erythrose, pyruvaldehyde, acetic acid, formic acid, and 5-HMF. The yields and kinetic rates were very similar to those obtained during glucose gasification at the same conditions [36]. Aida et al. [38] also studied SCWG of fructose, at temperatures up to 400°C and residence times from 0.14 to 0.78 s. In agreement with Kabyemela et al. [37], liquid yields of glyceraldehyde, dihydroxyacetone, pyruvaldehyde, lactic acid, and 5-HMF were reported.

Nanda et al. [7] gasified fructose at higher temperatures from 550 to 700°C, residence times of 30–75 s, and initial concentrations of 4–10 wt%. Broadly, higher temperatures and lower concentrations increased gasification efficiency. Higher concentrations of phenolic compounds were observed at 700°C, possibly indicating that char-forming pathways are enhanced at higher temperatures. Finally, residence times past 60 s increased CH_4 yields due to the consumption of H_2 via methanation reactions.

3.3 Alcohols

Several studies have investigated methanol decomposition in SCW, due to its chemical simplicity and prevalence as a refractory intermediate gasification product. Boukis et al. [39] reformed methanol in a continuous SCWR a T = 400–600°C, for residence times of 3–100 s, and initial methanol concentrations from 5 to 64 wt %. Gaseous yields of H_2, CO, CO_2, and trace CH_4 were detected, with higher temperatures and residence times increasing conversion rates. Bennekom et al. [40] also gasified methanol in a continuous reactor at T = 450–650°C and residence times of 6–173 s. H_2, CO, and CO_2 were reported as main products, with trace yields of methane, formaldehyde, and formic acid. The results show that methanol reforms to H_2 and CO in SCW, with CO converted to CO_2 by the WGS reaction.

Limited investigations are available related to the conversion of ethanol to gaseous products in SCW. Schanzenbacher et al. [41] subjected ethanol to SCW in a continuous reactor at temperatures from 433 to 494°C, for residence times from 2 to 12 s. Maximum conversion of 16.5% was observed, with acetaldehyde identified as the only reaction by-product. Pinkard et al. [26] gasified ethanol in a continuous SCWR at 560°C for residence times of 3–8 s, identifying reaction products as H_2, CO, CO_2, ethylene, ethane, and acetaldehyde.

Glycerol has been extensively studied due to its abundance as a by-product of biodiesel production. Reforming of this low-cost and widely available feedstock could be a source of inexpensive renewable H_2. Buhler et al. [42] reported intermediate products from glycerol gasification to include methanol, acetaldehyde, allyl alcohol, propionaldehyde, acrolein, ethanol, formaldehyde, and standard gaseous products. Non-Arrhenius decomposition behavior was observed, which was attributed to competing ionic and free radical reaction pathways. May et al. [43] reported acetaldehyde, hydroxyacetone, and acetic acid as major products from glycerol gasification in a continuous SCWR. Guo et al. [44] gasified glycerol at T = 445–600° C, reporting high conversion rates after 9 s at 600°C. The activation energy for decomposition was determined and is presented in **Table 1**. Bennekom et al. reported the activation energy for glycerol decomposition, for temperatures of 450–650°C, residence times of 6–173 s, and initial concentrations from 5 to 20 wt%.

3.4 Amino acids

Hydrothermal decomposition of amino acids has been studied in the context of valorizing protein-rich wastes from agriculture and seafood processing. Glycine, alanine, and leucine have all been identified as intermediate products from fish waste decomposition in subcritical water [45]. Islam et al. [46] also identified glycine and alanine as intermediate compounds from the decomposition of more complex amino acids, making them ideal model compounds for protein-rich waste.

Samanmulya and Matsumura [47] gasified glycine from 500 to 650°C with 1.0, 3.0, and 5.0 wt% initial concentrations. Higher temperatures and lower initial concentrations increased gasification efficiency; higher yields of char and tar were thought to decrease carbon conversion at higher concentrations.

Sato et al. [48] measured the decomposition of glycine at subcritical temperatures of 200–340°C. Gaseous yields were low, but liquid products were identified as ammonia, methylamine, glycolic acid, and formic acid. At subcritical conditions, ionic mechanisms likely facilitated the decomposition reactions. Klinger et al. [15] gasified glycine at subcritical and supercritical temperatures from 250 to 450°C. Strong temperature dependence was observed, with key liquid products identified as methylamine, diketopiperazine, glycolic acid, and formaldehyde. The decomposition rates for both studies are presented in **Table 1**. The slight discrepancy in observed products and reported reaction parameters is likely due to the different experimental temperature ranges; at supercritical temperatures, ionic mechanisms are suppressed, while free radical mechanisms are enhanced.

Samanmulya et al. [49] gasified alanine from 500 to 650°C with 1.0, 2.0, and 3.0 wt% initial concentrations. Results showed no change in gasification efficiency for varied initial concentrations, a positive indication of first-order reaction kinetics. Carbon conversion rates were found to be similar to those determined for glycine gasification, indicating that both amino acids likely react via similar free radical reaction mechanisms in high-temperature SCW. However, the gaseous yield from glycine was found to be rich in H_2 and CO_2, while the gaseous yield from alanine contained much more CO and CH_4. This can be attributed to the methyl (CH_3) group present in the alanine molecule, which likely reacts to form CH_4.

Sato et al. [50] investigated the decomposition of alanine at subcritical temperatures from 200 to 340°C major reaction products were identified as ammonia, ethylamine, carbonic acid, lactic acid, and pyruvic acid, which is in good agreement with the work of Klinger et al. [15], which identified major reaction products as lactic acid, ethylamine, acetaldehyde, and CO_2 from alanine decomposition at 250–450°C. Arrhenius rate parameters from both studies are presented in **Table 1** and show good agreement, likely due to the similar reactor configuration and temperature range used in the two studies.

Several studies also investigated the gasification kinetics of more complex amino acids, such as valine, leucine, proline, serine, and aspartic acid to determine the effect of the amino acid functional group on conversion rates and final product yields. Arrhenius parameters for the decomposition of serine and aspartic acid are presented in **Table 1**. The activation energies determined for all amino acids are within a similar range, showing that functional group has only a minor effect on the overall conversion rates of amino acids in SCW.

3.5 Other relevant model compounds

Some intermediate SCWG products are common across a range of feedstocks, notably aldehydes and organic acids. Formic acid has been identified as an intermediate product from SCWG of glucose [35], fructose [37], and glycine [48].

Yu and Savage [52] studied formic acid gasification in a continuous SCWR to understand the decomposition pathways and rates. Temperatures were varied between 320 and 500°C, pressures between 18 and 30.7 MPa, and residence times between 1.4 and 80 s. Major products were consistently identified as H_2 and CO_2, with minor yields of CO. Overall, this suggests the dominance of a decarboxylation pathway with a minor dehydration pathway. Arrhenius parameters for formic acid decomposition are presented in **Table 1**. Zhang et al. [53] also gasified formic acid in a continuous SCWR at temperatures between 550 and 650°C for residence times between 16 and 46 s. Again, H_2 and CO_2 were present as dominant reaction products, with minor CO yields. Trace yields of formaldehyde and methanol were also reported at the highest tested temperatures.

Pinkard et al. [5] gasified formic acid in a continuous SCWR at T = 300–430°C and residence times in the range of 4–65 s. In situ Raman spectroscopy was used to calculate kinetic rates and Arrhenius parameters. It was found that the transition across the critical point increased the reaction rate, favoring the production of H_2 and CO_2 via the decarboxylation reaction pathway. The notable change in reaction rate across the critical point indicates the importance of both ionic and free radical reactions to the overall decomposition of formic acid in SCW. Arrhenius parameters for formic acid decomposition in subcritical and supercritical water are presented in **Table 1**.

3.6 Gasification catalysts

It is beyond the scope of this chapter to review the body of work investigating various catalysts for SCWG. We direct the reader to the comprehensive reviews on the subject [2, 3]. However, it is worth mentioning some broad findings from this extensive area of research.

Numerous alkali metal compounds (some naturally present in biomass) have been investigated as suitable gasification catalysts. Broadly, these tend to enhance decomposition and increase gaseous product yields. A major disadvantage to adding salts is the persistent corrosion issue, as salts are insoluble in SCW, leading to precipitation of a molten or solid salt layer on the reactor walls. This scaling rapidly corrodes reactor components, increasing the frequency of component replacement.

Noble metals are effective gasification catalysts, with nickel and ruthenium generally accepted as the most promising and effective across a wide range of compounds and reaction regimes. Typically, metal catalysts are impregnated in a support compound (e.g., activated carbon, γ-Al_2O_3), after which they are crushed and loaded into a packed bed reactor. Both nickel and ruthenium are effective at cleaving C-C bonds, reducing char formation. However, issues exist with catalyst stability, longevity, and process economics. Nickel suffers from sintering and deactivation as carbon layers tend to accumulate on the catalytic surface, while ruthenium catalysts are expensive and can be poisoned by the presence of sulfur. More research is needed toward an economically viable catalyst for SCWG.

4. Discussion and opportunities for further exploration

In general, conversion rates for gasification of organic compounds in SCW can be improved by increasing the temperature and the residence time and by decreasing the initial feedstock concentration. Amino acids, carbohydrates, and simple organic acids are the compound classes with the fastest decomposition rates, with aromatic compounds and alcohols being the most recalcitrant compounds. Arrhenius plots with all mentioned compounds are presented in **Figures 2** and **3** for comparison between compound classes.

Additional studies are needed to understand chemical reactions in supercritical water. It is likely that key functional groups will behave similarly under SCW conditions and further experimentation and interpretation are needed to describe the reactions routes and rates. In situ product identification has the potential to provide accurate data for characterization of decomposition pathways. Special attention should be given to char-forming compounds, in order to understand the mechanisms leading to char formation and the conditions required to promote gasification. Additionally, the role of both homogeneous and heterogeneous catalysts in affecting reaction rates, pathways, and mechanisms should be explored and quantified.

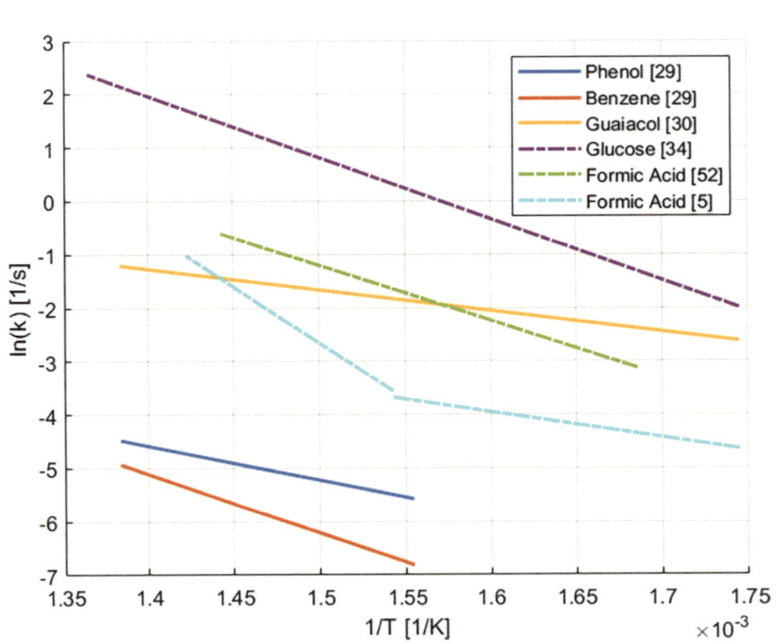

Figure 2.
Decomposition rates of model compounds in sub- and supercritical water.

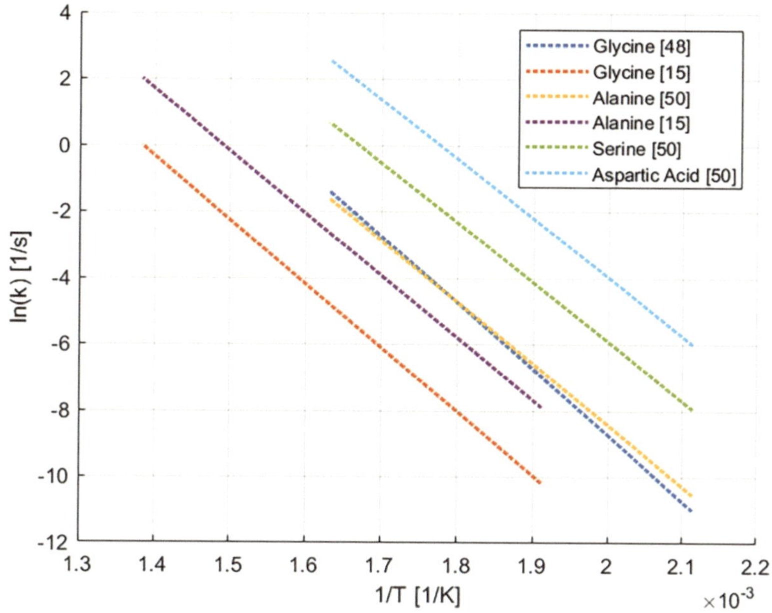

Figure 3.
Decomposition rates of model amino acids in sub- and supercritical water.

5. Conclusions

Supercritical water gasification promises to revolutionize the processing of waste streams to value-adding gaseous fuels. Technical barriers remain between the current state-of-the-art and widespread industrial adoption of the technology, several of which can be addressed through studying the chemistry of model compounds in supercritical water. The knowledge gained in these studies can be applied toward developing reaction pathways and mechanisms, lending insight toward reaction behavior of more complex feedstocks.

Acknowledgements

The authors would like to recognize funding provided by the DOD Defense Threat Reduction Agency (DTRA), Grant HDTRA1-17-1-0001. Special thanks to the University of Washington for providing resources toward the completion of this work.

Additional thanks to David Gorman, Kartik Tiwari, Elizabeth Rasmussen, Vedant Maheshwari, Anmol Purohit, Stuart Moore, Eric Molnar, Justin Davis, and other members of the Novosselov Research Group at the University of Washington who contributed to advancing knowledge of chemical reactions in supercritical water.

Author details

Brian Pinkard, John Kramlich, Per Reinhall and Igor Novosselov*
University of Washington, Seattle, WA, USA

*Address all correspondence to: ivn@uw.edu

References

[1] Savage PE. Organic chemical reactions in supercritical water. Chemical Reviews. 1999;**99**:603-621. DOI: 10.1021/cr9700989

[2] Guo Y, Wang SZ, Xu DH, Gong YM, Ma HH, Tang XY. Review of catalytic supercritical water gasification for hydrogen production from biomass. Renewable and Sustainable Energy Reviews. 2010;**14**:334-343. DOI: 10.1016/j.rser.2009.08.012

[3] Pinkard BR, Gorman DJ, Tiwari K, Rasmussen EG, Kramlich JC, Reinhall PG, et al. Supercritical water gasification: Practical design strategies and operational challenges for lab-scale continuous flow reactors. Heliyon. 2019; **5**:e01269. DOI: 10.1016/j.heliyon.2019.e01269

[4] DiLeo GJ, Savage PE. Catalysis during methanol gasification in supercritical water. The Journal of Supercritical Fluids. 2006;**39**:228-232. DOI: 10.1016/j.supflu.2006.01.004

[5] Pinkard BR, Gorman DJ, Rasmussen EG, Kramlich JC, Reinhall PG, Novosselov IV. Kinetics of formic acid decomposition in subcritical and supercritical water—A Raman spectroscopic study. International Journal of Hydrogen Energy. 2019;**44**: 31745-31756. DOI: 10.1016/j.ijhydene.2019.10.070

[6] Caputo G, Rubio P, Scargiali F, Marotta G, Brucato A. Experimental and fluid dynamic study of continuous supercritical water gasification of glucose. The Journal of Supercritical Fluids. 2016;**107**:450-461. DOI: 10.1016/j.supflu.2015.09.022

[7] Nanda S, Reddy SN, Hunter HN, Dalai AK, Kozinski JA. Supercritical water gasification of fructose as a model compound for waste fruits and vegetables. The Journal of Supercritical Fluids. 2015;**104**:112-121. DOI: 10.1016/j.supflu.2015.05.009

[8] Castello D, Kruse A, Fiori L. Low temperature supercritical water gasification of biomass constituents: Glucose/phenol mixtures. Biomass & Bioenergy. 2015;**73**:84-94. DOI: 10.1016/j.biombioe.2014.12.010

[9] Lee IG, Kim MS, Ihm SK. Gasification of glucose in supercritical water. Industrial & Engineering Chemistry Research. 2002;**41**:1182-1188. DOI: 10.1021/ie010066i

[10] Molino A, Migliori M, Macri D, Valerio V, Villone A, Nanna F, et al. Glucose gasification in super-critical water conditions for both syngas production and green chemicals with a continuous process. Renewable Energy. 2016;**91**:451-455. DOI: 10.1016/j.renene.2016.01.065

[11] Shen Z, Yang D, Wang S, Wang W, Li Y. Experimental and numerical analysis of heat transfer to water at supercritical pressures. International Journal of Heat and Mass Transfer. 2017; **108**:1676-1688. DOI: 10.1016/j.ijheatmasstransfer.2016.12.081

[12] Pioro IL, Duffey RB. Heat-transfer enhancement at supercritical pressures. In: Pioro IL, Duffey RB, editors. Heat Transfer & Hydraulic Resistance at Supercritical Pressures in Power Engineering Applications. New York: ASME Press; 2007. DOI: 10.1115/1.802523.ch9

[13] Hanush RG, Rice SF, Hunter TB, Aiken JD. Operation and Performance of the Supercritical Fluids Reactor (SFR). No. SAND-96-8203. Albuquerque, NM: Sandia National Labs; 1995

[14] Aida TM, Sato Y, Watanabe M, Tajima K, Nonaka T, Hattori H, et al.

Dehydration of d-glucose in high temperature water at pressures up to 80 MPa. The Journal of Supercritical Fluids. 2007;**40**:381-388. DOI: 10.1016/j.supflu.2006.07.027

[15] Klinger D, Berg J, Vogel H. Hydrothermal reactions of alanine and glycine in sub- and supercritical water. The Journal of Supercritical Fluids. 2007;**43**:112-119. DOI: 10.1016/j.supflu.2007.04.008

[16] Marrone PA, Hong GT. Corrosion control methods in supercritical water oxidation and gasification processes. The Journal of Supercritical Fluids. 2009;**51**:83-103. DOI: 10.1016/j.supflu.2009.08.001

[17] Boyd W, Pray H. Corrosion of stainless steels in supercritical water. Corrosion. 1957;**13**:33-42. DOI: 10.5006/0010-9312-22.10.280

[18] Kritzer P. Corrosion in high-temperature and supercritical water and aqueous solutions: A review. The Journal of Supercritical Fluids. 2004;**29**: 1-29. DOI: 10.1016/S0896-8446(03)00031-7

[19] Pinkard BR, Gorman DJ, Tiwari K, Kramlich JC, Reinhall PG, Novosselov IV. Review of gasification of organic compounds in continuous-flow, supercritical water reactors. Industrial & Engineering Chemistry Research. 2018;**57**:3471-3481. DOI: 10.1021/acs.iecr.8b00068

[20] Chuntanapum A, Matsumura Y. Char formation mechanism in supercritical water gasification of model compounds. Industrial & Engineering Chemistry Research. 2010;**49**: 4055-4062. DOI: 10.1021/ie901346h

[21] Matsumura Y, Goto S, Takase Y, Inoue S, Inoue T, Kawai Y, et al. Suppression of radical char production in supercritical water gasification by addition of organic acid radical scavenger. Energy & Fuels. 2018;**32**: 9568-9571. DOI: 10.1021/acs.energyfuels.8b02063

[22] Huelsman CM, Savage PE. Reaction pathways and kinetic modeling for phenol gasification in supercritical water. The Journal of Supercritical Fluids. 2013;**81**:200-209. DOI: 10.1016/j.supflu.2013.05.012

[23] Peterson AA, Vogel F, Lachance RP, Froling M, Antal MJ Jr, Tester JW. Thermochemical biofuel production in hydrothermal media: A review of sub- and supercritical water technologies. Energy & Environmental Science. 2008; **1**:32-65. DOI: 10.1039/B810100K

[24] Tiwari K, Pinkard BR, Gorman DJ, Davis J, Kramlich JC, Reinhall PG, et al. Computational modeling of mixing and gasification in continuous-flow supercritical water reactor. In: Proceedings of the 12th International Symposium on Supercritical Fluids (ISSF '18); 22–25 April. Antibes: Valence: ISASF; 2018. p. 2018

[25] McCreery RL. Raman Spectroscopy for Chemical Analysis. Vol. 225. Hoboken: John Wiley & Sons; 2005. 437 p. DOI: 10.1002/0471721646

[26] Pinkard BR, Rasmussen EG, Kramlich JC, Reinhall PG, Novosselov IV. Supercritical water gasification of ethanol for fuel gas production. In: Proceedings of the ASME 2019 13th International Conference on Energy Sustainability (ASME ES '19); 15–17 July. Bellevue. New York: ASME; 2019. p. 2019

[27] Rice SF, Hunter TB, Ryden AC, Hanush RG. Raman spectroscopic measurement of oxidation in supercritical water. 1. Conversion of methanol to formaldehyde. Industrial & Engineering Chemistry Research. 1996; **35**:2161-2171. DOI: 10.1021/ie950510f

[28] Hunter TB, Rice SF, Hanush RG. Raman spectroscopic measurement of oxidation in supercritical water. 2. Conversion of isopropyl alcohol to acetone. Industrial & Engineering Chemistry Research. 1996;**35**: 3984-3990. DOI: 10.1021/ie9505118

[29] Yong TLK, Matsumura Y. Kinetics analysis of phenol and benzene decomposition in supercritical water. The Journal of Supercritical Fluids. 2014;**87**:73-82. DOI: 10.1016/j.supflu.2013.12.018

[30] Yong TLK, Matsumura Y. Kinetic analysis of Guaiacol conversion in sub- and supercritical water. Industrial & Engineering Chemistry Research. 2013; **52**:9048-9059. DOI: 10.1021/ie4009748

[31] Yu D, Aihara M, Antal MJ Jr. Hydrogen production by steam reforming glucose in supercritical water. Energy & Fuels. 1993;**7**:574-577. DOI: 10.1021/ef00041a002

[32] Amin S, Reid RC, Modell M. Reforming and decomposition of glucose in an aqueous phase. In: Proceedings of the Intersociety Conference on Environmental Systems. San Francisco; 1975. ASME Paper 75-ENAS-21

[33] Kabyemela BM, Adschiri T, Malaluan RM, Arai K. Kinetics of glucose epimerization and decomposition in subcritical and supercritical water. Industrial & Engineering Chemistry Research. 1997; **36**:1552-1558. DOI: 10.1021/ie960250h

[34] Promdej C, Matsumura Y. Temperature effect on hydrothermal decomposition of glucose in sub- and supercritical water. Industrial & Engineering Chemistry Research. 2011; **50**:8492-8497. DOI: 10.1021/ie200298c

[35] Goodwin AK, Rorrer GL. Conversion of glucose to hydrogen-rich gas by supercritical water in a microchannel reactor. Industrial & Engineering Chemistry Research. 2008; **47**:4106-4114. DOI: 10.1021/ie701725p

[36] Hendry D, Venkitasamy C, Wilkinson N, Jacoby W. Exploration of the effect of process variables on the production of high-value fuel gas from glucose via supercritical water gasification. Bioresource Technology. 2011;**102**:3480-3487. DOI: 10.1016/j.biortech.2010.11.003

[37] Kabyemela BM, Adschiri T, Malaluan RM, Arai K. Glucose and fructose decomposition in subcritical and supercritical water: Detailed reaction pathway, mechanisms, and kinetics. Industrial & Engineering Chemistry Research. 1999;**38**:2888-2895. DOI: 10.1021/ie9806390

[38] Aida TM, Tajima K, Watanabe M, Saito Y, Kuroda K, Nonaka T, et al. Reactions of D-fructose in water at temperatures up to 400°C and pressures up to 100 MPa. The Journal of Supercritical Fluids. 2007;**42**:110-119. DOI: 10.1016/j.supflu.2006.12.017

[39] Boukis N, Diem V, Habicht W, Dinjus E. Methanol reforming in supercritical water. Industrial & Engineering Chemistry Research. 2003; **42**:728-735. DOI: 10.1021/ie020557i

[40] Bennekom JGV, Venderbosch RH, Assink D, Heeres HJ. Reforming of methanol and glycerol in supercritical water. The Journal of Supercritical Fluids. 2011;**58**:99-113. DOI: 10.1016/j.supflu.2011.05.005

[41] Schanzenbacher J, Taylor JD, Tester JW. Ethanol oxidation and hydrolysis rates in supercritical water. The Journal of Supercritical Fluids. 2002;**22**:139-147. DOI: 10.1016/S0896-8446(01)00119-X

[42] Buhler W, Dinjus E, Ederer HJ, Kruse A, Mas C. Ionic reactions and

pyrolysis of glycerol as competing reaction pathways in near- and supercritical water. The Journal of Supercritical Fluids. 2002;**22**:37-53. DOI: 10.1016/S0896-8446(01)00105-X

[43] May A, Salvado J, Torras C, Montane D. Catalytic gasification of glycerol in supercritical water. Chemical Engineering Journal. 2010;**160**:751-759. DOI: 10.1016/j.cej.2010.04.005

[44] Guo S, Guo L, Cao C, Yin J, Lu Y, Zhang X. Hydrogen production from glycerol by supercritical water gasification in a continuous flow tubular reactor. International Journal of Hydrogen Energy. 2012;**37**:5559-5568. DOI: 10.1016/j.ijhydene.2011.12.135

[45] Kang K, Quitain AT, Daimon H, Noda R, Goto N, Hu HY, et al. Optimization of amino acids production from waste fish entrails by hydrolysis in sub and supercritical water. The Canadian Journal of Chemical Engineering. 2009;**79**:65-70. DOI: 10.1002/cjce.5450790110

[46] Islam MN, Kaneko T, Kobayashi K. Reaction of amino acids in supercritical water-flow reactor simulating submarine hydrothermal systems. Bulletin of the Chemical Society of Japan. 2003;**76**:1171-1178. DOI: 10.1246/bcsj.76.1171

[47] Samanmulya T, Matsumura Y. Effect of activated carbon catalytic on supercritical water gasification of glycine as a model compound of protein. Journal of the Japan Institute of Energy. 2013;**92**:894-899. DOI: 10.3775/jie.92.894

[48] Sato N, Daimon H, Fujie K. Decomposition of glycine in high temperature and high pressure water. Kagaku Kogaku Ronbunshu. 2002;**28**:113-117. DOI: 10.1252/kakoronbunshu.28.113

[49] Samanmulya T, Inoue S, Inoue T, Kawai Y, Kubota H, Munetsuna H, et al. Gasification characteristics of alanine in supercritical water. Journal of the Japan Petroleum Institute. 2014;**57**:225-229

[50] Sato N, Quitain AT, Kang K, Daimon H, Fujie K. Reaction kinetics of amino acid decomposition in high-temperature and high-pressure water. Industrial & Engineering Chemistry Research. 2004;**43**:3217-3222. DOI: 10.1021/ie020733n

[51] Weiss-Hortala E, Kruse A, Ceccarelli C, Barna R. Influence of phenol on glucose degradation during supercritical water gasification. The Journal of Supercritical Fluids. 2010;**53**:42-47. DOI: 10.1016/j.supflu.2010.01.004

[52] Yu J, Savage PE. Decomposition of formic acid under hydrothermal conditions. Industrial & Engineering Chemistry Research. 1998;**37**:2-10. DOI: 10.1021/ie970182e

[53] Zhang Y, Zhang J, Zhao L, Sheng C. Decomposition of formic acid in supercritical water. Energy & Fuels. 2010;**24**:95-99. DOI: 10.1021/ef9005093

Chapter 6

Supercritical Water Oxidation for Environmentally Friendly Treatment of Organic Wastes

Yanhui Li and Shuzhong Wang

Abstract

Supercritical water oxidation is a promising, environment-friendly technology to efficiently deal with a wide variety of organic wastes such as wastewaters, industrial and municipal sludge, etc. As for the two key problems, i.e., corrosion and salt plugging, generally encountered in supercritical water oxidation plants, this chapter firstly reported the related mechanism analysis, solutions, research status, and development trends, respectively. From the perspectives of corrosion prevention and control, safety and automatic control, economic improvements, and development of novel reactors, a number of advanced technologies and equipment such as on-line desalination in supercritical water, new operation scheme assisted secondary traditional treatment, produced-gas recovery and oxygen reuse and novel TWM reactor, etc., were introduced systematically. Finally, this chapter summarizes the implementation status of industrial plants and the technological features of several firms being active in the construction of full-scale supercritical water oxidation plants. This chapter will provide very valuable information for the researchers and engineers who are interested in supercritical water oxidation for the harmless treatment of organic pollutants.

Keywords: supercritical water oxidation, waste treatment, corrosion, salt precipitation, technical status, industrial plants

1. Introduction

Supercritical water (SCW) refers to water whose temperature and pressure are above its critical point (374.15°C, 22.1 MPa). The physical properties of SCW such as density, viscosity, dielectric constant, hydrogen bonding, and ionic product is vastly different from those of normal water and steam. The density of SCW is about 1/3 of the normal state, the dielectric constant is reduced from 87.7 at 0°C to 2–30 of SCW, and the ion product is reduced 10^3 to 10^9 times than that of normal water. Supercritical water has a low dielectric constant, and thus is extremely soluble to nonpolar compounds such as organic matter and oxygen. In addition, SCW has a significantly lower viscosity than that of the normal water, the diffusion coefficient is improved, the mass transfer performance is enhanced, and thus it can be completely miscible with nonpolar gases and hydrocarbons. In view of the special properties of SCW, nonpolar molecules such as O_2 and organic matter have strong solubility in supercritical aqueous systems, which can be mutually dissolved in any ratio to form a homogeneous phase, and the mass transfer resistance at the interface is eliminated. There, SCW can be used as an excellent reaction medium for organic matter and oxygen [1].

Supercritical water oxidation (SCWO), which was firstly proposed by Modell in MIT in the middle of 1980s, is an effective and advanced oxidation technology to destruct organic matters by taking advantage of the unique properties of SCW under the typical operation conditions of 450–600°C, 24–28 MPa. A schematic of a typical SCWO process is displayed in **Figure 1**. In SCWO reactor, organic wastes can be thoroughly oxidized and decomposed into harmlessly small molecules, such as CO_2, N_2, water, etc., under excess oxidants in single-phase SCW. Hetero-atoms in organic matters are mineralized into corresponding acids or inorganic salts, and the formation of nitrous oxides is inhibited owing to the low reaction temperature. SCWO is particularly suitable for disposing organic wastewaters with high toxicity, high concentration, and bio-refractory components [2–4]. It can also recover energy and achieve heat self-sufficiency easily to ensure an economic advantage [5]. When a mass concentration of organic matters in feedstock is in the range of 3–4%, the whole reaction process does not require an extra energy input commonly. Furthermore, compared with incineration, SCWO does not have the problems of high cost, public resentment, and secondary pollutants like dioxins [5]. Hence, SCWO has attracted much attention in the past three decades. Brunner [1] has summarized SCWO results of real waste materials including textile wastewater, wastewater from terephthalic acids, food wastes, municipal excess sludge, and alcohol distillery wastewater. Furthermore, Veriansyah and Kim [5] have also systematically introduced SCWO experiments of toxic organic wastes such as pesticide, bacteria and dioxins, chlorophenol and chlorobenzene, pharmaceutical and biopharmaceutical wastes, solid rocket propellants, military wastes, etc. Some other feedstocks, including landfill leachate [6–9], industrial dyeing wastewater [10, 11], polychlorinated biphenyls (PCBs) [12], chlorinated wastes, oily wastes, etc., have also been disposed via SCWO efficiently.

Nowadays, SCWO plants have been commercialized by several famous companies and universities [14–16], such as General Atomics, Chematur AB, HydroProcessing, Supercritical Fluids International (SCFI), SuperWater Solution, Xi'an Jiaotong University (XJTU), etc. In 2001, two commercial-scale SCWO

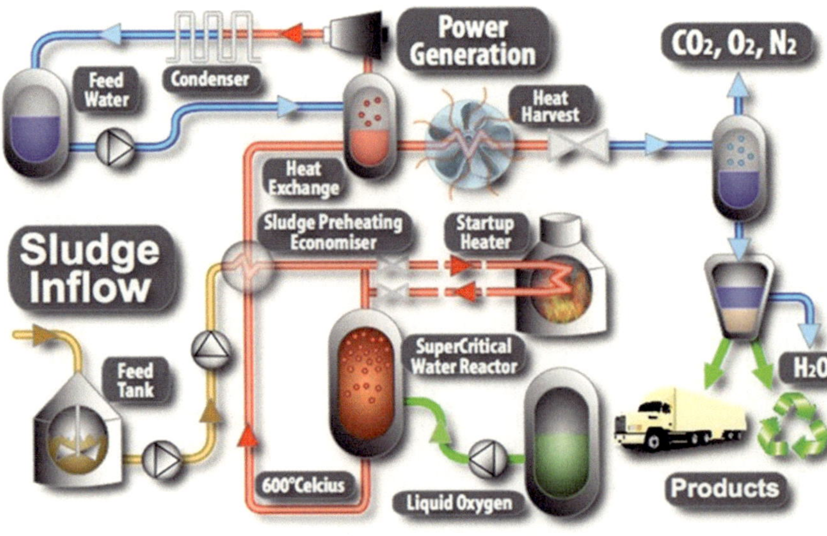

Figure 1.
A schematic of a typical SCWO process for sludge treatment [13].

plants were installed in Harlingen (Texas, USA) to treat sewage sludge with the solid-containing rates of 7–8 wt% [14]. Another SCWO plant with a treatment capacity of 7 m^3/h has also been built by Chematur Engineering AB (Sweden) to deal with sewage sludge [17]. From 2009 to 2011, SuperWater Solution installed and successfully tested a 5 dry t/d SCWO system for the Iron Bridge Regional Water Reclamation Facility in Orlando, FL, USA [16]. In 2007, the SCFI Group acquired patented SCWO technology (AquaCritox®) from the Sweden Chematur Engineering AB. Further work directed by the SCFI group has emphasized the reduction of costs. In 2015, the world's first 100 t/d SCWO plants for industrial refractory wastewater treatment was designed and constructed under the guidance of the authors. Besides, there were some other pilots or full-scale SCWO plants for the harmless treatment of sewage sludge and organic wastewaters. Nevertheless, the three key problems concerning corrosion, plugging triggered by salt precipitation, and high running cost still exist [18], even make some commercial scale SCWO plants inactive.

2. Key problems and status

The harsh conditions involved in SCWO process, such as high temperature, high pressure, excessive oxygen, corrosive ions, etc., easily induce severe reactor corrosion problems, meaning a shorter reactor life. On the other hand, inorganic salts will precipitate from SCW due to its extremely low dielectric constant, which may result in reactor plugging triggered by deposited salts and further cause expensive and frequent shutdowns of the whole SCWO plant. These two aspects, to a certain degree, are still not effectively solved and seriously hinder the extensive commercialization of SCWO.

2.1 Corrosion

Corrosion is a key obstacle to limit the commercial application of SCWO, which not only shortens reactor life but also induces a bad treatment effect of feedstock due to the formations of corrosion products. Harsh operation conditions (high concentration of oxidants, extreme pH values, high temperature, and pressure) together with reaction intermediate/ultimate products (high concentrations of ionic species, free radicals, and acids) result in severe corrosion problem in SCWO reactors. Corrosion mainly occurs on reactor's inner wall; it also appears in heat exchanger and cooler on inlet and outlet pipelines of the reactor [19–21].

Materials serving for SCWO include stainless steels, nickel-based alloys, titanium, tantalum, noble metal ceramics, etc. [21–25]. A series of investigations on the corrosion resistances of these materials under supercritical and/or subcritical conditions [20, 24, 26–30], reflected that no one kind of material can withstand corrosion at all conditions, but some exhibit perfect corrosion resistance under specific conditions, as given in **Table 1**. Thus, appropriate reaction conditions such as heteroatom types in feedstock, reaction temperature, and pressure should be optimized in order to minimize corrosion rate for a chosen reactor construction material. Generally speaking, nickel-based alloys show a relatively good corrosion resistance among all the acids listed in **Table 1** under supercritical conditions. Titanium is fit to be employed under subcritical conditions, and is a potential proper liner of preheater and cooler being installed before and after reactor, respectively.

Corrosion in SCWO circumstance also depends on parameters concerning both materials (alloy composition, surface condition, material purity, and heat

Materials	$T < T_c^a$; high density		$T > T_c^a$; low density	
	Good resistance	Poor resistance	Good resistance	Poor resistance
Nickel-based alloys	H_3PO_4, HF, alkaline solutions	HCl, HBr, H_2SO_4, HNO_3	All acids	$[H_3PO_4] > 0.1$ mol/kg, NaOH
Titanium	All acids	F^-	HCl	H_2SO_4, H_3PO_4

[a]*The critical temperature of water.*

Table 1.
Corrosion resistance of alloys against various media at subcritical and supercritical temperatures [19].

treatment) and the aqueous solutions (chemical dissolution process, electrochemical dissolution process, and influence of anions) [31]. Under the same condition, various alloys always display different corrosion resistances, which mainly depend on the intrinsic content of alloying elements, especially iron, chromium, nickel, and molybdenum. Iron is usually used to improve economy of iron-based alloys, nickel-based alloys, and titanium-based alloys. In SCW, Fe shows the highest oxidation rates to form stable oxide among the interested metal elements (Fe, Ni, Ti, Mo, and Cr). The corrosion rate of Fe in the oxidizing HCl solution is three times higher than that in SCW without HCl, indicating that the oxidizing acidic chlorinated solution can promote dealloying Fe of alloys in high density systems [32]. Fe always has a faster diffusion velocity than other elements so that the Fe oxides make up the majority of the outer layer of corrosion scales formed on alloys at temperature higher than 400°C [33]. The common Fe oxides includes Fe_2O_3, Fe_3O_4, and α(γ)-FeOOH under the SCWO condition. The mass loss of Fe is higher than that of Mo at 300–350°C and lower than that of Mo at 400–450°C in SCW with NaCl [34].

The Cr is considered as the most important alloying element to improve corrosion resistance [35]. The Cr_2O_3 presented the lowest solubility and the best oxidation resistance among the oxides of chromium, iron, and nickel [29, 32, 35–38]. The higher chromium-containing alloys presented the lower thicknesses of the formed oxide scales. The $NiCr_{25}$ alloy performed the best corrosion resistance among the interested binary Ni-Cr alloys (0–25 mass% Cr) in an aqueous solution resulting from the oxidation of CH_2Cl_2 at 40 MPa and temperatures of 100–415°C [39]. The cracking susceptibility of nickel-based alloys also decreased with an increase in Cr content of the substrate exposed to SCW containing HCl or NaOH [40]. The oxygen affinity of chromium is higher than those of iron and nickel, and thus Cr generally may be selectively oxidized to $Cr(OH)_3$ and Cr_2O_3. The Cr(III) of Cr_2O_3 can be transformed into Cr(VI) by further oxidation in the oxidizing high temperature water, and unfortunately the Cr(VI) possibly loss by the dissolution. A more continuous and stable Cr_2O_3 layer was formed at higher temperatures (450 and 500°C) than at 400°C in oxidizing SCW [41]. A series of results indicate that alloys generally suffer the more serious corrosion in high temperature subcritical water rather than in SCW at low density. Mo improves the resistance to pitting corrosion in subcritical salt solution obviously [42]. In nickel-based alloys, the synergistic effect of Mo and Cr on improving the corrosion resistance is remarkable [43]. The nickel-based alloys often show a severe depletion of nickel in high-density aqueous systems [32, 37], while NiO, which is an effectively protective oxide, covers the alloy substrate in SCW environments at low density [44].

For a specific reactor material, the corrosion behavior is commonly influenced by the dissociations of acids, salts, and bases, the solubility of corrosion products and gases, and the stability of protective oxide scales. All these characteristics are affected by density and ionic product of the solution [11, 20]. In order to decrease corrosion rate, it is better to adjust the solution density to be below 200 kg/m^3 [31], which may result from the variation in corrosion mechanisms at high- or low density aqueous systems. Lots of studies have been carried out on the mechanism of corrosion scales grown on metal materials in SCW [29, 45, 46]. Two typical mechanisms, such as solid-state growth mechanism and mixed model (formation of the inner layer by solid-state growth process and formation of the outer layer by a metal dissolution-oxide precipitation mechanism) depending on the water density below or above 100–200 kg/m^3, have been proposed [24, 29, 45, 47, 48]. Solid-state growth mechanism has been used successfully in corrosion problems of various materials such as alloys and ceramics in high temperature gaseous environments such as single gas (O_2, N_2, and CO_2), air, high temperature steam, and a series of mixture gases. According to the diffusion rate of metals involved in the substrate, some alloying elements with relative slow rate are retained and enriched in the inner layer, while the outer layer grows outwards resulting from that the metal ions, especially irons, transport along oxide grain boundaries outwards and then react with gas species at scales/environment interface [29, 38, 45]. However, the mixed model, obtaining satisfying application in condensed aqueous systems, emphasizes that the outer layer is formed by a metal dissolution-oxide precipitation mechanism, and the precipitated metal ions are released from the corroding metal itself or originate elsewhere in the system [27]. The dissolved metal cations combine with anions such as OH^- in aqueous environments to form oxides or hydroxides; they then precipitate on specimens surface to form and/or thicken the outer layer [49]. Li et al. [30] reported that the corrosion of Inconel 600 and Incoloy 825 in low density SCW containing sulfides also follows the solid-growth mechanism.

2.2 Salt precipitation

Salt management was identified as a critical issue for the success of SCWO technology. Except for the inorganic salts initially present in the feedstocks, a majority of inorganic salts derive from the SCWO treatment of high-concentration, stubborn organic matters including hetero-atoms such as chlorine, sulfur, and phosphorus [50]. Solubility of salts is reduced evidently in SCW, which is usually lower than 100 mg/L, and thus they are prone to deposition. In addition to corrosion issues, plugging triggered by salts precipitation is another main obstacle to hinder SCWO commercialization, which is induced by sticky salt agglomeration and deposition on the internal surface of reactor.

When plugging takes place, the SCWO plant has to be shut down, washed, and restarted, which will directly decrease the reliability of the plant and increase its running cost. Inorganic salts, whether soluble or not, may come from feedstock or reaction byproducts, and their viscosities decide the tendency of depositing on the reactor wall. In general, different salts own different deposition characteristics [51]. Salt deposition principles and the corresponding solutions have been systemically presented in the previous literatures [50, 52, 53]. In a commercialized SCWO plant, reactor plugging can be avoided by particular reactor designs, special instruments, and operation means [18]. Possible solutions are to adopt reverse flow tank reactor, reverse flow tubular reactor, transpiring wall reactor, centrifuge reactor, mechanical brushing, rotating scraper, reactor flushing, crossflow filtration, density separation, additives, high flow velocity, homogeneous precipitation, and extreme pressure [17]. In fact, precipitated salts are relatively difficult to be removed out of SCWO plant during the operation process; thus, most SCWO plants do not have salt removal function. Salts in SCWO are classified into Type 1 or Type 2, according to their solubility [54], Type 1 salt has a high solubility in the range of water's critical temperature, while Type 2 salt has a low solubility in this region. Type 2 salt can be separated by properties of SCW [50], and the suitable supercritical conditions should be controlled at approximately 400°C and 25 MPa according to the solubility of salts mentioned in the previous studies [52, 55]. Hydro cyclone or centrifugal reactor helps to remove soluble salts under the above-mentioned supercritical conditions, but severe wearing problem is inevitable. Of course, low salt concentration in feedstock is helpful for avoiding reactor plugging. Appropriate reactor configuration and pre-desalination before reactor may solve reactor plugging problem effectively and economically.

In addition, when the preheating temperature of feedstock is in the range of 200–450°C [56], some undesired intermediate products like tar and char will be generated, which may also plug preheating pipeline. To overcome this problem, the following options are taken into consideration, which include enlarging the inner diameter of the preheating pipeline, adding a small amount of oxidant into the preheating pipeline to inhibit the formations of tar and char, increasing the heating rate of feedstock, and preheating feedstock up to a lower temperature even not preheating [57]. However, the improvement of the heating rate is limited by setting space of heating part, heating methods, and withstanding temperature of preheater wall. The above fourth approach means that a large amount of heat will not be recovered into the SCWO system, which will undoubtedly reduce the energy efficiency and increase the running cost.

Salt deposition also accelerates catalyst inactivation rate and reactor corrosion rate, and reduces heat transfer coefficient [22, 58]. The potential catalysts for SCWO reactions may be poisoned and/or polluted quickly by precipitated salts in SCW [59]. It is also difficult to replace catalyst in traditional reactor configuration. That may be the reason why Savage has reported that no catalyst is implemented commercially for catalytic SCWO for organic waste treatment [60]. Thereby, it is important to separate these precipitated salts before they contact heterogeneous catalyst. Comparing with salt of ion form in subcritical water, salt in the form of molecule in SCW is relatively less corrosive [19, 21–23, 61]. Salt precipitated out from SCW mainly results in chemical corrosion through oxidation reaction. Salt in subcritical water or high density SCW mostly promotes electrochemical corrosion [58, 62], and may result in inter-granular corrosion starting from the edge of metal grain [49]. Nowadays, two contradictions of preventing salt deposition and minimizing reactor corrosion rate are displayed as follows: first, preventing salt deposition needs high SCW density because it will exhibit a relatively improved solvent property for precipitated salts. However, minimizing corrosion rate requires low SCW density to decrease the content of ionic salts [31]. Kritzer et al. [63] controlled SCW density at less than 250 kg/m^3 in order to reduce reactor corrosion rate. Second, adding alkali compounds independently or in feedstock before reactor is helpful for inhibiting reactor corrosion, but the possibility of reactor plugging increases because of salt deposition [19]. That is why some alkali compounds are delivered into reaction system from reactor outlet.

Salt deposition control achieves significant improvement through the innovation of research methods, the studies of salt deposition and separation performance, the evolution of salt control techniques, and new reactor configuration designs. However, several challenges such as the transport properties, deposition mechanisms, and practical salt control techniques, still restrict the technology application. Operation parameters play important roles in salt deposition processes, and various ions in the mixture may exhibit mutual influence or common effect. Thus, the prediction of salt deposition is quite difficult and needs extensive information on salts phase behavior in SCW. Moreover, the mechanisms of salt deposition in the multi-component system have not been recognized deeply, and the entire evolution process of salts from microscopic nucleation to macroscopic deposition in SCW requires to be further revealed. Hence, it is necessary to gain more phase behavior data for analysis in depth. The combination of the experiment and numerical simulation is a desired method for the investigation of salt deposition.

In future, salt deposition control in a SCWO system requires global considerations and designs instead of previous single or local protection. Certainly, reactor configuration design has the priority to the whole SCWO system due to its crucial roles in salt control. Wise reactor concept combines with appropriate operation techniques before the reactor, in the reactor and/or after the reactor, forming multiple control and security units. Notably, the determination of salt control schemes should also consider the characteristics of wastewater and each technique, as well as the economic costs. With the solution of salt deposition problems and the development of reactor corrosion resistant materials, SCWO will be able to achieve large-scale commercial applications.

3. Advanced technologies and equipment

Since the occurrence of SCWO technology firstly proposed by Modell at the Massachusetts institute of technology in the early 1980s, a number of researcher and industrial engineers have been devoted to related research on prevention and control of corrosion and salt precipitation, enhancement of SCWO economy, safety, automation control, etc. A series of effective solutions, technologies, and equipment were proposed, and some have been applied to the actual industrial SCWO plants, which are not focused in this chapter. Here, several latest or typical advanced technologies and equipment are introduced in detail.

3.1 Development of novel reactor

Beyond all doubt, reactor is the most important core equipment of SCWO plants. The produced salts due to SCWO reaction together with those initially contained in feedstock may result in reactor plugging and increase corrosion rate of equipment. In terms of whether or not the precipitated solid-state salt is promised to accumulate on reactor's internal surface, the precipitated salt and the internal surface of reactor have two kinds of relations in reactor [53]. For the former case, salt removal can be accomplished by special instruments such as mechanical brushing, rotating scraper, reactor flushing, etc. For the latter case, it can be achieved by particular reactor designs and operation means. They include reverse flow tank reactor with brine pool, reversible flow tubular reactor, transpiring wall reactor, centrifuge reactor, crossflow filtration, density separation, additive, high velocity flow, and homogeneous precipitation. Moreover, extreme pressure is provided to avoid salt precipitation in SCW. Philip et al. [53] have objectively reviewed the properties of the above specific reactor configurations and operational approaches in commercial applications. Furthermore, there are some other special reactor configurations such as cool wall reactor reported by Cocero and Martinez [64] and two pipes reactor introduced by Baur et al. [65]. Calzavara et al. [66] set a stirrer in their reactors. Príkopský [67] installed a protective metal sleeve replaced easily to prevent salt from depositing on the internal surface of their axial reactor in SCWO. There is no doubt that no one reactor design or operation mean has been proven to be clearly superior to the others in all aspects.

Next, three kinds of typical reactor configurations are given to introduce the development of the novel reactors. MODAR reactor or its variation belongs to a reverse flow tank reactor with brine pool, which separates/removes salt by SCW properties, as shown in **Figure 2**. Precipitated salts under supercritical condition fall down into the subcritical zone at the bottom of reactor by gravity, inertia, and forced convection. MODAR reactor has overcome the three disadvantages of tubular reactor, which are too long structure size, not suitable for dealing with salt-containing feedstock and easy to be plugged. However, due to the low falling velocity of small particle salt and the disturbance at the vertical direction, salt deposition is easy to take place on the internal surface of MODAR reactor in supercritical zone. Moreover, MODAR reactor may be eroded severely because its internal surface directly contacts corrosive reaction fluid.

Cool wall reactor [64] and transpiring wall reactor [68, 69] can help to solve salt deposition problems existing in MODAR reactor. These two kinds of reactors are divided into pressure-bearing wall and non-load-bearing wall. For cool wall reactor, precipitated salts are re-dissolved in a subcritical water film on the internal surface of no-load-bearing wall. The film is formed by low temperature feedstock cooling the no-load-bearing wall. However, severe corrosion still takes place because of the contact between corrosive reaction fluid and internal surface of no-load-bearing wall. When scale on the internal and external surfaces of no-load-bearing wall is left uncontrolled, the cool wall reactor will not continue to work normally since heat transfer efficiency reduces drastically. Transpiring wall reactor can not only prevent salt deposition but also decrease reactor corrosion rate through a porous transpiring wall element. Clean water flows across the porous wall to form a protective film to continuously dilute corrosive species, re-dissolve precipitated salt particles, and/or sweep them away from the internal surface of reactor. The transpiring wall (no-load-bearing wall) can be made of porous ceramic, sintered metal, or many thin and porous metal layers/platelets bonded together. The operational characteristics of transpiring wall reactor have been tested for a long time by some researchers. It is found that the concentrations of iron, nickel, chromium ions in reactor effluent were very low, so proving that the

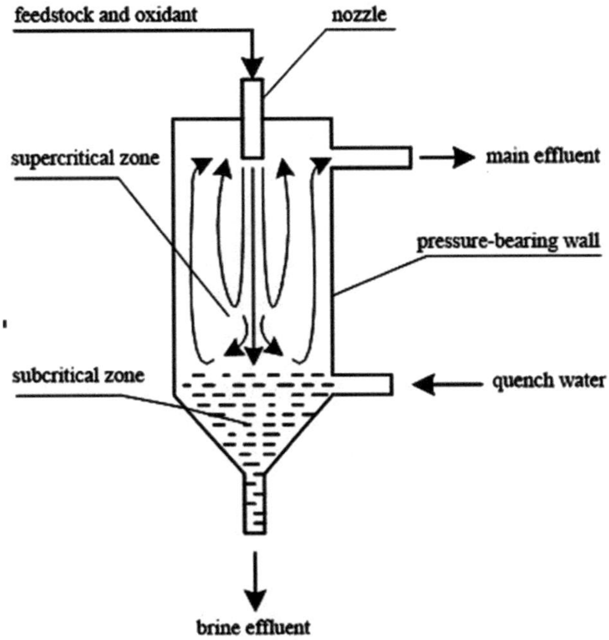

Figure 2.
A schematic diagram of reverse flow tank reactor [53].

corrosion rate of transpiring wall reactor was low. Fauvel et al. [70] also verified that the transpiring wall reactor had a good corrosion resistance property. Thus it can be claimed that the transpiring wall reactor is the best reactor configuration for minimizing reactor corrosion rate at present. Nonetheless, plenty of energy is required to preheat the transpiration water entering into the reactor. The pressure difference of two sides of transpiring wall should be adjusted at a suitable range since it has a bad mechanical property.

On the basis of summarizing characteristics of previously main reactor configurations, a novel concept reactor is designed and manufactured by the authors, as shown in **Figure 3**, which is the reactor installed into the first pilot-scale SCWO plant constructed in China finished by the authors [71]. The novel reactor owns characteristics of transpiring wall reactor and MODAR reactor for the first time, which is simply nominated as "TWM reactor." Similar to MODAR reactor, TWM reactor is also divided into supercritical zone above and subcritical zone below. Precipitated salts under supercritical condition fall down into the subcritical zone (formed by pumping into quench water) and are re-dissolved. After removing salts, clean fluid flows toward the top outlet of the reactor by the differences of density and flow resistance. The valves equipped on pipe lines of top and bottom outlets of reactor are used to regulate the flow rate of fluid at top outlet. Dirty fluid flows out TWM reactor from its bottom outlet, which contains a large amount of salts and solids. The flow pattern of TWM reactor is helpful for eliminating natural convection effects which were mentioned by Welling's group for transpiring wall reactor [55]. The MODAR reactor wall which is replaced by a porous transpiring wall and a pressure-bearing wall is set to surround it to form fundamental configuration of TWM reactor. Clean and cool transpiration water is pumped into TWM reactor and forms a protective film on the internal surface of transpiring wall. TWM reactor configuration is expected to prevent salt deposition and decrease corrosion rate.

Center pipe of MODAR reactor is comparatively thin and long, while transpiring wall reactor has no center pipe. For TWM reactor, the mixing of oxygen and sewage sludge preheating is accomplished by a specially simplified mixer equipped on the pipe line of reactor inlet. Center pipe is inserted into the reactor deeply and its diameter is big enough to avoid plugging. Reactants in center pipe are further preheated by high temperature fluid around or by reaction heat yielded through partial oxidation. Ideally, the mixture of oxygen and sewage sludge just reaches supercritical temperature at the outlet of center pipe. It is better to maintain subcritical temperature in center pipe because salt will not deposit on the internal surface of center pipe. The center pipe is set on a fixed platform by screw thread, which is easy to be replaced when plugged or eroded severely. Generally, the traditional transpiring wall reactor is a tubular reactor with a hydrothermal burner [69, 73, 74]. However, no hydrothermal burner is set in TWM reactor in order to decrease design temperature of reactor and operation cost of sewage sludge SCWO treatment.

3.2 Corrosion prevention and control

Marrone and Hong [75] have reported in detail some corrosion control approaches, which includes utilization of high corrosion resistance materials, liners, coatings, transpiring wall/film-cooled wall reactors, adsorption/reaction on fluidized solid phase, avoidance of corrosive feeds, pretreatment to remove corrosive species, vortex/circulating flow reactors, pre-neutralization, cold feed injection, feed dilution with noncorrosive wastes, effluent dilution/cooling, and optimization of operation conditions. The following text will exhibit several latest solutions of corrosion issues, which also have been verified by industrial applications.

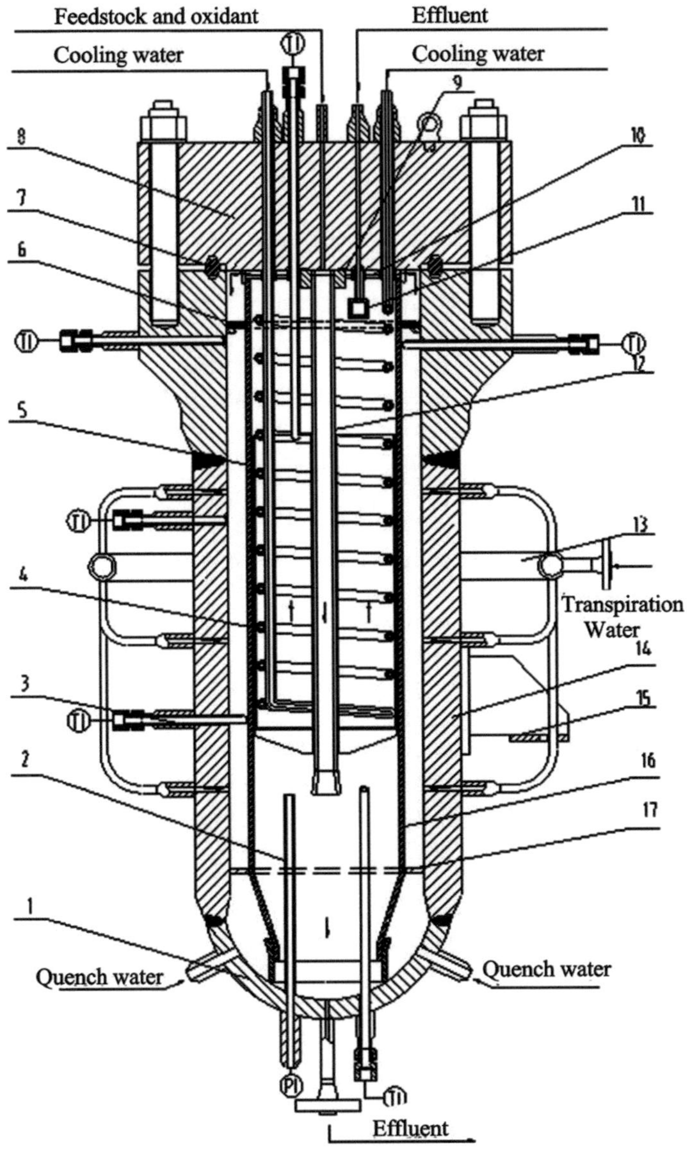

Figure 3.
A schematic diagram of TWM reactor for treating sewage sludge by SCWO [72]. 1—spherical head; 2—pressure measurement tube; 3—temperature measurement tube; 4—cooling spiral coil; 5—catalyst box; 6—support ring; 7—eight-square cushion; 8—top cover; 9—fixed platform; 10—cooling cover; 11—filter; 12—center pipe; 13—annular water box; 14—pressure-bearing wall; 15—ear base; 16—transpiring wall; 17—locating ring; TI—temperature indication; PI—pressure indicate.

3.2.1 On-line desalination in SCW

The existence of aggressive anions in SCWO environment is one of the key factors that causes severe corrosion problem in the SCWO system. Therefore, some methods should be used to remove the inorganic salts in the feed for SCWO treatment. Unfortunately, the conventional desalination methods (such as electrical dialysis, reverse osmosis, ion exchange, electrical adsorption) are difficult to be applied in the SCWO system for an organic wastewater with a high salinity.

An on-line desalination process in SCW, which is able to effectively separate the inorganic salts from the feeds, was firstly proposed by the authors, as exhibited in **Figure 4**. It is noteworthy that this diagram is just a part of a whole SCWO system, and the on-line desalination units can be added into any SCWO systems, which has no effects on the other parts of the system. The mixed wastewater is pressurized by a high-pressure metering pump and then sent to the heating furnace for preheating. A heat exchanging coil tube of the heating furnace in the system is arranged in two sections (including a high temperature section and a low temperature section). Fluid at an outlet of the low temperature section (namely the middle outlet of the heating furnace) reaches a supercritical temperature. After entering the hydrocyclone separator, most solid salts with a particle size more than 10 μm in reaction fluid are separated by centrifuging. After salts removal, the fluid enters the high temperature section of the heating furnace through the top outlet of the hydrocyclone, so as to ensure a heat exchanging coefficient of the heat exchange coil at the high temperature section, and effectively prevent pipes and reactors following the hydrocyclone from plugging. At the same time, solid inorganic salts separated at the bottom of the hydrocyclone are spirally sent to the buffer oxidizer by a motor installed on the hydrocyclone. When the buffer oxidizer is full of the solid inorganic salts, a cut-off valve at the top inlet of the buffer oxidizer is closed, the motor installed on a top portion of the hydrocyclone is closed, a bottom cut-off valve of the buffer oxidizer is slowly opened, and a spirally transporting motor installed on the buffer oxidizer is started, for transporting the solid inorganic salts in the buffer oxidizer to the expansion device. A fluid containing the solid inorganic salts expands in the expansion device. Steam generated enters the storage tank. Heat generated is recycled. The separated solid inorganic salts enter the salt storage and are taken out for further landfill at a certain period. In addition, after being spirally sent into the buffer oxidizer, the fluid with a high salinity separated by the hydrocyclone reacts with oxygen previously injected through an oxygen pipe, which harmlessly removes organic pollutants. After the supersaturated fluid with a high salinity enters the buffer oxidizer, supercritical fluid at the top of the buffer oxidizer, which hardly contains solid inorganic salts, enters the top outlet of the hydrocyclone. Meanwhile, the solid inorganic salt will deposit at the bottom of the buffer oxidizer.

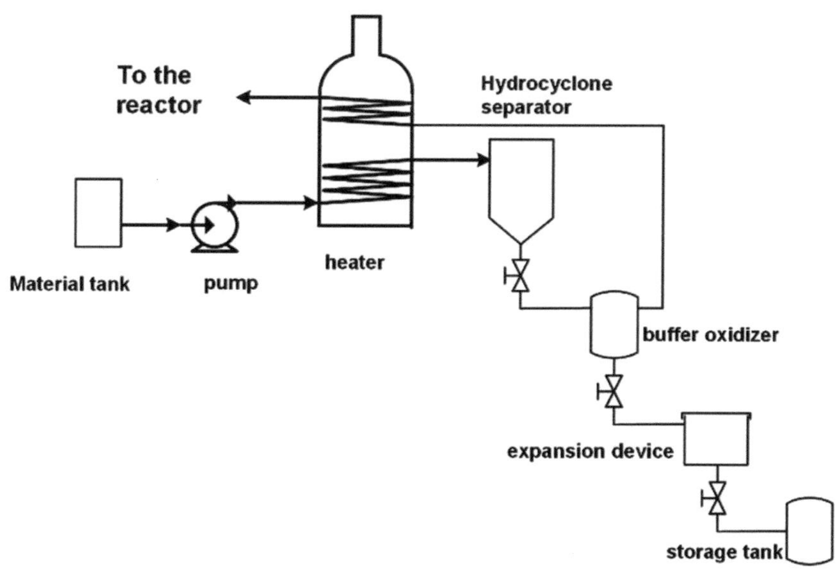

Figure 4.
A schematic diagram of the on-line desalination units.

3.2.2 A new scheme for mild SCWO condition and high removal

A high temperature and a high oxygen concentration are needed in order to achieve a satisfied removal efficiency of organic matters in a SCWO system. However, the corrosion of structural materials is also apparently promoted by the increasing temperature and oxygen concentration. Therefore, a new operation scheme should be used to achieve a high organic matters removal efficiency, but not at the expense of the facilities materials corrosion. The core concept of the new operation scheme is to reduce the operation temperature and oxygen concentration in a reactor. As a result, the chemical oxygen demand (COD) removal efficiency cannot reach the standard value, and the concentration of ammonia nitrogen in the effluence was also unsatisfied. To work around this problem, some secondary treating methods were used for thorough removal of ammonia nitrogen and COD, including an ammonia stilling process, a phosphorous removal process, and a biochemistry process, which are all conventional water treatment techniques and can be operated easily with a relatively low operation cost.

3.2.3 Multipoint injection mixer

In a reactor, the heated feed and oxygen first meet in a mixer, where the local oxygen concentration is much higher than that in other areas in the reactor. There exist two technical drawbacks. First, drastic oxidation reaction occurs once the sludge and oxygen are mixed, producing a dangerous increase in temperature in the reactor. Second, the local high oxygen concentration and local high temperature in the mixer and the forepart of the reactor cause severe corrosion of structural materials, leading to unpredictable corrosion effects, such as pitting corrosion and stress corrosion cracking. A new multipoint mixer is developed to optimize the oxygen injection process in the reactor. As seen in **Figure 5**, this mixer consists of six oxidant injection holes in the core tube, and helical mixing elements are used to enhance mixing. The size number of the holes are determined by computational fluid dynamics (CFD) simulation to ensure that the input rate and flow distribution of oxygen are appropriate. Through this method, severe corrosion occurrence and hot-spot formation are avoided.

Additionally, it worthy to say that the design of appropriate reactors is also a very effective approach to overcome reactor corrosion and plugging issues. The detail of a novel design of the anti-corrosion reactors is available in Section 3.1.

3.3 Safety and automatic control technology

According to the schematic diagram of a typical SCWO process (see **Figure 6**) proposed by the authors, some safety and automatic control technologies are introduced, as follows.

3.3.1 Pretreatment of feedstock: thermal hydrolysis

Thermal hydrolysis pretreatment (THP) is widely used for sludge pretreatment, which has been proven to be an efficient method to reduce the viscosity of sludge [76]. To some extent, the use of thermal hydrolysis pretreatment is a safety control technology. The power of the diaphragm pump must be increased to transport viscous sludge, which will significantly increase the risk of overload of the pump's motor. Additionally, a large heat exchange area is needed if high viscosity sludge is fed into the heat exchanger, which substantially increases the risk of plugging. As shown in **Figure 6**, the feed is heated to 85°C in the preparation tank (4), and the thermal hydrolysis of the sludge occurs in the sludge buffer tank (8). The reaction time and reaction temperature of the sludge in the tank (8) are 180 min and 170°C, respectively. Thus, the viscosity is as low as 33 mPa·s and to a certain degree, avoids the overload of the pump and the plugging of the heat exchanger.

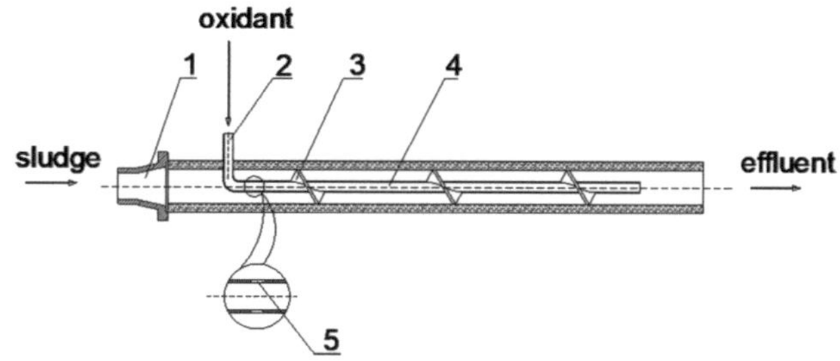

Figure 5.
A schematic diagram of the novel mixer: 1—adapter flange; 2—core tube; 3—helical mixing element; 4 and 5—injection holes.

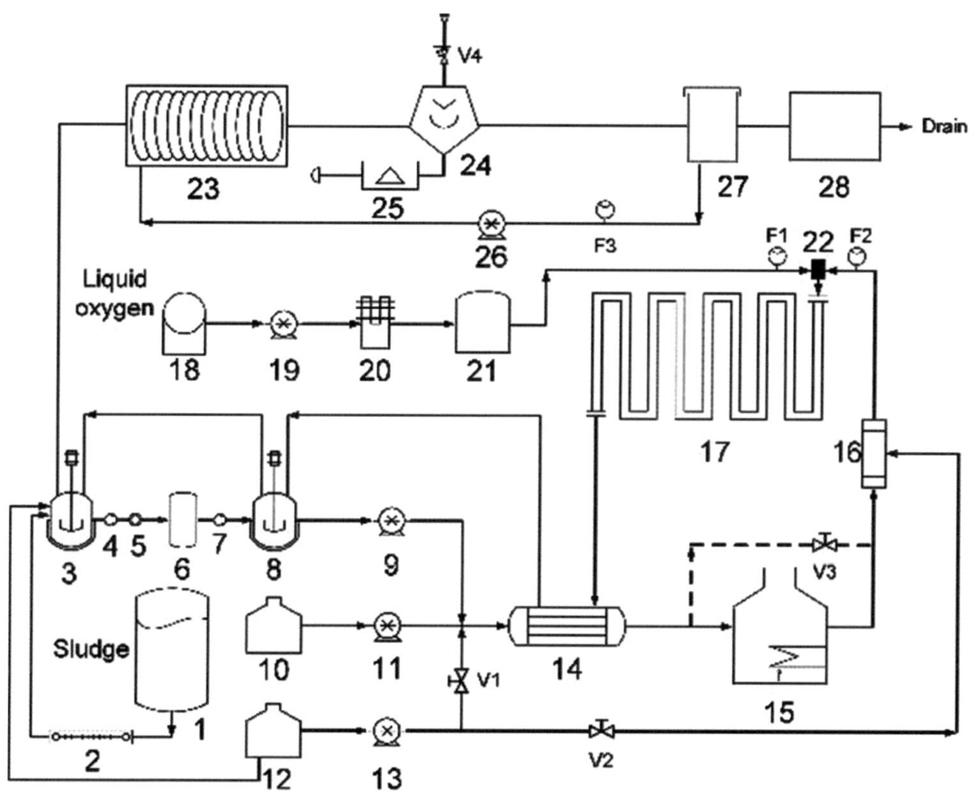

Figure 6.
A schematic diagram of the commercial plant for mixed sludge SCWO treatment: 1—sludge tank; 2—screw conveyer; 3—preparation tank; 4—screw pump; 5—homogeneous emulsification pump; 6—sludge tank for sampling; 7—screw pump; 8—sludge buffer tank; 9—diaphragm pump; 10—methanol tank; 11—plunger pump; 12—water tank; 13—plunger pump; 14—heat exchanger; 15—heating furnace; 16—multiple nozzle desuperheater; 17—tube reactor; 18—liquid oxygen tank; 19—liquid oxygen pump; 20—vaporizer; 21—oxygen buffer; 22—mixer; 23—pressure relief device; 24—three phase separator; 25—centrifugal dehydrator; 26—plunger pump; 27—effluent buffer tank; 28—secondary treatment unit; (F1-F3)—flowmeter; (V1-V5)—electric valve.

3.3.2 System start-up with auxiliary methanol

In traditional SCWO plants, during start-up process, water is heated in a heating module to reach supercritical temperature (generally 400°C) and then the feed is switched to organic wastes. Once the process has started, the effluent is used to preheat the cold sludge in the heat exchanger, and the heat released by the oxidation reaction is enough to preheat the cold feed, which makes the process self-sufficient and permits the heating module to be switched off. Generally, the organic waste to be treated often contains some soluble salts that have low solubility in SCW. In the pipes of the heating module, the pipe walls are heated directly and are the first to reach supercritical temperature, forming a high-temperature microenvironment. Many studies have shown that inorganic salts tend to crystalize in the overheated microenvironment and stick to the inner walls of pipes, leading to clogging and interruption of the continuous process.

Here, the proposed start-up strategy can effectively avoid or relieve the risk of plugging and overheat in the heater. As shown in **Figure 6**, not the sludge but the methanol solution in the methanol tank (10) is fed into the system and is heated by the heating furnace (15). When the final preheating temperature achieves to approximately 400°C, the first step of start-up is finished. Then, the heating furnace is shut down, and oxidant is injected into the tube reactor (17) to react with methanol. A large amount of heat is released during the oxidation reaction, which maintains the energy balance of the whole system. After a few minutes, the electric valve (V3) is opened, and the heating furnace (15) is bypassed. Then, the feed is switched to sludge in the sludge buffer tank (8), which has already been heated to 170°C. Through this method, no sludge passes the heating furnace, so effectively avoiding plugging in the heating facility. Additionally, only pure water, whose corrosivity is obviously lower than that of sludge at supercritical temperature and pressure, passes through the heating furnace. As a result, the risk of corrosion is also reduced.

3.3.3 Thermal control of the SCWO reactor

The temperature of the SCWO reactor is one of the key parameters which affects the COD removal efficiency of organic matter. Thus, various methods were designed to realize precise thermal control of the SCWO reactor in this commercial plant. There are several temperature measurement points in the reactor. If the temperature of the reactor increases, as shown in **Figure 6**, the electric valve (V5) is slightly opened to allow some of the high-temperature effluents to bypass the heat exchanger (14), decreasing the temperature of the outlet of the heat exchanger (14). However, the opening of the electric valve (V5) should be limited to prevent a large amount of high-temperature effluent from directly passing through the coil heat exchanger in the sludge buffer tank (8) and causing unpredictable damage. If the temperature of the reactor increases sharply in a short time, the electric valve (V2) will be opened, and cooling will be injected into the system via the multiple nozzle desuperheater (16). If the temperature of the reactor decreases, however, the electric valve (V5) will be slightly closed to allow more hot effluent to pass through the heat exchanger (14). Furthermore, if the heat released by the SCWO reaction is not sufficient, methanol in the tank (10) will be pumped into the reactor (17) and mixed with the feed in order to enhance the SCWO reaction.

3.3.4 Depressurization of fluids containing solids

SCWO plants are high temperature and high pressure systems, meaning that the reaction effluent is still required to be cooled and depressed. As mentioned, cooling the high temperature effluent can be achieved by the indirect heat transfer system. However, the conventional method normally used to maintain and decrease the pressure of SCWO system by adapting a pressure regulating valve. It means that the pressure drop of the whole system is achieved on one point, which will lead to extremely high velocities of feedstock and noise issues here, make the valve occur failure easily, and eventually affect the operation reliability of the SCWO system. In addition, if the feedstock contains solid particles, the existence of inert inorganic salt will result in severe abrasion or clogging around the regulating valve during the depressurization process of the SCWO system. Therefore, it is not advisable to achieve a complete depressurization in a single step by a regulating valve during the SCWO process, especially for the particles-containing feedstocks. Capillary pressure reduction is achieved by the frictional drag generating from the fluid flowing through the small diameter and high length capillary, realizing the step-down of the system step by step and effectively avoiding the problems about high velocities in tubes and corrosion damage of valves compared with depressurization by a single valve. What is more, adjusting the addition of choke water flowing into the capillary could accurately control the pressure of system, which ensures the reliability of the system operation.

3.4 Economic improvements

3.4.1 Catalytic SCWO

High reaction temperature and pressure result in a relatively expensive running cost, which also further affects the development of SCWO. The effective means may be to load proper catalyst to moderate the harsh reaction conditions, to minimize oxidant consumption, and to recover heat of reactor effluent as well as increase byproduct income. First, catalyst can improve reaction conditions, but it is also easy to be inactive due to the critical reaction conditions. The employed homogeneous and heterogeneous catalysts are mainly heteropolyacids, alkali carbonates, carbons, transition metal oxides, and bulk MnO_2 [60]. Besides, they also especially pointed out that the bulk MnO_2 has high activity, hydrothermal stability, and ability of maintenance, and resistance to metal leaching under supercritical reaction conditions.

3.4.2 Utilization of liquid oxygen

Liquid oxygen, gaseous oxygen, hydrogen peroxide, and air can be applied as oxidants in SCWO. Unfortunately, hydrogen peroxide is not commercially applicable because it is very expensive. Nitrogen in air does not react with feedstock, and costs a lot to separate them from gaseous products. Liquid oxygen is chosen as the oxidant of this plant due to its relatively cheaper price compared with the same amount of gaseous oxygen. Oxygen is regarded as one of the major costs, whereas high oxygen excess coefficient is not necessary because it is not realistic for a prospective commercial application. In this plant, oxygen excess coefficient is anticipated to control at about 1.1 to ensure complete decomposition of organic matters, and to avoid recovering excessive oxygen as well as decrease the running cost. The system uses liquid oxygen cold energy to realize the centralized liquefying and recycling of carbon dioxide, and greatly improves the system economy. In addition, the system is pre-desalted by the system of liquid oxygen cooling, reducing the subsequent desalting load and corrosivity of the system, reducing the grade of the equipment material, reducing the investment of the equipment, and recovering and selling valuable inorganic salts to improve the system economy.

3.4.3 Gas recovery and oxygen reuse

The SCWO system innovatively adopts the process of combining gas-phase product carbon dioxide recovery with oxygen recovery. By developing regular exhaust and compressor variable flow operation technology, high rate oxidation and low rate oxygen consumption are realized in the reaction process, that is, the reactor operates with high oxidation coefficient, and the gas phase after reaction is produced. The excess oxidants in the material are separated and recycled under variable flow conditions.

3.4.4 Waste heat recovery

Energy recovery is often achieved by preheating feedstock through reactor effluent of high temperature via heat exchangers [77]. Surplus energy is able to be utilized to produce steam for electricity generation or to form hot water for industrial application. For a larger scale SCWO system, energy recovery may potentially employ a supercritical steam turbine to generate electricity through the direct expansion of products [78]. However, it is not possible to design and construct a supercritical steam turbine in a SCWO plant presently due to the critical operation conditions and the very difficult separation of solids under SCWO conditions [79]. It is supposed that SCWO would be developed as a multifunctional technology for generating heat and electricity on the basis of removing organic pollutants in future. There is no doubt that sufficient energy recovery and heat income must be adequately considered in order to decrease the running cost, otherwise an external or internal heat source is required. In this apparatus, heat is recovered enough by preheating feedstock to realize a low running cost. In addition, major byproducts such as hot water and CO_2 can be collected for sale or reuse.

4. Industrial applications

Since the early 2000s, the significant maturation of SCWO technology on the basis of better understanding of the core reaction principles and effective solutions for mitigating the corrosion and salt precipitation problems in the past two decades started a shift in emphasis of SCWO activity from fundamental studies to industrial applications. For several feedstocks such as sewage sludge, industrial wastewaters, explosives, and other hazardous wastes, some full-scale SCWO treatment plants have been built under the guidance or promotion of companies, universities and colleges, and scientific research institutes. A series of associated SCWO vendors, such as MODAR, MODEC, Oxidyne, EcoWaste Technologies, and Foster Wheeler, have left this field, but the associated technologies still survived in some sense through acquisition or licenses by the former. For instance, the company Chematur Engineering located in Sweden acquired the assets of EcoWaste Technologies in 1999, and in 2007, licensed the SCWO patents to SCFI Group Ltd. Nowadays, a number of additional firms such as SCFI Group, InnovEOX, SuperWater Solution, XJTU, etc. are active in SCWO field (**Table 2**).

4.1 HydroProcessing

HydroProcessing constructed the first commercial SCWO sludge processing equipment for the Harlingen wastewater treatment plant, in Texas, USA [14]. In the HydroProcessing system, a hydrocyclone is applied to separate the flow after the reactor and to maintain the solids in the underflow. The overflow and subsequent

Company/University	HydroProcessing	Chematur Engineering AB	SuperWater Solutions	Xi'an Jiaotong University
Country	USA	Sweden	USA	China
Capacity	150 t/d	168 m^3/d	35 dry ton/d	3 t/d
Solid content (wt.%)	6–9	15	10	8
COD (mg/L)	80,000–120,000	110,000	Not available	5750
Temperature (°C)	592	510–580	600	400
Pressure (MPa)	23.47	25	26	25
Reactor	Tubular	Tubular	Tubular	TWM
Capital cost	3 million USD	5 million GBP	33.7 million USD	0.58 million USD
Operation cost	100 USD/dry ton	105 USD/dry ton	268 USD/dry ton	76.56 USD/dry ton

Table 2.
Operation and economic parameters of typical SCWO systems for municipal sludge treatment.

underflow are used to heat the feed. The heated feed mixes with oxygen and they enter the gas-fired heater together for further heating before entering a tubular reactor. The operation results show that the destruction efficiencies are 99.93–99.96, 49.6–84.1, and 92.89–98.90% for COD, ammonia, and total solid in the overflow, respectively. Meanwhile, the destruction efficiencies of sludge are 99.92–99.93, 49.6–86.4, and 62.72–88.94% for COD, ammonia, and total solid in the underflow, respectively. Due to the high ammonia concentration of the effluent, it can be sent to municipal wastewater plants or to industrial plants as a nutrient. A net operation and maintenance cost of approximately 100 USD/dry ton sludge was obtained [80].

4.2 Chematur AB and SCFI

For demonstration purposes, Chematur AB has built a 6 t/d unit in Karlskoga, Sweden, for the treatment of both undigested and digested sludge, which has been in operation since 1998 [15]. Chematur AB licensed the SCWO process to the Shinko Plantech Co. of Japan, and they have built a pilot SCWO unit with a capacity of approximately 26.4 t/d for the treatment of sludge in Kobe, Japan [17]. The developed SCWO process, named as the AquaCritox process, is presented in **Figure** 7. This system consists of a series of homogeneous sludge equipment including a feed tank with a paddle mixer, a macerator, an ultra turrax, and a grinder. The first three facilities along with a mono pump constitute a re-circulation loop, which provides size reduction and homogenization during a batch operation. After homogenization, sludge enters a hose diaphragm piston pump pressurized to approximately 25 MPa and is pumped to a double-pipe economizer heated by the reactor effluent. Fouling problems in the economizer are avoided due to the high velocity and the reduced feed particle size. After heat exchange, sludge enters a heater for further temperature increase. The removal rate of COD is greater than 99.9% when T > 520°C. A temperature higher than 540°C is needed to completely destroy total nitrogen (TN). They also evaluated the economic value of a unit treating 168 m^3 of sludge per day, being approximately 105 USD/dry ton sludge [17].

In 2007, the SCFI Group acquired patented SCWO technology (AquaCritox®) from Chematur Engineering AB of Sweden. Further work directed by the SCFI group has emphasized the reduction of costs. The residual inorganic fraction of

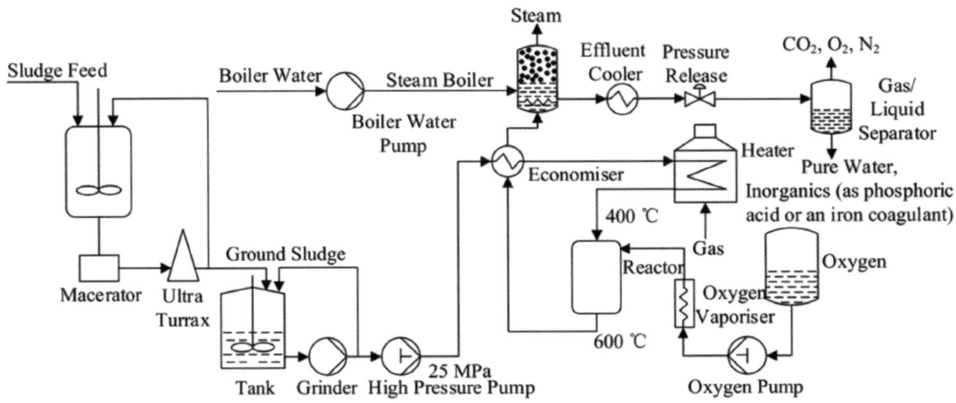

Figure7.
Flow sheet of the Chematur AB system [15].

the sludge can be recovered as phosphoric acid and iron coagulant if required. The orthophosphates from the supernatant, representing 78% of the total influent phosphorus, can also be recovered by using fluidized bed crystallization technologies [81].

4.3 SuperWater Solution

From 2009 to 2011, SuperWater Solution installed and successfully tested a 5 dry t/d SCWO system for the Iron Bridge Regional Water Reclamation Facility in Orlando, FL, USA [16]. They also assessed the economic value of 268 USD/dry ton for a unit with a capacity of 35 dry tons of sludge per day. For the employed SCWO process. It is interesting that the SuperWater Solution system has an oxygen recycling process based on the difference in liquefaction temperatures between O_2 and CO_2. Through the oxygen recycling process, the sludge can be oxidized at a high n value with low oxygen consumption, thus increasing the effluent quality and decreasing operation costs.

4.4 Energy and Environment Institute in XJTU of China

Since the early 2000s, Energy and Environment Institute led by one of the authors (Shuzhong Wang) in XJTU, based on the supercritical water technologies, has committed to the related research on harmless disposal and energy-oriented utilization of high concentration organic wastewaters and municipal/industrial sludge, and has accumulated abundant experience in water quality evaluation and products characterization. The research areas are involved in organic molecule reaction mechanisms and kinetics, catalytic characteristics and mechanisms, mechanisms and theories of alloy corrosion, salt crystallization kinetics, fuel and auxiliary heating mechanisms, etc. [6, 11, 21, 22, 28–30, 45, 50, 71, 82–85].

The stepwise engineering amplifications have been performed from initial experimental facility to test equipment (4.5 L/h), to small pilot plant (0.5 t/d). To further verify the industrialization feasibility of SCWO technology in China, in 2009, the authors successfully designed and constructed the first industrial scale plant of China (see **Figure 8**), actually of which the processing capacity of 3 ton/day, filling the China blank in the SCWO field [71]. This SCWO plant installed with the newly developed "TWM reactor" reaches a COD removal rate of more than 99.5% and surplus heat utilization efficiency of >80%. A brine pool is formed in the bottom of the reactor. Fluid in the pool with a high concentration of salts then enters

Figure 8.
First industrial scale SCWO plant in China [71].

the salt-removing device and is heated to a supercritical state. The salts are precipitated again and finally stored in the salt-collecting tank. The equipment investment is 0.58 million USD, and the operating cost is 76.56 USD/dry ton for sludge, which is much lower than the cost of incineration.

Depending on the accumulated research, design, and construction experiences, for overcoming or improving the potential higher investment, severe corrosion and blockage risk of the heat exchanging units, an indirect heat transfer SCWO process with a capacity of 50 t/d for petrochemical sludge treatment was developed, which is currently in the detailed design phase of the process package. The indirect heat transfer SCWO system is shown in **Figure 9**.

SCWO reactions happen at high temperature and pressure, meaning that the heating and pressuring of feedstock and the opposite process for the reaction effluents are necessary and unavoidable. In a conventional SCWO process, heat exchangers are usually used to cool the reaction effluents by untreated feedstock as cooling medium; meanwhile, the feedstock is preheated to the expected temperature, and then enters the reactor and reacts with the oxidants. But the critical aqueous solutions zones at temperature of 320–410°C corresponding exactly to the heat exchanger, which owns high-density, large dielectric constant and high solubility of inorganic salts, are aggressive to the piping and equipment materials generally inducing serious corrosion issues by improper controls, posing a potential threat to the reliable operation of the SCWO system. In addition, various organic wastes generally contains considerable amounts of heteroatom such as chlorine, sulfur, phosphorus, etc., which are transformed into corresponding inorganic acids after SCWO. Please note that the oxidant supplied into the system is often excessive to ensure the complete degradation of organic matters, causing a high level of residual oxidant in the reaction products. Therefore, reaction effluents are more corrosive than the initial feedstock. For the double-pipe heat exchanger used in conventional SCWO process, the cooling medium in the inner tube is feedstock while the reaction products as thermal fluid exists in the outer tube, causing internal side of the shell tube and both internal and external surfaces of the inner tubes facing serious corrosion problems and requiring expensive corrosion-resistant alloys, which greatly increases the investment of the equipment. Further, when the reaction effluents in the outer tube contains a large amount of inorganic salts or insoluble solids, plugging problems of the shell tube is extremely likely to occur due to its relatively complex multi-turn flow channel. In the indirect heat transfer SCWO system (see **Figure 9**), the feedstock being pre-grinded to 50 μm by a low pressure pump (2) is compressed by a high pressure feedstock

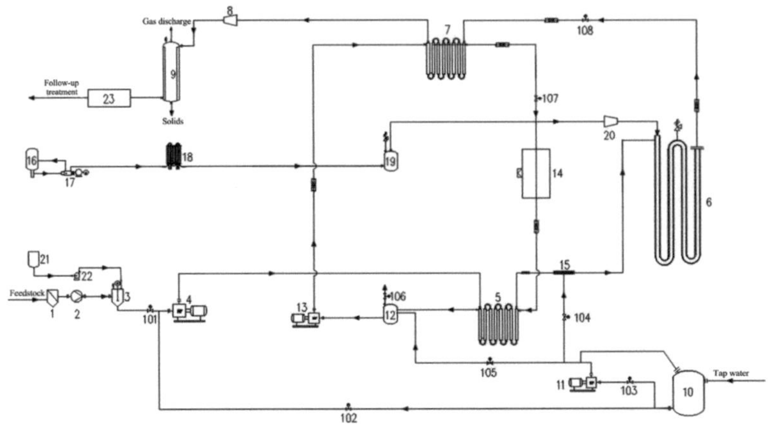

Figure 9.
A Schematic diagram of the indirect heat transfer SCWO comprehensive system: 1—filter; 2—low pressure pump; 3—feedstock tank; 4—high pressure feedstock pump; 5—preheater; 6—reactor; 7—heat exchanger; 8—capillary depressurization device; 9—three phase separator; 10—water storage tank; 11—high pressure water pump; 12—water buffer tank; 13—booster pump; 14—heater; 15—spray desuperheater; 16—liquid oxygen tank; 17—low temperature oxygen pump; 18—vaporizer; 19—oxygen buffer tank; 20—oxygen pressure regulator; 21—caustic tank; 22—low pressure caustic pump; (101—108)—valve; 23—biochemical treatment unit.

pump (4), preheated in the preheater (5) by high temperature intermediate medium, and then made to enter into reactor (6) when reaching the expected temperature. Liquid oxygen is pressured and its flow rate are regulated by the low temperature oxygen pump (17), subsequently gasified in the vaporizer (18), and then gaseous oxygen is stored in the oxygen buffer tank (19), ultimately entering the reactor (6) after being depressurized by the oxygen pressure regulator (20) and reacting with the feedstocks. Meanwhile, the SCWO reaction releases a certain amount of heat and reaction effluent with high-temperature and high-pressure flows into the inner tube of heat exchanger (6) to heat the intermediate medium in the outer tube, and the effluents are also cooled through the intermediate medium at the same time. Then the cooled effluents are reduced to an appropriate pressure by the capillary depressurization device (8). Gaseous products, liquid products, and solid products are similarly separated in the low pressure three-phase separator (9). The intermediate medium from the water buffer tank (12) pressurized by high pressure booster pump (13), firstly absorbs the heat of the reaction effluent through the regenerator (7); flows through the heater (14); and then, if necessary to further increase the temperature, eventually enters the preheater (5) for preheating the feedstock. Therefore, the corrosive fluid only passes through the inner tube of the preheater and the regenerator, and both outer tubes of the heat exchanger are clean water. Therefore, only inner tubes of the preheater and regenerator require the use of high-end corrosion-resistant alloys, and the outer tubes can use cheap carbon steel or low alloy steel, which will greatly reduce the investment cost of heat exchanger for the SCWO process. In addition, the clean water in the outer tube also dramatically avoids the blockage risk compared with dirty fluid containing insoluble solids.

Additionally, in view of specific characteristics of various municipal/industrial sludge and organic wastewaters such as printing and dyeing wastewater, phenol-containing wastewater, pharmaceutical/pesticide-production wastewater, etc., combining with their individual industrial production processes and customer specific requirements, the Energy and Environment Institute in XJTU has carried out the process integration innovation of large-scale supercritical water treatment system to maximize the efficiency, economy, and safety, and has finished a series of mature, reliable SCWO process packages [82].

5. Conclusions

In virtue of the special physicochemical properties of supercritical water, supercritical water oxidation (SCWO) can efficiently and thoroughly degrade a wide variety of organic pollutants into harmless small molecules, such as CO_2, N_2, H_2O, etc., which has been widely regarded as the most promising, environment-friendly treatment technology for organic wastes. Not a specific material can withstand all kinds of SCWO conditions; however, the suitable material can be selected for a certain condition considering corrosion resistance, strength, and economy. Abundant corrosion control approaches can mitigate the equipment corrosion effectively. Salts precipitation, possibly resulting in plugging of pipes and instruments and accelerating corrosion, is another main obstacle to hinder SCWO commercialization. Fortunately, pre-desalination, on-line salt separation, novel reactor configuration such as TWM reactor, etc., all exhibit better performances in preventing and controlling salt deposition and corrosion issues occurred in reactors. The global considerations and designs instead of previous single or local protection will be in focus in the future. On the basis of decades of implementation experience of full-scale SCWO plants, considering the incessant emergence of more advanced technologies and equipment which aims to overcome the abovementioned two key obstacles and to further elevate the system economy, safety, and automatic control level, the industrial SCWO plants will achieve the harmless disposal of various organic wastes in a more economical, safe, credible way.

Acknowledgements

This chapter and involved contents were supported by the Projects from National Natural Science Foundation of China (21576219, 51871179, and 51876174), China Postdoctoral Science Foundation [2019TQ0248], the National High Technology Research and Development Program of China (2006AA06Z313), and the National Key Research and Development Program of China (2016YFC0801904, 2017YFB060360).

Author details

Yanhui Li and Shuzhong Wang*
Key Laboratory of Thermo-Fluid Science and Engineering of MOE, School of Energy and Power Engineering of Xi'an Jiaotong University, Xi'an, Shaanxi, China

*Address all correspondence to: szwang@aliyun.com

References

[1] Brunner G. Near and supercritical water. Part II: Oxidative processes. The Journal of Supercritical Fluids. 2009;**47**:382-390

[2] Gong YM, Guo Y, Wang SZ, Song WH, Xu DH. Supercritical water oxidation of quinazoline: Reaction kinetics and modeling. Water Research. 2017;**110**:56-65

[3] Xu DH, Wang SZ, Zhang J, Tang XY, Guo Y, Huang CB. Supercritical water oxidation of a pesticide wastewater. Chemical Engineering Research & Design. 2015;**94**:396-406

[4] Zhang J, Wang SZ, Guo Y, Xu DH, Li XD, Tang XY. Co-oxidation effects of methanol on acetic acid and phenol in supercritical water. Industrial & Engineering Chemistry Research. 2013;**52**:10609-10618

[5] Veriansyah B, Kim J-D. Supercritical water oxidation for the destruction of toxic organic wastewaters: A review. Journal of Environmental Sciences. 2007;**19**:513-522

[6] Gong YM, Guo Y, Sheehan JD, Chen ZF, Wang SZ. Oxidative degradation of landfill leachate by catalysis of $CeMnOx/TiO_2$ in supercritical water: Mechanism and kinetic study. Chemical Engineering Journal. 2018;**331**:578-586

[7] Wang SZ, Guo Y, Chen CM, Zhang J, Gong YM, Wang YZ. Supercritical water oxidation of landfill leachate. Waste Management. 2011;**31**:2027-2035

[8] Gong YM, Wang SZ, Li YH. The treatment of concentrated landfill leachate from membrane-based processes by supercritical water oxidation. Applied Mechanics and Materials. 2014;**522-524**:560-564

[9] Gong YM, Wang SZ, Qian LL, Li YH. Supercritical water oxidation of fulvic acid as a model organic compound in landfill leachate. Applied Mechanics and Materials. 2014;**522-524**:600-604

[10] Zhang J, Wang SZ, Guo Y, Xu DH, Gong YM, Tang XY. Supercritical water oxidation of polyvinyl alcohol and desizing wastewater: Influence of NaOH on the organic decomposition. Journal of Environmental Sciences-China. 2013;**25**:1583-1591

[11] Zhang J, Wang S, Li Y, Lu J, Chen S, Luo X. Supercritical water oxidation treatment of textile sludge. Environmental Technology. 2016:1-12

[12] Marulanda V, Bolaños G. Supercritical water oxidation of a heavily PCB-contaminated mineral transformer oil: Laboratory-scale data and economic assessment. The Journal of Supercritical Fluids. 2010;**54**:258-265

[13] SCFI. What Is Super Critical Water Oxidation; 2012

[14] Griffith JW, Raymond DH. The first commercial supercritical water oxidation sludge processing plant. Waste Management. 2002;**22**:453-459

[15] Patterson DA, Stenmark L, Hogan F. Pilot-scale supercritical water oxidation of sewage sludge. In: Proceedings of the 6th European Biosolids and Organic Residuals Conference. 2001

[16] Sloan DS, Pelletier RA, Modell M. Sludge management in the city of Orlando—It's supercritical! Florida Water Resources Journal. 2008:46-54

[17] Gidner A, Stenmark L. Supercritical water oxidation of sewage sludge—State of the art. In: Chematur Engineering AB. 2001

[18] Marrone PA. Supercritical water oxidation-current status of full-scale commercial activity for waste destruction. Journal of Supercritical Fluids. 2013;**79**:283-288

[19] Kritzer P, Dinjus E. An assessment of supercritical water oxidation (SCWO)—Existing problems, possible solutions and new reactor concepts. Chemical Engineering Journal. 2001;**83**:207-214

[20] Li Y, Xu T, Wang S, Yang J, Li J, Xu T, et al. Characterization of oxide scales formed on heating equipment in supercritical water gasification process for producing hydrogen. International Journal of Hydrogen Energy. 2019. In press. DOI: 10.1016/j.ijhydene.2019.01.284

[21] Xingying T, Shuzhong W, Lili Q, Mengmeng R, Panpan S, Yanhui L, et al. Corrosion properties of candidate materials in supercritical water oxidation process. Journal of Advanced Oxidation Technologies. 2016;**19**:141-157

[22] Tang XY, Wang SZ, Qian LL, Li YH, Lin ZH, Xu DH, et al. Corrosion behavior of nickel base alloys, stainless steel and titanium alloy in supercritical water containing chloride, phosphate and oxygen. Chemical Engineering Research & Design. 2015;**100**:530-541

[23] Tang XY, Wang SZ, Xu DH, Gong YM, Zhang J, Wang YZ. Corrosion behavior of Ni-based alloys in supercritical water containing high concentrations of salt and oxygen. Industrial & Engineering Chemistry Research. 2013;**52**:18241-18250

[24] Li YH, Wang SZ, Yang JQ, Xu DH, Guo Y, Qian LL, et al. Corrosion characteristics of a nickel-base alloy C-276 in harsh environments. International Journal of Hydrogen Energy. 2017;**42**:19829-19835

[25] Li YH, Wang SZ, Li XD, Lu JM. Corrosion of an austenitic heat-resistant steel HR3C in high-temperature steam and supercritical water. Advanced Materials Research. 2014;**908**:67-71

[26] Asselin E, Alfantazi A, Rogak S. Corrosion of nickel–chromium alloys, stainless steel and niobium at supercritical water oxidation conditions. Corrosion Science. 2010;**52**:118-124

[27] Yang J, Li YH, Xu AN, Fekete B, Macdonald DD. The electrochemical properties of alloy 690 in simulated pressurized water reactor primary water: Effect of temperature. Journal of Nuclear Materials. 2019;**518**:305-315

[28] Yang JQ, Wang SZ, Tang XY, Wang YZ, Li YH. Effect of low oxygen concentration on the oxidation behavior of Ni-based alloys 625 and 825 in supercritical water. The Journal of Supercritical Fluids. 2018;**131**:1-10

[29] Li YH, Wang SZ, Sun PP, Xu DH, Ren MM, Guo Y, et al. Early oxidation mechanism of austenitic stainless steel TP347H in supercritical water. Corrosion Science. 2017;**128**:241-252

[30] Li YH, Wang SZ, Tang XY, Xu DH, Guo Y, Zhang J, et al. Effects of sulfides on the corrosion behavior of Inconel 600 and Incoloy 825 in supercritical water. Oxidation of Metals. 2015;**84**:509-526

[31] Kritzer P. Corrosion in high-temperature and supercritical water and aqueous solutions: A review. Journal of Supercritical Fluids. 2004;**29**:1-29

[32] Son M, Kurata Y, Ikushima Y. Corrosion behavior of metals in SCW environments containing salts and oxygen. Corrosion. Denver, Colorado: NACE International; 2002. pp. 12

[33] Was GS, Jiao Z, Teysseyre SS. Corrosion of austenitic alloys in supercritical water. Corrosion. 2005;**62**:989-1005

[34] Ikushima Y, Kurata Y, Hatakeda K, Kim H, Son M. Corrosion behavior of metals in supercritical water solutions in the presence of salts. Corrosion. Houston, Texas: NACE International; 2001. pp. 13

[35] Daigo Y, Watanabe Y, Sugahara K, Isobe T. Compatibility of nickel-based alloys with supercritical water applications: Aging effects on corrosion resistance and mechanical properties. Corrosion. 2006;**62**:174-181

[36] Sue K, Arai K, Watanabe Y, Tsujinaka N, Adschiri T. Relationship between corrosion rate and metal oxide solubility in supercritical water. Corrosion. 2002;**2002**

[37] Kim H, Mitton DB, Latanision R. Stress corrosion cracking of alloy 625 in pH 2 aqueous solution at high temperature and pressure. Corrosion. 2011

[38] Li YH, Wang SZ, Sun PP, Tong ZX, Xu DH, Guo Y, et al. Early oxidation of Super304H stainless steel and its scales stability in supercritical water environments. International Journal of Hydrogen Energy. 2016;**41**:15764-15771

[39] Schroer C, Konys J, Novotny J, Hausselt J. Aspects of the corrosion kinetics for binary nickel-chromium alloys under conditions found in supercritical water oxidation plants processing chlorinated substances. Corrosion. 2006;**62**:444-459

[40] Fujisawa R, Sakaihara M, Kurata Y, Watanabe Y. Corrosion behaviour of nickel base alloys and 316 stainless steel in supercritical water under alkaline conditions. Corrosion Engineering, Science and Technology. 2005;**40**:244-248

[41] Sun M, Wu X, Zhang Z, Han E-H. Analyses of oxide films grown on alloy 625 in oxidizing supercritical water. The Journal of Supercritical Fluids. 2008;**47**:309-317

[42] Tang XY, Wang SZ, Qian LL, Lu JM. Selection of material with high salt concentration in preheating section of supercritical water oxidation system. In: Zhang H, Jin D, Zhao XJ, editors. Advanced Research on Material Engineering, Chemistry and Environment. Stafa-Zurich: Trans Tech Publications Ltd; 2013. pp. 440-443

[43] Fujisawa R, Nishimura K, Sakaihara M, Watanabe Y, Kurata Y, Nishida T. Corrosion behavior of ni base alloys and 316 stainless steel in less oxidizing or reducing SCW containing HCl. Corrosion. 2005

[44] Kritzer P, Boukis N, Dinjus E. Corrosion of alloy 625 in aqueous solutions containing chloride and oxygen. Corrosion. 1998;**54**:824-834

[45] Li YH, Wang SZ, Sun P, Yang JQ, Tang XY, Xu DH, et al. Investigation on early formation and evolution of oxide scales on ferritic–martensitic steels in supercritical water. Corrosion Science. 2018;**135**:136-146

[46] Li YH, Xu TT, Wang SZ, Fekete B, Yang J, Yang JQ, et al. Modelling and analysis of the corrosion characteristics of ferritic-martensitic steels in supercritical water. Materials. 2019;**12**:409

[47] Viswanathan R, Sarver J, Tanzosh JM. Boiler materials for ultra-supercritical coal power plants-stearnside oxidation. Journal of Materials Engineering and Performance. 2006;**15**:255-274

[48] Bischoff J, Motta AT. Oxidation behavior of ferritic-martensitic and ODS steels in supercritical water. Journal of Nuclear Materials. 2012;**424**:261-276

[49] Li YH, Wang SZ, Wang LS, Yu PF, Yang JQ. Corrosion behavior of austenitic steel 304 in nearcriticalaqueous solutions.

Advances in Engineering Research. 2015;**45**:189-192

[50] Xu DH, Huang CB, Wang SZ, Lin GK, Guo Y. Salt deposition problems in supercritical water oxidation. Chemical Engineering Journal. 2015;**279**:1010-1022

[51] Kawasaki S-I, Oe T, Itoh S, Suzuki A, Sue K, Arai K. Flow characteristics of aqueous salt solutions for applications in supercritical water oxidation. The Journal of Supercritical Fluids. 2007;**42**:241-254

[52] Hodes M, Marrone PA, Hong GT, Smith KA, Tester JW. Salt precipitation and scale control in supercritical water oxidation—Part a: Fundamentals and research. Journal of Supercritical Fluids. 2004;**29**:265-288

[53] Marrone PA, Hodes M, Smith KA, Tester JW. Salt precipitation and scale control in supercritical water oxidation—Part B: Commercial/full-scale applications. The Journal of Supercritical Fluids. 2004;**29**:289-312

[54] Schubert M, Regler JW, Vogel F. Continuous salt precipitation and separation from supercritical water. Part 1: Type 1 salts. Journal of Supercritical Fluids. 2010;**52**:99-112

[55] Prikopsky K, Wellig B, von Rohr PR. SCWO of salt containing artificial wastewater using a transpiring-wall reactor: Experimental results. The Journal of Supercritical Fluids. 2007;**40**:246-257

[56] Matsumura Y, Minowa T, Potic B, Kersten SRA, Prins W, van Swaaij WPM, et al. Biomass gasification in near- and super-critical water: Status and prospects. Biomass and Bioenergy. 2005;**29**:269-292

[57] Portela JR, López J, Nebot E, Martínez De La Ossa E. Elimination of cutting oil wastes by promoted hydrothermal oxidation. Journal of Hazardous Materials. 2001;**88**:95-106

[58] Li YH, Wang SZ, Yang JQ, Zhang Y, Xu DH, Guo Y. Effect of salt deposits on corrosion behavior of Ni-based alloys in supercritical water oxidation of high salinity organic wastewater. Journal of Environmental Engineering. 2019;**145**:04019080

[59] Waldner MH, Krumeich F, Vogel F. Synthetic natural gas by hydrothermal gasification of biomass: Selection procedure towards a stable catalyst and its sodium sulfate tolerance. The Journal of Supercritical Fluids. 2007;**43**:91-105

[60] Savage PE. A perspective on catalysis in sub- and supercritical water. Journal of Supercritical Fluids. 2009;**47**:407-414

[61] Yang JQ, Wang SZ, Xu DH, Guo Y, Yang C, Li YH. Effect of ammonium chloride on corrosion behavior of Ni-based alloys and stainless steel in supercritical water gasification process. International Journal of Hydrogen Energy. 2017;**42**:19788-19797

[62] Wang J, Qian S, Li Y, Macdonald DD, Jiang Y, Li J. Passivity breakdown on 436 ferritic stainless steel in solutions containing chloride. Journal of Materials Science & Technology. 2019;**35**:637-643

[63] Kritzer P, Boukis N, Dinjus E. Factors controlling corrosion in high-temperature aqueous solutions: A contribution to the dissociation and solubility data influencing corrosion processes. The Journal of Supercritical Fluids. 1999;**15**:205-227

[64] Cocero MJ, Martinez JL. Cool wall reactor for supercritical water oxidation—Modelling and operation results. Journal of Supercritical Fluids. 2004;**31**:41-55

[65] Baur S, Schmidt H, Krämer A, Gerber J. The destruction of industrial

aqueous waste containing biocides in supercritical water—Development of the SUWOX process for the technical application. The Journal of Supercritical Fluids. 2005;**33**:149-157

[66] Calzavara Y, Joussot-Dubien C, Turc HA, Fauvel E, Sarrade S. A new reactor concept for hydrothermal oxidation. The Journal of Supercritical Fluids. 2004;**31**:195-206

[67] Príkopský K. Characterization of Continuous Diffusion Flames in Supercritical Water Oxidation. Zürich: Swiss Federal Institute of Technology; 2007

[68] Wellig B. Transpiring wall reactor for the supercritical water oxidation. ProQuest Dissertations Publishing; 2003

[69] Li Y, Wang S, Ren M, Zhang J, Xu D, Qian L, et al. Recent advances on research and application on supercritical hydrothermal combustion technology. HuagongJinzhan/Chemical Industry and Engineering Progress. 2016;**35**:1942-1955

[70] Fauvel E, Joussot-Dubien C, Tanneur V, Moussiere S, Guichardon P, Charbit G, et al. A porous reactor for supercritical water oxidation: Experimental results on salty compounds and corrosive solvents oxidation. Industrial & Engineering Chemistry Research. 2005;**44**:8968-8971

[71] Xu D, Wang S, Tang X, Gong Y, Guo Y, Wang Y, et al. Design of the first pilot scale plant of China for supercritical water oxidation of sewage sludge. Chemical Engineering Research & Design. 2012;**90**:288-297

[72] Xu DH, Wang SZ, Gong YM, Guo Y, Tang XY, Ma HH. A novel concept reactor design for preventing salt deposition in supercritical water. Chemical Engineering Research & Design. 2010;**88**:1515-1522

[73] Li YH, Wang SZ, Qian LL, Wen S. An assessment of supercritical hydrothermal combustion (SCHC) for organic wastes destruction. Advanced Materials Research. 2014;**955-959**:1777-1782

[74] Qian LL, Wang SZ, Li YH. Review of supercritical water oxidation in hydrothermal flames. Advanced Materials Research. 2014;**908**:239-242

[75] Marrone PA, Hong GT. Corrosion control methods in supercritical water oxidation and gasification processes. The Journal of Supercritical Fluids. 2009;**51**:83-103

[76] Qian LL, Wang SZ, Savage PE. Hydrothermal liquefaction of sewage sludge under isothermal and fast conditions. Bioresource Technology. 2017;**232**:27-34

[77] Dolores Bermejo M, Rincon D, Martin A, Jose Cocero M. Experimental performance and modeling of a newcooled-wall reactor for the supercritical water oxidation. Industrial & Engineering Chemistry Research. 2009;**48**:6262-6272

[78] Bermejo MD, Cocero MJ. Destruction of an industrial wastewater by supercritical water oxidation in a transpiring wall reactor. Journal of Hazardous Materials. 2006; **137**:965-971

[79] Marias F, Mancini F, Cansell F, Mercadier J. Energy recovery in supercritical water oxydation process. Environmental Engineering Science. 2008;**25**:123-130

[80] Svanström M, Fröling M, Modell M, Peters WA, Tester J. Environmental assessment of supercritical water oxidation of sewage sludge. Resources, Conservation and Recycling. 2004;**41**:321-338

[81] O'Callaghan P. Phosphorus recovery from sewage sludge using the

aquacritoxsupercritical water oxidation process. O_2 Environmental and SCFI. 2010:1-8

[82] Yang JQ, Wang SZ, Li YH, Zhang Y, Xu DH. Novel design concept for a commercial-scale plant for supercritical water oxidation of industrial and sewage sludge. Journal of Environmental Management. 2019;**233**:131-140

[83] Zhang YS, Wang SZ, Song WH, Yang JQ, Xu TT, Li JN, et al. Characteristics of sodium sulfate deposition in hydrogen production from supercritical water gasification: A review. International Journal of Hydrogen Energy. 2019. In press. DOI: 10.1016/j.ijhydene.2019.05.125

[84] Zhang J, Wang SZ, Ren MM, Lu JL, Chen SL, Zhang HM. Effect mechanism of auxiliary fuel in supercritical water: A review. Industrial & Engineering Chemistry Research. 2019;**58**:1480-1494

[85] Qian LL, Wang SZ, Ren MM, Wang S. Co-oxidation effects and mechanisms between sludge and alcohols (methanol, ethanol and isopropanol) in supercritical water. Chemical Engineering Journal. 2019;**366**:223-234

Chapter 7

Synthesis and Functionalization of Nanoparticles in Supercritical CO_2

Ludmila Otilia Cinteza and Daniela Bala

Abstract

A review of recent results on fabrication of inorganic and organic nanoparticles in supercritical carbon dioxide will be presented, with particular emphasis on the metallic and polymeric nanoparticles used in biomedicine. The use of the water-in-$scCO_2$ microemulsion in the synthesis of metal nanoparticles will be also discussed. The recent progress in preparation of polymeric nanoparticles with desired size and porosity obtained through processing methods in $scCO_2$ as drug delivery systems will be described. The efficiency of the drug encapsulation in organic and inorganic nanoparticles using supercritical CO_2 as dissolving media is another topic of interest. Various methods to achieve surface functionalization of nanoparticles in supercritical and subcritical CO_2 will be evaluated, considering the challenges and limitations in efficiency, scalability, and development of new applications.

Keywords: supercritical CO_2, nanoparticles, drug delivery, biomedicine

1. Introduction

The pharmaceutical industry has a major problem concerning the production of active pharmaceutical ingredients, which have a low water solubility and bioavailability. Therefore an appropriate technology for producing these active components is needed with certain properties like particle size (smaller than 1000 nm, typically under 500 nm), solubility, efficacy, state transition (polymorphism and crystallization), cost-effectiveness, etc.

In many nanomedical applications used in nanomedicine, processes based on supercritical fluids (SCFs) can be applied because they allow controlled fabrication of biological active nanostructured microparticles, nanoparticles, and nanoporous/nanostructured materials. Supercritical carbon dioxide ($scCO_2$) as a green solvent that possesses many beneficial properties (it is nonflammable, nontoxic, biocompatible, cost-effective, and abundant) has gained huge interest in the food and pharmaceutical industries; it is considered environmentally benign and one of the few solvents not regulated as a volatile organic compound (VOC) by the US Environmental Protection Agency.

There are several methods for the manufacturing of solid particles scaled from micron to nanosize, divided into bottom-up, top-down, and combination approaches of these [1].

Bottom-up techniques produce nanosized particles by precipitation from a supersaturated drug solution. Precipitation by addition of liquid antisolvent is made by simple mixing methods (using a static mixer) or by modified mixing methods (sonoprecipitation or high gravity controlled precipitation). Other bottom-up techniques involve supercritical fluids, rapid expansion of supercritical solution (RESS), and supercritical antisolvent technique (SAS) or solvent removal by nano-spray dryer and spray freezing into liquid techniques. These methods have some disadvantages which include the size of particles that cannot be properly controlled in the non-sized range, but using of some additives (excipients, surfactants, etc.) can produce the stabilization nanocrystals or nanoparticles regarding the morphological properties and the crystallized polymorphic form [2]. Bottom-up techniques showed some advantages because they are low-energy processes and less expensive than the other methods, and obtained particles have narrow size distribution. In order to obtain smaller particles, these methods have been used in combination with top-down techniques [1].

Top-down techniques are used for particle size reduction of drugs to the nanometer size range by application of friction, involving high-energy processes such as media milling (wet bead milling) and high-pressure homogenization techniques (Dissocubes homogenization and NanoPure technology). These methods have some disadvantages which include a limited control of crystal size and surface properties and thermal or mechanical degradation generated by intensive energy of mixing [3]. These techniques are used in the last decade, but few nanocrystals under 100 nm have been obtained. Drug particles with smaller size than 100 nm have novel physical properties and better permeation through different biological barriers [4] and improved bioavailability of poorly aqueous soluble drugs, having different routes of administration such as oral, ocular, dermal, buccal, and pulmonary.

Supercritical fluids and their mixtures have specific properties like very fast mass transfer, near zero surface tension, and effective solvent elimination. The liquid-like and/or gas-like properties of SCFs and the possibility to modify several process parameters (temperature, pressure, and surface tension) can be benefits to produce several medical products at nanoscale. Several SCF-based processes are applied to nanomedicine applications: supercritical antisolvent precipitation (SAS), rapid expansion of supercritical solutions, supercritical emulsion extraction (SEE), supercritical assisted phase separation, supercritical gel drying, supercritical assisted liposome formation (SuperLip), supercritical assisted atomization (SAA), electrospinning in $scCO_2$, supercritical assisted injection in a liquid antisolvent (SAILA), and depressurization of an expanded solution into aqueous media (DESAM) [5].

$scCO_2$ method	Nanoparticles	Temperature (K)	Pressure (MPa)	References
RESS	Drug micro−/nanoparticles Polymeric micro−/nanoparticles Drug encapsulated in polymeric microparticles	313–373	8–25	[6, 7]
SAS	Drug micro−/nanoparticles Inorganic nanoparticles	303–323	9–15 MPa	[8]
SEE	Drug encapsulated in lipid or polymeric nanoparticles	308–323	8–10 MPa	[5, 9, 10]
SAA	Drug nanoparticles	343–353	7–9	[11]
SAILA	Drug nanoparticles	343–353	7–10	[5]
RESOLV	Drug nanoparticles Polymer-stabilized drug NPs	308–318	10–20	[12]

Table 1.
Reaction conditions (temperature and pressure) in $scCO_2$-assisted synthesis of nanoparticles.

Numerous methods for nanoparticle fabrication and functionalization using supercritical carbon dioxide have been proposed, in various reaction conditions. In **Table 1**, some examples of temperature and pressure range frequently used for the preparation of nanoparticulated materials are summarized, with particular emphasis on drug or drug delivery systems.

The selection of the temperature and pressure condition is crucial for the particle size and drug encapsulation efficiency, and the specific values are chosen according to the characteristics of the active substance, such as solubility in $scCO_2$ phase, thermal stability, etc.

In this chapter, we present several applications of supercritical carbon dioxide ($scCO_2$) for preparation of polymeric nanoparticles as drug delivery systems, active principle encapsulation, water-in-$scCO_2$ microemulsion in the synthesis of metal nanoparticles, and various methods for surface functionalization of nanoparticles in supercritical and subcritical CO_2.

2. Applications of $scCO_2$ for preparation of polymeric nanoparticles as drug delivery systems

2.1 Supercritical fluids and their properties

SCFs can bring their contributions in different fields for certain applications as chromatography, fluid extractions, and micro- and nanoparticle formation. There are several compounds that can be used as supercritical fluids. Among them, hydrocarbons are toxic and inflammable, water has high critical parameters, but carbon dioxide has appropriate critical temperature and pressure, being suitable to be used as a green solvent.

2.2 Rapid expansion of supercritical solution

Methods using supercritical fluids (SCFs) can produce particles with narrow size distribution. Because of their special properties, SCFs can be applied to micronization of several types of compounds: drugs, biopolymers, polymers, food, coloring maters, explosives, etc. Pharmaceutical micronizations using supercritical fluids have some advantages due to the absence of organic solvent, and the particle size distribution could be controlled by process parameters. During the micronizations by SCFs methods, the dissolution rate is increased. The use of $scCO_2$ provides several advantages in comparison with the previous conventional techniques. The $scCO_2$ can be used as a solvent, antisolvent, and extracting agent for the organic phase of oil-in-water emulsions or at particle formation from gas saturated solution and improve the spraying process in different techniques [13].

The RESS process is based on the saturation of the supercritical medium with a solute (polymer) followed by a rapidly depressurization of the solution through a heated nozzle at high speed. During the pressure drop, the system passes from supercritical to atmospheric conditions, the solvent power decreases, and a fast nucleation of the solute in the form of very small particles with uniform size takes place. The properties of the obtained particles are influenced by its solubility in $scCO_2$; state parameters from precipitation vessel (temperature and pressure); size, length, and shape of the nozzle; distance of the jet stream; and impact angle onto the surface. A schematic representation of this process is presented in **Figure 1**.

The RESS technique has several advantages like the simple control of process parameters, the absence of organic solvents, and easy implementation on lab-scale when a single nozzle is used. RESS has also disadvantages as difficulty in scaling-up,

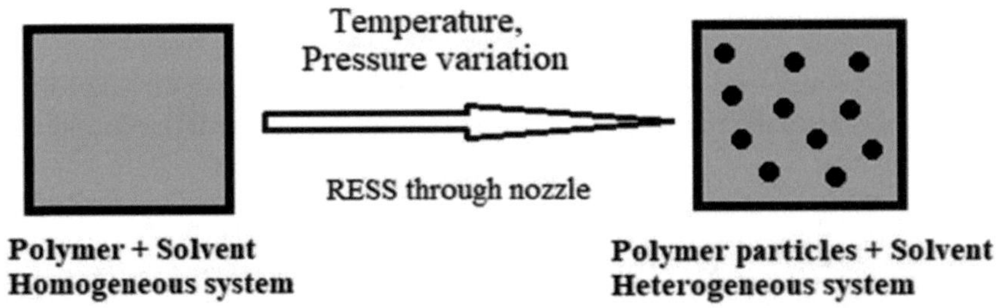

Figure 1.
Schematic representations of the RESS process.

the possible particle aggregation and/or nozzle blockage, using an important amount of carbon dioxide, and low solubility of most pharmaceutical compounds in supercritical CO_2. But, the $scCO_2$ solvent power can be increased by using of a co-solvent.

2.3 Rapid expansion of supercritical solution into a liquid solvent (RESOLV)

RESOLV represents a variation of RESS. This technique can reduce the particle aggregation during the jet expansion. In this process the supercritical solution is depressurized through a nozzle into a collection chamber containing an aqueous solution at room temperature. Different types of water-soluble polymers or surfactants can added to the aqueous medium in order to stabilize the obtained nanoparticle suspension [12]. Biocompatible polymer nanoparticles composed of biodegradable polymers such as polylactic acid (PLA), poly(lactic-coglycolic) acid (PLGA), and poly(ε-caprolactone) (PCL) are drawing a considerable interest in the scientific community because they can be used in medicine as biodegradable support materials and drug delivery vehicles. There are studies focused on the generation of polysaccharide particles. Polysaccharides are biobased polymers used in a lot of domains such as nutrition, energy, health care, and materials science, with large applications in the industry. An example of this family is chitosan, an amino-polysaccharide derived from chitin, being the second most biosynthesized polymer after cellulose. Because chitosan is biocompatible and biodegradable (mucoadhesive with antibacterial and cytocompatible), it can be used in pharmaceutics and biomedical applications, cosmetics, food packaging, agriculture, water treatment, etc. Another polysaccharide is alginate that can be used in several areas like drug delivery, tissue engineering, and wound dressing. For these purposes and due to its essential functional groups (hydroxyl and carboxyl), alginates can be transformed to hydrogels, porous scaffolds, and micro- and nanoparticles [14].

2.4 Nanoparticles and porous scaffolds obtained by high-pressure CO_2 processing

In order to obtain microparticles, an experimental apparatus equipped with a high-pressure cell having a capillary nozzle was used. The scheme of this handmade experimental setup is presented in **Figure 2**.

Biopolymeric microparticles were prepared in our laboratory by rapid expansion of high-pressure CO_2-chitosan solution in sodium bis-(2-ethylhexyl) sulfosuccinate solution. At pressures higher than 2 MPa, ultrafine particles were formed, while under this value, wires were obtained (**Figure 3**). The Chi/AOT ultrafine particles are instantaneously formed when 2 mL Chi solution 2% (wt/wt), preheated 1 hour at 40°C, is sprayed using CO_2 at different pressures into 20 mL AOT 0.03 M aqueous solution through a stainless steel capillary nozzle of 30 mm length and 0.4 mm

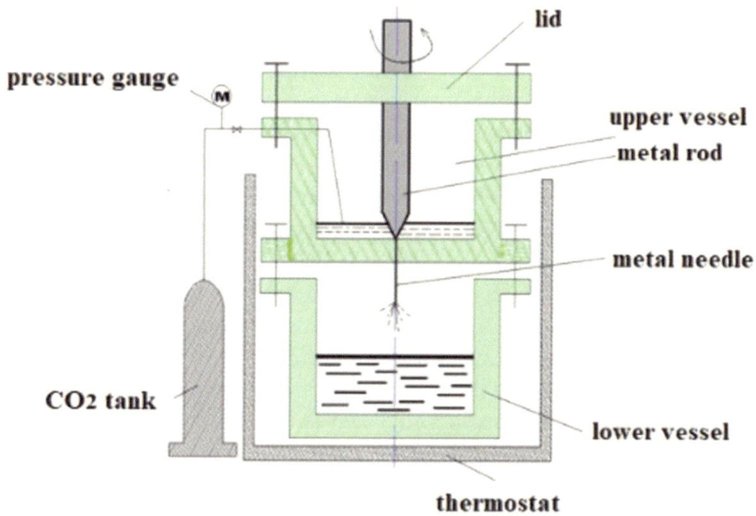

Figure 2.
Experimental setup for RESS technique.

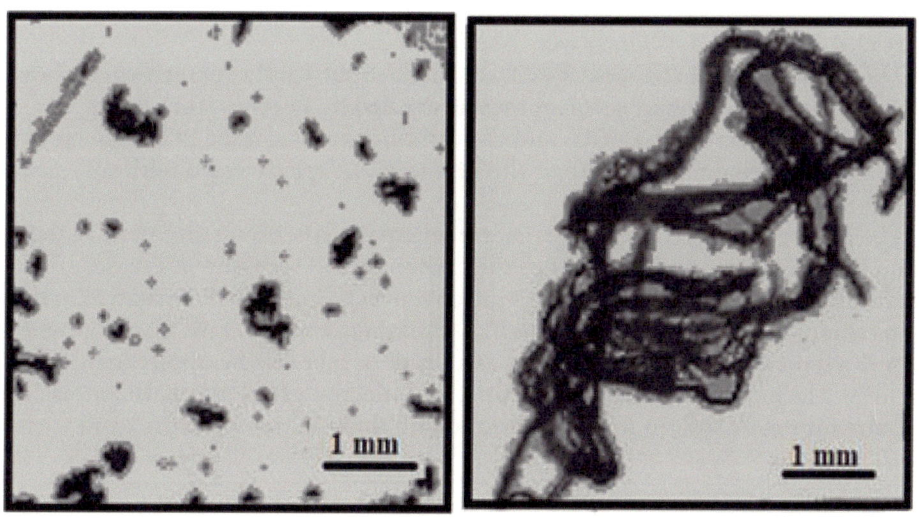

Figure 3.
Optical images of CO_2-chitosan (chi)/bis-(2-ethylhexyl) sulfosuccinate (AOT) microparticles and wires.

diameter. The pre-expansion pressure was 1–5 MPa, and the distance from the nozzle tip to surfactant solution interface was of about 20 mm. We observe that with increasing the spraying pressure of polymer, the size of the particles decreases. The microparticles obtained at high pressure are quasi-spherical in aqueous medium and irregular with many pores and a rough surface after freeze-drying. The morphology of synthesized particles recommends them for possible applications in adsorption of organic and inorganic substances from aqueous medium. The Chi/AOT microparticles were an effective adsorbent for removal of phenol and o-cresol from aqueous solution [15].

Porous alginate matrices were obtained using sub- and supercritical carbon dioxide. Calcium alginate matrices had uniform porous texture generated by high-pressure CO_2 as foaming agent without co-solvents. Sodium alginate solutions were processed in high-pressure CO_2, with freezing. After depressurization, the frozen samples were ionically cross-linked with calcium ions with and without glycerol. The effects of the presence of glycerol as plasticizer, carbon dioxide pressure, temperature, and processing time (20 minutes and 5 hours) on the structure of the obtained calcium alginate matrices were investigated. The porosity of alginate matrices increased with CO_2 pressure, processing time, and glycerol adding. The plasticizer, glycerol, improves mechanical properties and texture for scaffolds.

Poly(ε-caprolactone) is another polymer with potential application in the biomedical field due to its properties such as good solubility, low melting point (59–64°C), and very good blending compatibility. PCL is suitable for controlled delivery of drug because it has high permeability for several drugs and excellent biocompatibility and it can be completely eliminated from the body. Due to the low melting point and good rheological and mechanical properties, PCL can be used as a biomaterial in cardiovascular and bone tissue engineering.

Microparticles of poly(ε-caprolactam) were obtained by rapid expansion in water of the PCL polymeric solution (with a cosolvent, methanol or dimethylformamide) saturated with high-pressure carbon dioxide, in experimental setup presented in **Figure 3**. The solution was denoted by P1 – PCL in methanol (1 g/mL) and P2 – PCL in dimethylformamide (1 g/mL). The working temperature was 70°C and pressure 6.5 and 8.5 MPa. In **Table 1** and **Figure 4**, values for pressure are expressed in bar. Depending on the nature of the cosolvent, temperature, and pressure, poly(ε-caprolactam) particles of spherical shape and variable dimensions (particles diameter 1.4–6.7 μm) were obtained.

The morphological changes of PCL solutions in sub- and supercritical carbon dioxide and the addition of cosolvent were revealed by Boethius microscopy.

The interactions between CO_2 and the carbonyl groups in the PCL molecules led to the lower melting temperature of the P1 and P2 samples treated with sub- and supercritical CO_2 (**Table 2**).

The presence of high-pressure CO_2 influences the stretching vibration in the group C=O (carbonyl). For P1, once the carbon dioxide pressure increases, the maximum absorption of the carbonyl group moves to higher values than P2 where the maximum absorption decreases with increasing pressure. This displacement reaches a limit at the highest pressures as a result of increased carbon dioxide mobility. The integral area of the carbonyl peak decreases linearly with increasing pressure for the P1 system (data consistent with those in the literature) and varies

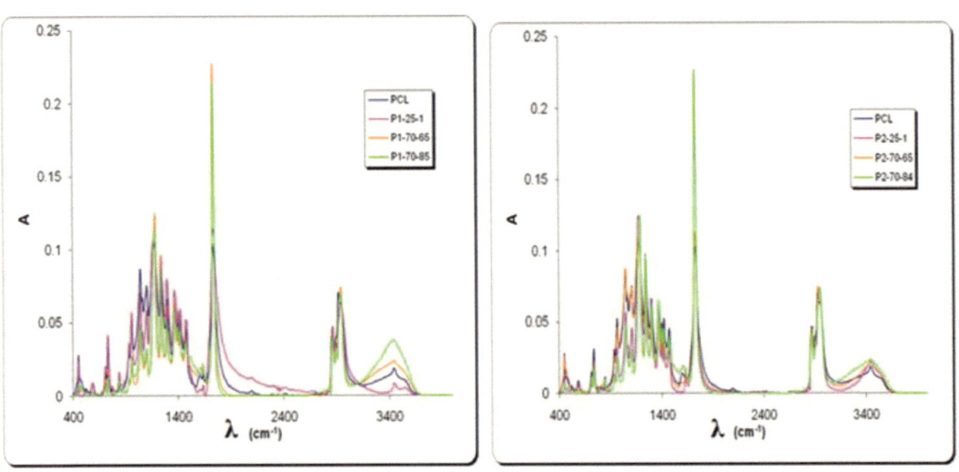

Figure 4.
FTIR spectra for PCL, P1, and P2 and different temperatures and pressures.

System	Melting point (°C)	Observations
PCL	60.3–61.9	White crystals
P1-25-1	55.8–57.5	White crystals
P2-25-1	57.1–58.9	White crystals
P1-70-85	55.5–57.6	Amorphous
P1-70-65	58.7–59.8	White crystals
P2-70-85	58.4–60.3	White crystals
P2-70-65	57.0–59.3	Amorphous

*Structural changes of PCL were evidenced by the FTIR technique (**Figure 4**).*

Table 2.
Melting point of high-pressure samples of PCL.

nonlinearly for the P2 system. It has been observed that with the decrease of carbon dioxide pressure, the frequency of the C-O-C stretching vibration moves to higher values. At higher temperatures (70°C) and higher pressures (6.5 and 8.5 MPa), the polymer melts and recrystallizes after expansion in a semicrystalline or amorphous state, resulting in melting temperatures below PCL. For the samples treated at high pressures, we observe a shift of the maximum attributed to the crystallinity to smaller wave numbers, so a decrease of the crystallinity produces a decrease of melting temperature for the PCL samples treated under high-temperature and high-pressure conditions.

3. Drug encapsulation using $scCO_2$

In the last decade, drugs loaded in porous biodegradable polymeric foams have found to have important applications in tissue engineering and delivery systems. These polymeric porous scaffolds with an open-pore structure can ensure and increase seeding, attachment, growth of cells, extracellular matrix production, vascularization, and tissue growth. Supercritical CO_2 is an excellent choice to produce impregnated polyester foams in a one-step process creating porosity, without residual solvent in the products. After polymer scaffold degradation, obtained tissue would not contain synthetic polymer. The rate degradation of the scaffold should be similar or slower than the rate of tissue formation; therefore, it is important to elucidate the mechanism of this degradation process.

Some examples of polymers used as porous scaffolds were poly(D,L-lactide) (PLA) and poly(D,L-lactide-co-glycolide) (PLGA), respectively, and poly(methyl methacrylate), PMMA, PMMA/poly(ε-caprolactone), PCL, etc. Such polymers are harmless to the growing cells and can be removed from the organism by normal metabolic pathways. They also can be used in other in vivo applications, such as resorbable sutures [16].

Encapsulation of drugs within colloidal-sized polymeric matrix is largely used to improve the sustained release, reduce the side effect of the drugs, and increase the bioavailability of the drug from the pharmaceutical formulations. Polymeric beads for the controlled release of the drugs are most often used as the best solution due to easy preparation procedure and high drug-loading efficiency.

The special properties of $scCO_2$ made it a good transport vector for solid matrix impregnation. This process depends on the partition of interest substance between the supercritical fluid phase and matrix (such as porous polymers) used for impregnation. First the substance is mixed with high-pressure carbon dioxide and then is placed in contact with matrix used as support. Because $scCO_2$ has a good diffusion capacity into porous scaffolds, inducing a swelling and/or plasticization, impregnation of a lot of materials could be possible. Among them were polymers as swellable matrices or silica as non-swellable matrices [17].

The controlled release of the very hydrophilic drugs is still a challenge due to the difficulties to encapsulate the drug in a suitable drug delivery system able to control the rate of the release and to ensure a minimum retention of the active substance inside the pharmaceutical vehicles, but with a good control of the particle size and surface properties, nanoparticles may be directed to specific sites for targeted drug delivery.

The encapsulation of active substances can be improved by impregnation in the presence of compressed fluids like carbon dioxide [18], at temperatures and pressures near or above critical point (CO_2, p_{cr} = 7.382 MPa; t_{cr} = 31.04°C).

In a typical experiment in our laboratory, 2-pyridinealdoxime methochloride (PAM) was chosen as a hydrophilic drug model, with a high solubility in water. PAM is an acetylcholinesterase reactivator used as an antidote in poisoning with organophosphoric substances characterized by high solubility in water. Due to its hydrophilicity, PAM is rapidly eliminated from the body. Consequently, PAM is relatively short acting, and repeated doses may be needed. The encapsulation of PAM in alginate beads (used as sustained drug release system) was tested.

Alginate is a water-soluble linear polysaccharide extracted from brown seaweed, and it is composed of alternating blocks of 1–4 linked α-L-guluronic (G) and β-D-mannuronic (M) acid fragments. Alginate is a biocompatible and a hydrophilic biopolymer. These properties and its relatively low cost have recommended it for pharmaceutical applications [19]. The anionic biopolymer has the ability to bind multivalent cations, leading to the formation of insoluble hydrogels with "egg box"-type structure [19]. In the formation of water-insoluble gels, a specific interaction occurs between calcium ions (Ca^{2+}) and -COO^- and -OH groups of the guluronic acid fragments in a simple procedure of dripping a sodium alginate aqueous solution into a calcium chloride solution [20]. The drug can be encapsulated in alginate beads by two methods. A first method consists in dripping the aqueous solution of alginate/PAM mixture in calcium chloride solution. The deficiency of this method is the elimination of PAM outside the beads during the ionically cross-linking of alginate with calcium ions. A second method consists in immersing and soaking the polymeric beads into the solution of drug. In this case the impregnation efficiency is low.

A certain number of calcium alginate beads were immersed in the 3×10^{-3} wt% solution of PAM. The impregnation of PAM in calcium alginate beads was carried out in a high-pressure cell, in the presence of compressed carbon dioxide. The impregnation was performed at different pressures (2.5, 5.0, 7.5 MPa) and temperatures (20, 40, 60°C) and also at atmospheric conditions (0.1 MPa, 20°C). The impregnation time was 30 minutes. The samples are designated p/t, where p and t represent the pressure and the temperature of impregnation.

The efficiency of PAM impregnation at various temperatures and pressures is presented in **Figure 5**.

The results showed that the efficiency of PAM impregnation increases with temperature at 2.5 MPa and the interaction between the cationic drug and active sites of biopolymer are favored by high pressure. Both state parameters influence the encapsulation of the ionic drug in the polymeric beads. The encapsulation of PAM in calcium alginate beads at 2.5 MPa depends on temperature. Adsorption or/and absorption of PAM are the main processes which take place between 20 and 40°C. At 60°C, the dissociation constants of polyelectrolyte and ionic drug increase and electrostatic interactions between the polymer and drug molecules are promoted. At higher pressure (5.0 and 7.5 MPa), the impregnation efficiency varies in the same way like ratio of CO_2 density to viscosity versus temperature and pressure.

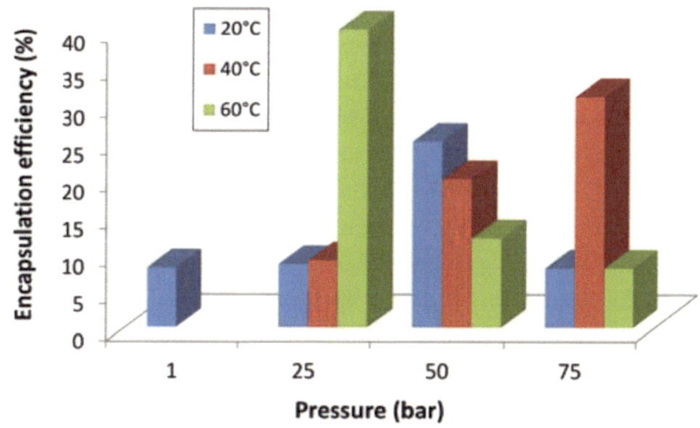

Figure 5.
Efficiency of encapsulation of PAM in alginate microbeads at different pressures and temperatures.

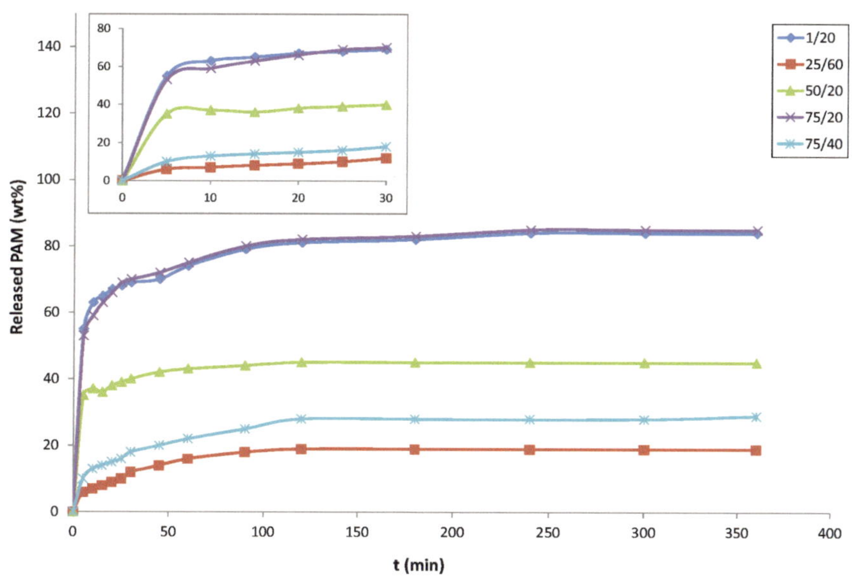

Figure 6.
PAM release profiles from impregnated alginate beads; inset, the release curves in the first 30 minutes.

3.1 In vitro release studies

The release curves for sample 1/20 and for samples with higher encapsulation efficiency are presented in **Figure 6**. The release curves for samples 0.1/20, 7.5/20, and 5.0/20 show a "burst effect"; the encapsulated PAM is released in the first 15 minutes. The sustained release of PAM is observed for samples 7.5/40 and 2.5/60. The maximum amount of released PAM is lower than 50% for samples 7.5/40, 2.5/60, and 5.0/20. The observed initial burst release profiles may be an indication that the impregnated drug was mainly located at the polymer surface for the sample impregnated with PAM at 20°C and different pressures. The kinetic data were fitted by first-order, Weibull, and Korsmeyer-Peppas kinetic equations. The experimental concentration values of the active substance released as a function of time were processed for establishing the release kinetics.

The kinetic parameters offer information about the encapsulation mechanism which can take place by adsorption on the surface of the beads or by interactions between drug and polymer inside the alginate beads. The PAM release curves show that the release mechanism depends on the drug impregnation conditions.

The encapsulation of PAM in alginate beads was improved by impregnation in the presence of compressed carbon dioxide at high temperature. The release of ionic drug from drug delivery beads obtained by impregnation of PAM in supercritical CO_2 (7.5 MPa, 40°C) is controlled by diffusion of active substances. The alginate beads with encapsulated PAM in subcritical conditions (7.5 MPa, 40°C) release PAM according to swelling and erosion of an amorphous biopolymer.

The release of PAM from alginate beads can be controlled by changing the initial conditions of impregnation in the presence of CO_2.

4. Synthesis of nanoparticles in $scCO_2$-based microemulsions

Nanoparticles with tunable size and high monodispersity present great interest for biomedical application, in particular to be used in bioimaging or as drug delivery systems. Synthesis of inorganic and organic NPs in $scCO_2$ as green alternative to the physical and chemical traditional methods suffer from the same difficulties in controlling the nanoparticle growth step and to ensure narrow size distribution of the final product. One of the most efficient routes developed to fabricate monodisperse nanoparticles is the synthesis performed in heterogeneous media, i.e., in microemulsions or in blue emulsions, in order to ensure the nucleation and growth of nanoparticles in the restricted domain of liquid nanoreactors (oil or aqueous droplets). Microemulsions, as liquid-liquid colloidal systems consisting of nanometric droplets with size ranging from few nanometers to 100, stabilized with surfactant and cosurfactant mixture, are considered the most suitable reaction media for the synthesis of organic and inorganic nanoparticles with tunable size and morphology. Oil-in-water microemulsions are usually used for the fabrication of hydrophobic polymeric nanoparticles, while water-in-oil ones are used as reaction media for inorganic nanoparticles, especially for metal NPs obtained from reduction of metallic salt. All advantages of the controlled synthesis in water-in-oil microemulsions could be preserved using water-in $scCO_2$ microemulsions, since the general structure of the colloidal system is the same (**Figure** 7).

However, the stabilization of miniemulsions and the formation of thermodynamically stable microemulsions in supercritical CO_2 are challenging tasks, due to the specific properties of carbon dioxide. Supercritical carbon dioxide possesses peculiar properties such as very high diffusivity and low viscosity, different from usual oily phases in the microemulsion composition; thus special requirements are

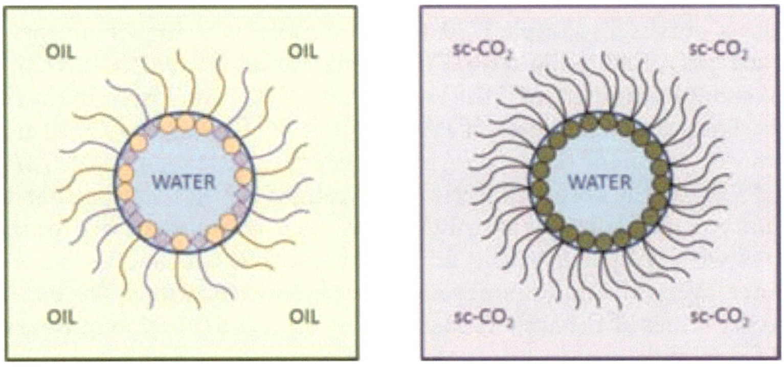

Figure 7.
The schematic view of structure and stabilization of water droplets in W/O and W/$scCO_2$ microemulsions.

needed for the surfactants that enable the formation of W-in-scCO_2 microemulsion: (i) enhanced absorbability at water/scCO_2 interface, (ii) enhanced surface activity in lowering water/scCO_2 interfacial tension, and (iii) strong repulsive interaction (steric effect) between CO_2-philic groups, which reduces the probability of droplet aggregation and leads to the stabilization of microemulsion [21].

Many researches have been performed addressing the problem of surfactant solubility in supercritical and dense CO_2, in order to evidence their ability to stabilize water-in-CO_2 microemulsions, and surfactants with special designed molecular structure were synthesized for this purpose [21].

The early studies [22] conclude that conventional theories in molecular design for surfactant with superior efficiency in lowering interfacial tension at W/O interfaces are inapplicable to the W-scCO_2 ones, because CO_2-philicity could not be directly compared with oleophilicity. Considerable efforts have been made to find the most favorable chemical structures able to ensure adequate solubility of the surfactant in supercritical CO_2, together with better adsorption at W-scCO_2 interface. Highly branched hydrocarbon surfactants were found to possess good solubility, but unfortunately they are not efficient enough as stabilizer, compared to AOT sodium salt, a commercially available cost-effective common surfactant widely used for the formulation of water-in-oil microemulsions. The typical surfactants exhibit very different Hildebrandt solubility parameters for carbon dioxide; thus their solubility in CO_2 is very low, explaining their lack in stabilizing supercritical CO_2 emulsions and microemulsions.

Based on the data evaluated [22], an interesting family of surfactants were synthesized as AOT analogues AO-VAc bearing vinyl-acetate oligomeric chains and AOK bearing *t*-butyl and carbonyl groups, grafted to promote CO_2-philicity. Those syntheses are very demanding, and the difficult fabrication of industrial scale quantities limits the extension of the new compounds in application such as preparation of W/scCO_2 microemulsions as reaction media. Sagisaka et al. [23] conducted a systematic study on the influence of the hydrophobic tail structure on the efficiency of fluorinated surfactants to stabilize water-in- supercritical CO_2 microemulsions. Since anionic hydrocarbon surfactant AOT sodium salt is the most successful surfactant in obtaining W/O microemulsions, fluorinated surfactants with similar structures were investigating, and the results suggest that fluorinated analog bis(1*H*,1*H*,2*H*,2*H*-heptadecafluorodecyl)-2 sulfosuccinate sodium salt is the most effective in the formulation of water-in-scCO_2 microemulsion, allowing a water solubility up to $W^c_0 = 32$ (W^c_0 is the maximum water-to-surfactant molar ratio in W/O or W/scCO_2 microemulsions).

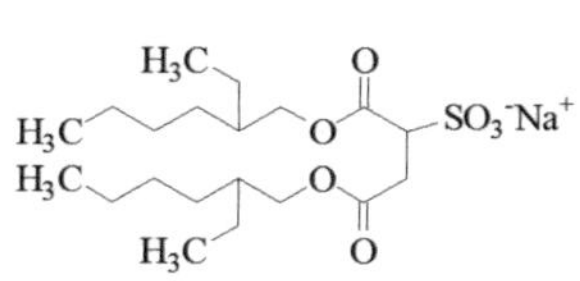

AOT

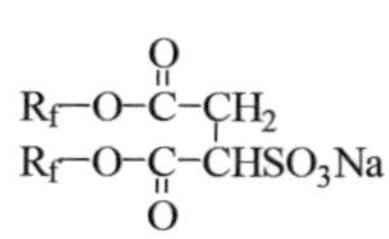

AOT analogue
Rf fluorinated tail

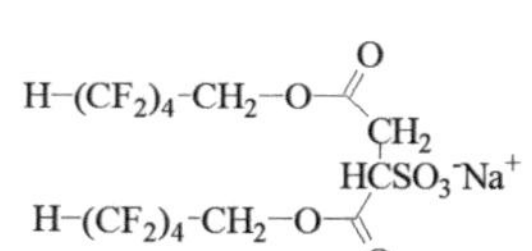

bis(2,2,3,3,4,4,5,5-octafluoro-1-pentanol) sulfosuccinate

In order to increase the region of microemulsion in the phase diagram of the water-supercritical CO_2-surfactant system, novel amphiphilic compounds were synthesized with a more flexible molecule due to the presence of linkers. A series of fluorocarbon derivatives having different fluorocarbon chain lengths $F(CF_2)_n$ with n = 4,6,8, and 10 and oxyethylene groups $(CH_2CH_2O)_{m/2}$ (where m = 2 and 4) introduced into the molecule as spacers were developed [23]. The analysis of phase diagram of quaternary systems water-scCO_2-*n*FS(EO)*m* surfactants suggests that an optimum length of the fluorocarbon tail is required for the formation of W/scCO_2 microemulsion with increased amount of water included. The maximum water/surfactant ratio W^{c}_{0} is obtained for the CO_2-philic tail with 12–14 Å length, corresponding to the surfactants with eight carbon atoms and two oxyethylene groups or six carbon atoms and four oxyethylene groups.

These fluorinated surfactants with specific chemical structure are quite difficult to obtain, and their use is restricted in the "green chemistry" methods due to the high toxicity. In this respect, other classes of surfactants were investigated, in particular polymeric derivatives. Block copolymers have been extensively studied regarding their ability to act as stabilizer of water droplets in scCO_2 since J. de Simone reports for the first time in 1994 the solubility of fluorinated polymers in liquid and scCO_2. Chemicals such as fluoropolymers, polysiloxanes, and some poly(vinyl esters) derivatives prove to be CO_2-philic polymers but with reduced surface activity. In the last decades, the progresses in the synthesis techniques allow the fabrication of a large variety of block copolymers with CO_2-philic moieties and specially designed structures, such as amphiphilic block, amphiphilic comb copolymers, or gradient copolymers bearing linear or branched chains, such as poly(1,1,2,2-tetrahydroperfluorooctyl methacrylate)-based copolymers with either poly(N,N-dimethylaminoethyl methacrylate) (PDMAEMA) or poly(oligo(ethylene glycol) methacrylate) (POEGMA) as hydrophilic blocks or 1,1,2,2-tetrahydroperfluorodecyl tetrahydroperfluorodecyl acrylate and vinyl benzoic phosphonic acid [24]. A remarkable strategy to enhance the surfactant efficiency in scCO_2 media is based on the synergistic effect obtained by ion-pairing cationic and anionic surfactants. Sagisaka et al. [25] report the properties of short tail, unbranched fluorinated surfactants with sulfate ionic groups and their cationic analogues synthesized by Verdia [26].

[$C_6F_{13}S$]	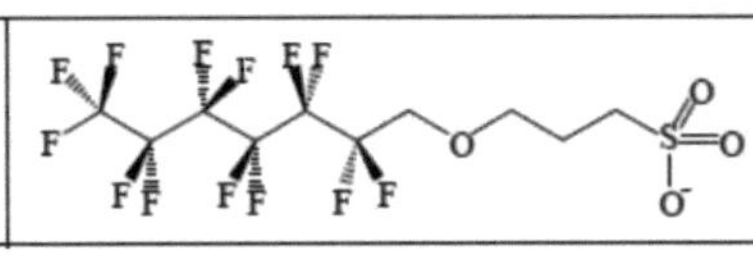	[C_6F_{13}mim]	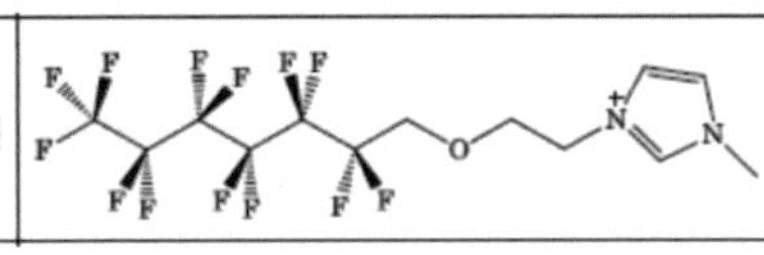

Individual compounds, for example, [$C_6F_{13}S$] and [C_6F_{13}mim], both exhibit very low solubility in CO_2 and produce limited water solubility; even in water they have modest surface activity. The anionic and cationic pair of surfactants with similar hydrophobic tail used together in adequate molar ratio produces a spectacular synergistic effect, which enable the formulation of water-in-scCO_2 microemulsion with significant water content (W_0 max up to 50).

Metal nanoparticles (Ag, Au, and Cu) are particularly interesting for many industrial applications, due to their size- and shape-dependent unique optic and electric properties. In the last decades, applications of metal nanoparticles in nanomedicine, in emerging fields such as in vivo bioimaging, innovative cancer therapy, and diagnostics evolve in solution with enormous positive impact in the health-care system. The large-scale applicability of these nanomaterials is still restricted due to the concern about their intrinsic biocompatibility and the difficulties to remove toxic residual precursors and solvents during the synthesis, in particular regarding synthesis in W/O microemulsion as heterogeneous reaction media. Thus, new methods to synthesize metal or metal oxide nanoparticles using "green chemistry" approach as alternative to classic technologies have been investigated. The first reported synthesis of silver nanoparticles in CO_2 microemulsion was reported in the late 1990s. The results were promising in terms of the quality of the nanopowder obtained, with additional advantage of fine-tuning the nanoparticle size. W/O microemulsions, as heterogeneous reaction media allow the synthesis of nanoparticles with desired average size through the specific mechanism of precipitation in arrested environment, are defined by the water droplets. The radius of the water nanoreactor could be tuned in a facile way by changing water/surfactant ratio (W_0^c) in the composition of microemulsion. Fernandez et al. [27] demonstrated that the use of $scCO_2$ microemulsion as reaction media exhibits additional advantage in continuous tuning of the Ag nanoparticle size. The nanoparticles were prepared in $scCO_2$ microemulsion with different water/surfactant molar ratios (W_0^c), that is, different water nanopool dimensions, stabilized with surfactant AOT fluorinated analogue at 10 mM concentration. The change of fluid phase density is obtained by simply adjusting temperature and pressure conditions. For the $scCO_2$ microemulsion with lower water content (W = 6) very fine tuning was achieved, producing silver nanoparticles with average size ranging from 1.9 to 9.3 nm for a variation of density from 0.96 up to 0.80 g/mL.

The same strategy is used to obtain tunable size of quantum dots, cadmium sulfide and zinc sulfide [28]. Similarly, Cu nanoparticles have been obtained [29] by reduction of the copper precursor (copper nitrate) dissolved in the water droplets of microemulsion after injection of CO_2-soluble reducing agents such as *N,N,N,N*-tetramethyl-*p*-phenylenediamine or sodium cyanoborohydride. The $scCO_2$ microemulsion used as reaction media was formulated with a mixed surfactant system consisting in 12.8 mM AOT and 25.3 mM fluorinated derivative (perfluoropolyether-phosphate) at a water-to-surfactant ratio $W([H2O]/[AOT]) = 12$. Nanoparticles with average size of 5–15 nm were obtained, according to the type of the reduction agent and rate of the reagent addition.

Polymeric nanoparticles could also be obtained through polymerization processes in $scCO_2$ microemulsion, using adapted strategy from emulsion/microemulsion polymerization with common oily phases. The preparation of polyamide nanoparticles was reported by Ohde et al. [30], using a $scCO_2$ microemulsion stabilized with a mixture of commercially available hydrocarbon and fluorinated surfactants (AOT and perfluoropolyether-phosphate PFPE–PO4). The NPs were prepared by using polymerization of acrylamide monomer in the presence of potassium persulfate, both dissolved in the water core of $scCO_2$ microemulsions with identic composition. The reaction is performed in homemade interconnected high-pressure vessels that allow the equilibration of separate $scCO_2$ microemulsions containing monomer and initiator and mixing them at the desired time. The polymerization was completed very quickly, in less than 1 minutes at 25°C, 20 MPa, and in less than 30 s at 50°C, 15 MPa.

Metallic nanoparticles deposited on the surface of the graphenes or carbon nanotubes are hybrid material with particular interest in various domains, for example, catalysis, energy saving, modified electrodes in biosensing, etc. The classic methods employed suffer from various drawbacks, from high amount of organic reagents to the lack of dimension control. Deposition of metal nanoparticles on the carbon nanotubes using $scCO_2$ microemulsion as reaction media was proposed as green alternative to wet chemistry or other technologies of deposition using physical methods.

Shimizu et al. [31] report the preparation of hybrid electrocatalyst consisting in Pt NPs deposited on multiwalled carbon nanotubes in three different media: water-in-supercritical CO_2 microemulsion, supercritical CO_2 fluid, and water-in-hexane microemulsion.

As it is shown in **Figure 8**, the reaction in $scCO_2$ microemulsion leads to a better dispersion of Pt NPs on the surface of nanotubes that is obtained in homogeneous $scCO_2$ phase and conventional W/O microemulsion.

Among the many techniques of processing and synthesis in fluid carbon dioxide phases, the use of water-in-$scCO_2$ microemulsions is one of the most effective in

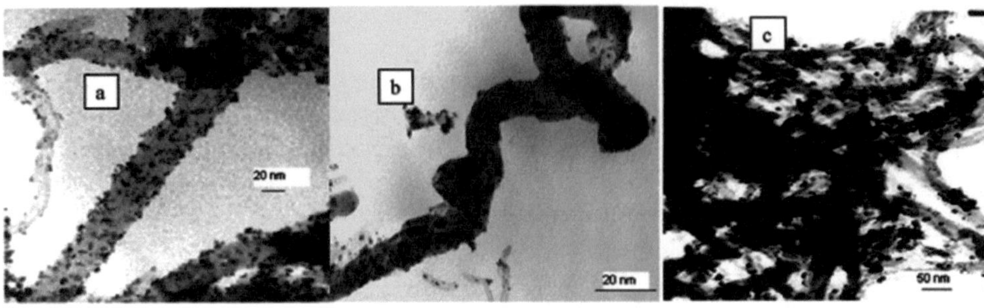

Figure 8.
TEM images of hybrid catalyst Pt-CNT synthesized in various media by (a) water-in-supercritical CO_2 microemulsion, (b) water-in-oil microemulsions, (c) direct deposition in supercritical CO_2 (adapted with permission from ref. [31]).

terms of control on the size, size distribution, and shape of the final product. The stabilization of such reaction media with low amount of affordable and nontoxic surfactants remains the major bottleneck and limits the extension of this method.

5. Functionalization of nanoparticles in $scCO_2$

The modification of nanoparticle surfaces, either in solutions or as layers deposited on various surfaces, is a key issue in many processes to fabricate functional materials, in many applications. Commonly used methods for the modification of nanoparticles such as quantum dots, graphene, metallic nanoparticles, etc. are based on solution processing involving organic solvents, which makes the whole process very difficult to implement as eco-friendly technology. For biomedical applications in particular, the proper functionalization of the surfaces in synthesis that involve green solvents is particularly encouraged. Industrial or large-scale processes for the modification of surfaces by vapor-related physical methods suffer from huge limitations due to the high-cost and energy-consuming devices. On the other hand, liquid-related chemical methods have an additional drawback: the solvent removal after the reaction is completed on the targeted surface. Due to the variety of roles that can be attributed to carbon dioxide, i.e., solvent, antisolvent, reaction medium, respectively, unique flexibility for surface engineering is obtained using $scCO_2$.

A variation of RESS method can be applied for coating processes, with the condition that coating material dissolves in $scCO_2$. The solution is introduced via a nozzle in another chamber that contains the solid support to be functionalized. During the rapid depressurization, the density of CO_2 changes significantly, and consequently a supersaturation followed by precipitation of the material forming a coating occurs. For example, superhydrophobic coatings on paper were obtained from alkyl ketene dimer (AKD) dissolved in $scCO_2$ by this modified RESS technique.

Surface modification of silica particles with organomodified silane derivatives for highly hydrophobic films has been reported by Purcar et al. [32]. The formation of silica nanoparticles using sol-gel synthesis in various conditions of temperature and pressure of CO_2 was investigated, both in subcritical and supercritical phases. Tetraethoxysilane (TEOS) was used as the particle precursor and octyltriethoxysilane (OTES) as the co-precursor for coating. The dynamic light scattering measurements show that the higher CO_2 pressure leads to larger diameters of coated particles. The agglomeration of the silica nanoparticles was found to be dependent

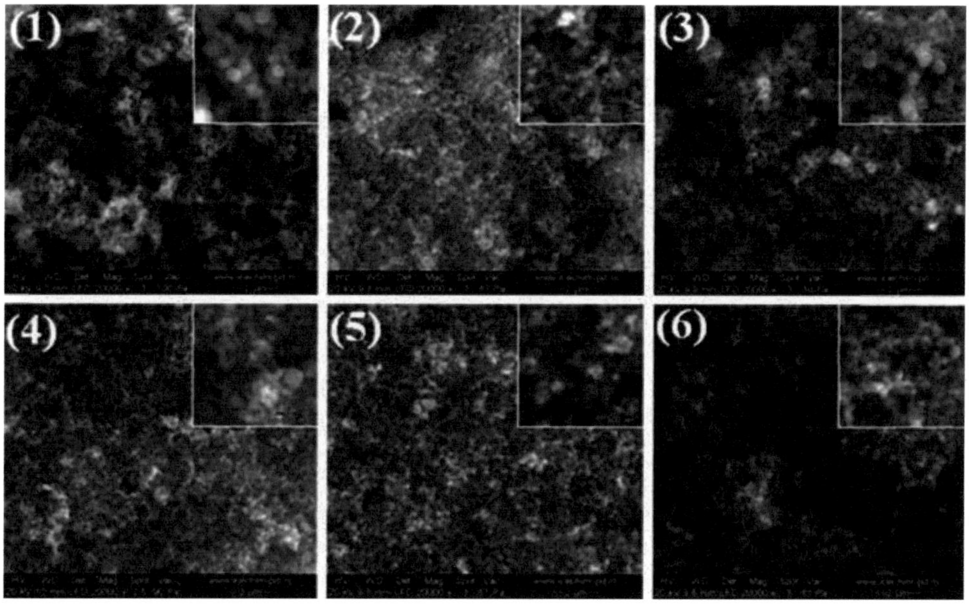

Figure 9.
SEM images of coated silica nanoparticles prepared in CO_2 in various conditions of temperature and pressure. (1 = 25°C, 6 MPa; 2 = 40°C, 6 MPa; 3 = 60°C, 6 MPa; 4 = 25°C, 7.5 MPa; 5 = 40°C, 7.5 MPa; 6 = 60°C, 7.5 MPa). Reprinted with permission from Ref. [32].

on the temperature and pressure of the CO_2 reaction media. The SEM images confirm the tendency of aggregation and differences in the size of silica nanoparticles coated with octyl-modified derivative (**Figure 9**).

The size of the resulting silica particles is around 100–250 nm when the CO_2 pressure is 6 MPa and decreases up to 150 nm when the CO_2 pressure increases to 7.5 MPa. It was observed that the higher CO_2 pressure gives narrower size distributions. The thin films obtained by the as-prepared silica nanodispersions deposited on glass substrate show high contact angle values, ranging from 127 to 145^0.

The NP processing using $scCO_2$ as solvent leads not only to the adequate modification of the surface but, sometimes, also to the decrease of the aggregates and better dispersibility and interaction with polymeric matrix. Wei et al. [33] reported recently the silane modification of graphene nanoparticles for the fabrication of composite polymeric films with superior barrier properties. The obtained graphenes functionalized with silane exhibit a fluffy appearance with small size of sheets. The graphene surface was silanized through the hydrolysis and condensation of different organomodified silane derivatives: aminopropyl triethoxysilane, glycidyloxypropyltrimethoxysilane, and methacryloxy propyl trimethoxysilane. The reaction was conducted in $scCO_2$ for 2 hours at 40°C and 20 MPa. The successful formation of SiO2 coating on graphene sheets was proven by FTIR spectra and thermogravimetric analysis and supports the enhanced dispersibility of modified GNs in polymeric membrane, together with a significant exfoliation.

Nanoparticle surface functionalization with polymeric shells is an important issue in preparation of biocompatible colloidal vectors in drug delivery that allow further attachment of bioactive ligands. The transfer of such synthesis in $scCO_2$ for the functionalization of nanoparticles used as drug delivery systems in pharmaceutical formulations benefits from the absence of many posttreatment steps that are generally required to remove remaining solvents or nondesirable by-products generated in wet chemistry methods. Marre et al. [34] demonstrated the feasibility

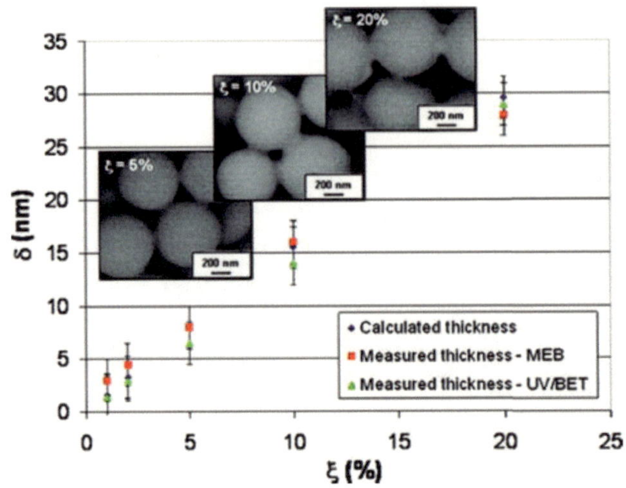

Figure 10.
The variation of polymeric shell thicknesses on nanoparticles as a function of the initial mass fraction of polymer in the initial solution (reprinted with permission from ref 34).

of a versatile method for coating silica particles used as model substrates, with either a hydrophilic (polyethylene glycol PEG) or hydrophobic polymer (hydroxyl-terminated polybutadiene, HTPB) using a supercritical antisolvent process (**Figure 10**).

The principle of the precipitation from compressed antisolvent PCA techniques is allowing the contact of the polymeric solution with the $scCO_2$ as antisolvent, when a succession of different phenomena governed by hydrodynamic, precipitation kinetics, thermodynamic, and mass-transfer effects result eventually in the precipitation of an organic shell on the nanoparticles dispersed previously in the polymeric solution. According to the polymer used (PEG or PBHT) and the reaction conditions, various sizes of the deposited polymeric shell have been obtained, from 3 to 28 nm, by simply modification of polymer concentration in the initial solution (**Figure 10**).

In order to effectively exploit the advantages of "green" processing in $scCO_2$, Roy et al. describe the successful grafting of a model drug (salicylic acid, SA) on TiO2 nanoparticles [35]. The chosen active compound could react with the nanoparticle surfaces through the carboxylic and hydroxyl groups. The functionalization of the TiO2 surface was performed in mild conditions, at 40°C and 16 MPa for a short period of time, up to 2 hours. The thermogravimetric analysis shows a calculated amount of drug SA deposited on TiO_2 nanopowder of approximately 8 wt%, and the SEM images of coated samples do not reveal rod-like crystals specific for solid salicylic acid, which confirm the absence of residual drug other than the coating deposited on nanoparticles.

In the last decades, methods to perform functionalization of nanoparticles using $scCO_2$ were reported, as viable alternative to classical synthesis and showing remarkable advantage of reduced impact on the environment.

6. Conclusion

The recent works in supercritical fluids demonstrate the multiplicity of advantages in synthesis of nanomaterials, either organic or inorganic. Complex synthesis of hybrid materials consisting in nanoparticles deposited on other structures (carbon nanotubes) or core-shell nanoparticles was also developed in $scCO_2$.

An important problem that requires further research, especially for fine-tuning the nanoparticle size, is design and fabrication of CO_2-philic surfactants that enable the formation of water-in-scCO_2 microemulsion as reaction media. The successful synthesis of nontoxic and cost-effective surfactants will open the opportunity to extend this method to many nanoparticles used as drug delivery vehicles, where the size and size distribution control are a key issue.

Another domain which is expected to benefit from further research in order to advance this technology is the development of novel strategy for surface functionalization, especially for scalable processes.

Acknowledgements

This work was supported by grants of the Romanian National Authority for Scientific Research and Innovation, CCCDI - UEFISCDI, project number PN-III-P1-1.2-PCCDI-2017-0743/P5, and PN-III-P1-1.2-PCCDI-2017-0428/P2 within PNCDI III.

Conflict of interest

The authors have no conflict of interest to declare.

Nomenclature

p	pressure, MPa
T	temperature, K
t	time, s
W	maximum water content in microemulsion, expressed as water-to-surfactant molar ratio [H_2O]/[surfactant]
δ	polymeric shell thicknesses on nanoparticles, nm
λ	wavelength, cm^{-1}
ξ	initial mass fraction of polymer, %
cr	critical
Acronyms	
AKD	alkyl ketene dimer
AOT	sodium bis-(2-ethylhexyl) sulfosuccinate
Chi	chitosan
DESAM	depressurization of an expanded solution into an aqueous medium
DSC	dynamic light scattering
FTIR	Fourier-transform infrared spectroscopy
GNs	graphene
NPs	nanoparticles
OTES	octyltriethoxysilane
PAM	2-Pyridinealdoxime methochloride
PBHT	polybutadiene hydroxy terminated
PCL	poly(ε-caprolactone)
PDMAEMA	poly(N,N-dimethylaminoethyl methacrylate)
PEG	polyethylene glycol
PLA	polylactic acid
PLGA	poly(lactic-coglycolic)
PMMA	poly(methyl methacrylate)
POEGMA	poly(oligo(ethylene glycol) methacrylate)

RESOLV	rapid expansion of supercritical solution into a liquid solvent
RESS	rapid expansion of supercritical solution
SAA	supercritical assisted atomization
SAILA	supercritical assisted injection in liquid antisolvent
SAS	supercritical antisolvent technique
$scCO_2$	supercritical carbon dioxide
SCFs	supercritical fluids
SEE	supercritical emulsion extraction
SEM	scanning electron microscopy
TEOS	tetraethoxysilane
VOC	volatile organic compound
W/O	Water-in-oil
W-scCO2	water-in-supercritical CO2

Author details

Ludmila Otilia Cinteza* and Daniela Bala
Department of Physical Chemistry, Faculty of Chemistry, University of Bucharest, Bucharest, Romania

*Address all correspondence to: ocinteza@yahoo.com

References

[1] Long B, Ryan KM, Padrela L. From batch to continuous—New opportunities for supercritical CO_2 technology in pharmaceutical manufacturing. European Journal of Pharmaceutical Sciences. 2019;**137**:104971. DOI: 10.1016/j.ejps.2019.104971

[2] Pfund LY, Price P, Frick JJ, Matzger AJ. Controlling pharmaceutical crystallization with designed polymeric Heteronuclei. Journal of the American Chemical Society. 2015;**137**:871-875. DOI: 10.1021/ja511106j

[3] Lehocký R, Pěček D, Štěpánek F. Scale-up from batch to flow-through wet milling process for injectable depot formulation. European Journal of Pharmaceutical Sciences. 2016;**95**: 122-129. DOI: 10.1016/j.ejps.2016.08.043

[4] Sinha B, Müller RH, Möschwitzer JP. Bottom-up approaches for preparing drug nanocrystals: Formulations and factors affecting particle size. International Journal of Pharmaceutics. 2013;**453**(1):126-141. DOI: 10.1016/j.ijpharm.2013.01.019

[5] Campardelli R, Baldino L, Reverchon E. Supercritical fluids applications in nanomedicine. The Journal of Supercritical Fluids. 2015;**101**:193-214. DOI: 10.1016/j.supflu.2015.01.030

[6] Elizondo E, Veciana J, Ventosa N. Nanostructuring molecular materials as particles and vesicles for drug delivery, using compressed and supercritical fluids. Nanomedicine. 2012;7(9): 1391-1408. DOI: 10.2217/nnm.12.110

[7] Samei M, Vatanara A, Fatemi S, Rouholamini Najafabadi A. Process variables in the formation of nanoparticles of megestrol acetate through rapid expansion of supercritical CO2. Journal of Supercritical Fluids. 2012;**70**:1-7. DOI: 10.1016/j.supflu.2012.05.011

[8] Kalani M, Yunus R. Application of supercritical antisolvent method in drug encapsulation: A review. International Journal of Nanomedicine. 2011;**6**:1429. DOI: 10.2147/IJN.S19021

[9] Campardelli R, Cherain M, Perfetti C, Iorio C, Scognamiglio M, Reverchon E, et al. Lipid nanoparticles production by supercritical fluid assisted emulsion–diffusion. Journal of Supercritical Fluids. 2013;**82**:34-40. DOI: 10.1016/j.supflu.2013.05.020

[10] Porta GD, Campardelli R, Falco N, Reverchon E. PLGA microdevices for retinoids sustained release produced by supercritical emulsion extraction: Continuous versus batch operation layouts. Journal of Pharmaceutical Sciences. 2011;**100**(10):4357-4367. DOI: 10.1002/jps.22647

[11] Porta GD, De Vittori C, Reverchon E. Supercritical assisted atomization: A novel technology for microparticles preparation of an asthma-controlling drug. AAPS PharmSciTech. 2005;**6**(3):421-428. DOI: 10.1208/pt060352

[12] Pathak P, Meziani MJ, Desai T, Sun Y-P. Nanosizing drug particles in supercritical fluid processing. Journal of the American Chemical Society. 2004;**126**(35):10842-10843. DOI: 10.1021/ja046914t

[13] Esfandiari N. Production of micro and nano particles of pharmaceutical by supercritical carbon dioxide. The Journal of Supercritical Fluids. 2015;**100**:129-141. DOI: 10.1016/j.supflu.2014.12.028

[14] Venkatesan J, Anil S, Singh SK, Kim S-K. Preparations and applications of alginate nanoparticles. In:

Venkatesan J, Anil S, Kim S-K, editors. Seaweed Polysaccharides. BV, Amsterdam: Elsevier; 2017. pp. 251-268. DOI: 10.1016/b978-0-12-809816-5.00013-x

[15] Peretz S, Florea-Spiroiu M, Anghel D-F, Bala D, Branzoi F, Moreno J-C. Chitosan/anionic surfactant microparticles synthesized by high pressure spraying method for removal of phenolic pollutants. Open Chemistry. 2012;**10**(6):1969-1979. DOI: 10.2478/s11532-012-0126-y

[16] Cabezas LI, Gracia I, de Lucas A, Rodríguez JF. Novel model for the description of the controlled release of 5-Fluorouracil from PLGA and PLA foamed scaffolds impregnated in supercritical CO_2. Industrial and Engineering Chemistry Research. 2014;**53**(40):15374-15382. DOI: 10.1021/ie404028t

[17] García-Casas I, Crampon C, Montes A, Pereyra C, Martínez de la Ossa EJ, Badens E. Supercritical CO_2 impregnation of silica microparticles with quercetin. The Journal of Supercritical Fluids. 2019;**143**:157-161. DOI: 10.1016/j.supflu.2018.07.019

[18] Manna L, Banchero M, Sola D, Ferri A, Ronchetti S, Sicardi S. Impregnation of PVP microparticles with ketoprofen in the presence of supercritical CO_2. The Journal of Supercritical Fluids. 2007;**42**(3):378-384. DOI: 10.1016/j.supflu.2006.12.002

[19] Chan LW, Jin Y, Heng PWS. Cross-linking mechanisms of calcium and zinc in production of alginate microspheres. International Journal of Pharmaceutics. 2002;**242**:255-258. DOI: 10.1016/S0378-5173(02)00169-2

[20] Peretz S, Cinteza O. Removal of some nitrophenol contaminants using alginate gel beads. Colloids and Surfaces A: Physicochemical and Engineering Aspects. 2008;**319**(1-3):165-172. DOI: 10.1016/j.colsurfa.2007.06.012

[21] Sagisaka M, Ogiwara S, Ono S, James C, Yoshizawa A, Mohamed A, et al. New class of amphiphiles designed for use in water-in-supercritical CO_2 microemulsions. Langmuir. 2016;**32**(47):12413-12422. DOI: 10.1021/acs.langmuir.6b01670

[22] Eastoe J, Gold S, Steytler DC. Surfactants for CO_2. Langmuir. 2006;**22**(24):9832-9842. DOI: 10.1021/la060764d

[23] Sagisaka M, Ono S, James C, Yoshizawa A, Mohamed A, Guittard F, et al. Effect of fluorocarbon and hydrocarbon chain lengths in hybrid surfactants for supercritical CO.$_2$. Langmuir. 2015;**31**(27):7479-7487. DOI: 10.1021/acs.langmuir.5b01737

[24] Ribaut T, Oberdisse J, Annighofer B, Stoychev I, Fournel B, Sarrade S, et al. SANS study of the self-organization of gradient copolymers with ligand groups in supercritical CO_2. Soft Matter. 2009;**5**(24):4962. DOI: 10.1039/B912268K

[25] Sagisaka M, Saito T, Yoshizawa A, Rogers SE, Guittard F, Hill C, et al. Water-in-CO_2 microemulsions stabilized by fluorinated Cation–anion surfactant pairs. Langmuir. 2019;**35**(9):3445-3454. DOI: 10.1021/acs.langmuir.8b03942

[26] Verdia P, Gunaratne HQN, Goh TY, Jacquemin J, Blesic M. A class of efficient short-chain fluorinated catanionic surfactants. Green Chemistry. 2016;**18**(5):1234-1239. DOI: 10.1039/C5GC02790J

[27] Fernandez CA, Wai CM. Continuous tuning of silver nanoparticle size in a water-in-supercritical carbon dioxide microemulsion. Small. 2006;**2**(11): 1266-1269. DOI: 10.1002/smll.200600222

[28] Fernandez CA, Wai CM. Continuous tuning of cadmium sulfide and zinc sulfide nanoparticle size in a water-in-supercritical carbon dioxide

microemulsion. Chemistry—a European Journal. 2007;**13**(20):5838-5844. DOI: 10.1002/chem.200601259

[29] Ohde H, Hunt F, Wai MC. Synthesis of silver and copper nanoparticles in a water-in-supercritical-carbon dioxide microemulsion. Chemistry of Materials. 2001;**13**(11):4130-4135. DOI: 10.1021/cm010030g

[30] Ohde H, Wai CM, Rodriguez JM. The synthesis of polyacrylamide nanoparticles in supercritical carbon dioxide. Colloid & Polymer Science. 2006;**285**:475-478. DOI: 10.1007/s00396-006-1582-8

[31] Shimizu K, Cheng IF, Wang JS, Yen CH, Yoon B, Wai CM. Water-in-supercritical CO_2 microemulsion for synthesis of carbon-nanotube-supported Pt Electrocatalyst for the oxygen reduction reaction. Energy & Fuels. 2008;**22**(4):2543-2549. DOI: 10.1021/ef800052b

[32] Purcar V, Cinteza O, Donescu D, Bala D, Ghiurea M, Petcu C, et al. Surface modification of silica particles assisted by CO_2. Journal of Supercritical Fluids. 2014;**87**:34-39. DOI: 10.1016/j.supflu.2013.12.025

[33] Wei J, Peng S, Xue B, Yang Z, Qin S, Yu J, et al. Effect of silane functionalized graphene prepared by a supercritical carbon dioxide process on the barrier properties of polyethylene terephthalate composite films. RSC Advances. 2019;**9**(38):21903-21910. DOI: 10.1039/C9RA02479D

[34] Marre S, Cansell F, Aymonier C. Tailor-made surface properties of particles with a hydrophilic or hydrophobic polymer Shell mediated by supercritical CO_2. Langmuir. 2008;**24**:252-258. DOI: 10.1021/la702154z

[35] Roy C, Vega-González A, García-González CA, Tassaing T, Domingo C, Subra-Paternault P. Assessment of $scCO_2$ techniques for surface modification of micro- and nanoparticles: Process design methodology based on solubility. Journal of Supercritical Fluids. 2010;**54**(3):362-368. DOI: 10.1016/j.supflu.2010.05.018

Chapter 8

Impregnation of Materials in Supercritical CO_2 to Impart Various Functionalities

Molla Tadesse Abate, Ada Ferri, Jinping Guan, Guoqiang Chen and Vincent Nierstrasz

Abstract

Supercritical CO_2 ($scCO_2$) impregnation has attracted growing interest due to its unique properties such as high diffusivity, low surface tension, and ease of solvent removal at the end of the process. In addition, $scCO_2$ is the most environmentally acceptable solvent possessing many advantages compared with the conventional aqueous and solvent-based processing. $scCO_2$ impregnation has a wide range of applications mainly used to incorporate various active principles such as pharmaceuticals, functional finishing agents, colorants, and other agents into a polymeric matrix. This chapter reviews some studies carried out so far about the application of $scCO_2$ as impregnation medium to develop various functional materials and it is intended to stimulate further research into the application of $scCO_2$ to textile functionalization. It mainly focuses on applications related to textiles and some polymeric films.

Keywords: supercritical CO_2, impregnation, functionalization, dyeing

1. Introduction

Supercritical fluid (SCF) is defined as a substance for which both its pressure and temperature are above the critical values simultaneously [1]. SCFs have been applied in many areas such as extraction, dyeing, impregnation, cleaning, polymerization, fractionation, formation of powdered polymers, and so on [2, 3]. Among the SCFs, carbon dioxide (CO_2) is the most popular as it offers several advantages including low toxicity, ready availability, low cost, non-flammability, environmental sustainability, and it is chemically inert under many conditions. In addition, CO_2 has an easily attainable critical temperature of 31°C and critical pressure of 7.4×10^6 Pa which are lower compared with other SCFs such as water (critical temperature > 374°C and pressure > 22×10^6 Pa) and other organic solvents. Moreover, at least 90% of the CO_2 introduced can be recovered and recycled at the end of the procedure, which is attractive from waste minimization viewpoint. This also reduces the production cost and avoids the undesirable solvent residue in the produced material [4].

Impregnation is the process of infusing or depositing solute molecules dissolved in a solvent into a polymer matrix to modify the property of the material by physically or chemically binding or absorbing impregnates to a bulk or surface [5].

The conventional aqueous or solvent-based impregnation processes have many drawbacks such as low diffusion rates, high temperature, limited penetration depth, very long contact time, use of hazardous solvents, consumption of high energy, water, solvents, and other additives. To solve these problems, several techniques have been developed, and it has been shown that supercritical CO_2 ($scCO_2$) is an attractive alternative to conventional organic solvents used in polymer impregnation [6].

$scCO_2$ has appeared to be the appropriate candidate to replace conventional impregnation using organic solvents due to several unique properties suitable for impregnation of polymeric materials. It has high diffusivity and low viscosity allowing faster penetration of molecules to the polymer matrix than in water. The absence of surface tension also improves the penetration of molecules into polymeric structures and avoids the unwanted distortion of delicate materials during processing. In addition, the possibility to recover high purity and dry product free from residual solvent is one key advantage especially important when considering the production of food and pharmaceuticals [6–8]. Furthermore, $scCO_2$ reduces the environmental pollution and the associated cost incurred for the removal of the residual solvent, cost of freshwater input, and wastewater treatment. Due to these important attributes, today, $scCO_2$-assisted impregnation has been used in many fields and it is a promising candidate to replace organic solvents in the future.

In this chapter, studies involving $scCO_2$ dyeing and impregnation processes to develop products for various functional applications are reviewed. The chapter focuses on studies related to $scCO_2$ impregnation of textile fibers and polymers and some polymeric films, made of similar polymers. The references used are not exhaustive, as many articles are published covering the same subject area, but only the most relevant ones for this chapter are presented.

2. Impregnation mechanism in $scCO_2$

Various studies utilizing $scCO_2$ as impregnation medium of textiles and polymers have been reported in the literature. Generally, $scCO_2$ impregnation of additives can be performed based on two mechanisms. The first mechanism works if the solute molecule is readily soluble in $scCO_2$ solvent. When polymers are introduced into $scCO_2$ bath containing solutes, the small CO_2 molecules penetrate to the free volume of the amorphous region and swell the material creating additional free volumes. This causes plasticization of the material due to a decrease in the glass transition temperature (T_g) [9]. Then, the dissolved solutes are transported to the fiber surface and subsequently penetrate and diffuse into the swollen polymer matrix. Finally, upon depressurization, the CO_2 molecules are removed by the shrinking polymer, and the impregnate molecules are trapped inside the polymer matrices [10]. The second mechanism applies for solute molecules, which are poorly soluble but having high affinity to the polymer. In this case, the solute molecules partition preferably toward the polymer matrix than the fluid because of their higher affinity to the polymer. This is the key mechanism by which polar dye molecules are incorporated into the polymer matrix in $scCO_2$ dyeing and impregnation of drug molecules into polymers [8]. Therefore, the impregnation process is feasible when the active principle (solute) is soluble in $scCO_2$ or the partition coefficient is favorable toward the polymer charging enough solute, and the polymer itself is well swollen by the $scCO_2$ solvent [6]. The general steps of impregnation of polymeric fibres in $scCO_2$ are illustrated in **Figure 1.**

Functional active principles such as functional dyes, antimicrobial agents, flame retardant, antioxidants, fragrances, pharmaceutical drugs, and others can be impregnated into a polymer by exposing the polymer to $scCO_2$ medium containing

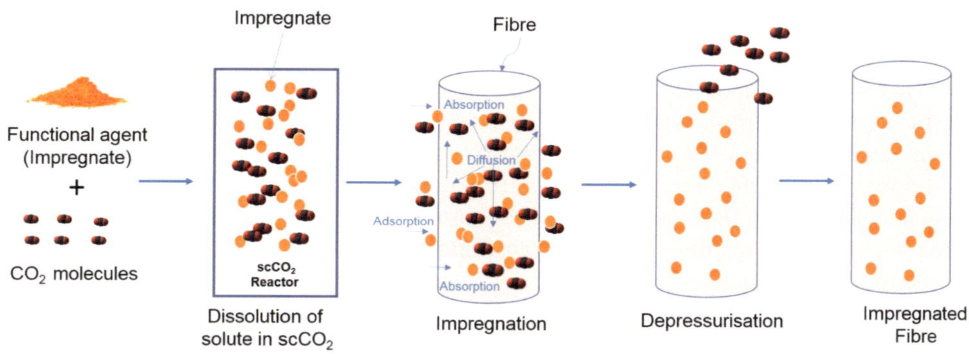

Figure 1.
Schematic of impregnation mechanism of polymeric fibres with functional agent in $scCO_2$.

these agents based on the mechanisms explained above [5]. It has been shown that pharmaceutical drugs can be impregnated into a swollen polymer matrix at operating temperature low enough to avoid thermal degradation of temperature-sensitive drugs. After impregnation and depressurization, the impregnated drug materials slowly diffuse out from the polymer matrix at a slower rate than the rate it was diffused into the polymer which can be used to form a novel controlled release of drugs [11]. The same principle works for deodorizing and antimicrobial agents as well.

3. Solubility of functional agents in $scCO_2$

The most important property for the design of processes in $scCO_2$ medium is the solubility of the compounds in $scCO_2$ fluid. For this reason, solubility data of many compounds including dyes are available in the literature [12–14]. The properties of the compounds such as molecular structure, size, and polarity are the main factors determining their solubility in $scCO_2$ solvent. The solvent character of $scCO_2$ is very much like a hydrocarbon solvent such as n-hexane [15], in which polar compounds are poorly soluble and nonpolar molecules such as disperse dyes have relatively higher solubility [16, 17]. To improve the solubility, polar co-solvents (also called entrainer or modifier), such as acetone or alcohols, are usually added to the $scCO_2$ bath. Furthermore, the solvent power of $scCO_2$ is a function of its density, and this density can be fine-tuned by changing the pressure and temperature of the system [18]. Disperse dyes are among the most investigated compounds, owing to acceptable solubility and suitable molecular size for dyeing polyester and other synthetic fibers in $scCO_2$ [13]. However, the solubility data of functional finishing agents commonly used for textile finishing are still scarce in the literature. According to reports, several non-ionic, low molecular mass organic materials are soluble in $scCO_2$, but only two classes of polymeric materials such as fluoropolymers and silicones showed appreciable solubility in $scCO_2$ at a readily accessible temperature and pressure [19]. Thus, future research should focus on studying the solubility of functional compounds in $scCO_2$.

4. Functionalization in $scCO_2$

In the conventional process, functional finishing agents are usually applied at the end of the process during dyeing or finishing stages. The common problems with these conventional finishing processes are the requirement of a higher amount of water, energy, and auxiliary chemicals, which generates toxic wastewater causing environmental pollution and increases production cost. Due to this, a new dyeing and impregnation process has been developed in which $scCO_2$ is used as a solvent and transport media owing to unique and important properties as explained earlier. In this section, attempts that have been made so far to functionalize different textile fibers and polymers using $scCO_2$ impregnation technique are reviewed. The functional finishing agents used in impregnating polymers are categorized as functional dyes based on natural and synthetic origin, silicon and fluoropolymer-based, natural functional compounds, and organometallic-based agents.

4.1 Functional dyes

One strategy that has been followed to functionalize textiles in $scCO_2$ is by using different dyes having additional functional property. In this method, the functional dyes are either prepared by modifying them to contain functional groups through molecular design or those dyes which inherently possessed the required functional property are directly used. In most of the cases, disperse dyes are modified to contain functional groups based on the needed functionality, and some of them are presented in this section. Fluorescence functional dyes, such as disperse fluorescent yellow 82 were used to dye polyester in $scCO_2$ with the aim to manufacture protective clothing [20]. Results showed that polyester fabric was successfully dyed in $scCO_2$ medium exhibiting better photostability and fastness properties, and no morphological change was detected. Abou Elmaaty et al. [21] synthesized new hydrazonopropanenitrile dyes and applied the new species to polyester fabric using $scCO_2$ for potential antimicrobial application. Efficient dyeing and excellent antimicrobial and fastness properties were obtained using $scCO_2$ dyeing procedure. A series of disperse azo dyes with potential antibacterial activity were also applied to nylon 6 fabric using $scCO_2$ technique and compared with aqueous dyeing [22]. The comparison showed that samples dyed under $scCO_2$ medium had excellent antibacterial efficiency and better color fastness properties compared with the conventional exhaust dyeing with the advantage of the elimination of auxiliary chemicals. Impregnation of polyester (PET) films and poly(hydroxybutyrate) (PHB) granules with curcumin natural dye in $scCO_2$ has been reported [23]. In this study, the impregnation process was successfully developed with different amounts of curcumin add-on depending on the dyeing conditions and no significant detrimental effect observed on the material properties. More recently, curcumin has been used to dye and functionalize polyester in $scCO_2$ in our research group [24]. Dyed samples exhibited excellent color strength and fastness properties with improved antibacterial, antioxidant, and UV protection properties. Thus, the strategy of utilizing functional dyes which are suitable for $scCO_2$ process is a promising approach toward the production of colored and functional material in a single step.

4.2 Silicon and fluoropolymer-based functional agents

As stated earlier, silicon and amorphous fluoropolymers are known to have appreciable solubility in $scCO_2$ solvent. Due to this, functional agents based on these compounds have been employed to functionalize various textiles, polymers, and films. Mohamed et al. used a modified dimethyl siloxane terminated with silanol groups (DMS) to functionalize cotton fabric in $scCO_2$ [25]. Different crosslinking agents were used for covalently bonding silicon and cellulose. The results confirm that $scCO_2$ medium provides good coating (thickness between 1 to 2×10^{-6} m) of the cotton surface with a 3D network of DMS compound and crosslinker. Chen et al. [26] synthesized CO_2-philic silicon-containing quaternary ammonium salt (QAS) and applied to cotton in $scCO_2$ to prepare antimicrobial fabric. The treated fabric exhibited potent antimicrobial activity with good durability against washing and UV irradiation. They also synthesized silicone-containing 2,2,6,6-tetramethyl-4-piperidinol (TMP)-based N-chloramine and applied to polyethylene (PE) fiber via $scCO_2$ impregnation technique. A uniform coating of TMP-based N-chloramine reaching up to 70×10^{-9} m was obtained using 28×10^6 Pa pressure. The obtained PE modified with TMP-based N-chloramine imparted powerful and durable biocidal activity [27]. The same research group synthesized a CO_2-philic biocidal fluorinated pyridinium silicon and applied to cotton yarn using $scCO_2$ impregnation medium. Up to 50×10^{-9} m thickness of biocidal layer with pyridinium groups segregated on the top surface was attained at 24×10^6 Pa and 50°C. The obtained material provided higher biocidal efficiency [28]. Furthermore, polyester fabric was treated with low molecular weight polytetrafluoroethylene in $scCO_2$ medium and a high degree of water repellency was consistently obtained [29, 30]. Xu and co-workers [31] prepared a water/oil repellent polyester fabric using a solution of organic fluorine in $scCO_2$. A uniformly distributed fluorine could be obtained with good water/oil repellency keeping good air permeability and improved strength. Recently, perfluoroalkyl methacrylate/hydroxyalkyl methacrylate and a crosslinking agent (diisocyanate) have used to treat nylon fabric in $scCO_2$ medium to fabricate a durable water and oil repellent coatings [32]. A uniform, highly repellent, and durable coating was obtained by $scCO_2$ treatment compared with a coating deposited from a liquid solvent. These studies show that silicon and fluoropolymer-based materials have been playing a key role in the application of $scCO_2$ processing method for functionalization of textiles and polymeric materials.

4.3 Natural functional compounds

Supercritical CO_2 has also been used to impregnate polymers with natural functional compounds to impart different functionalities. Zizovic and co-workers widely investigated the application of thymol to various textile-based substrates in $scCO_2$ to develop different functional materials. They studied the solubility of thymol in $scCO_2$ and its impregnation on cotton gauze [33], cellulose acetate [34, 35], corona modified polypropylene non-woven material [36], and polycaprolactone (PCL) and polycaprolactone hydroxyapatite (PCL-HA) composites. Thymol has been shown soluble in $scCO_2$ solvent, and impregnation process was successful. All the samples prepared using $scCO_2$ impregnation exhibited strong antimicrobial effect against a wide range of bacteria strain. The same research group also used $scCO_2$ impregnation medium for loading cellulose acetate beads with carvacrol in order to fabricate a biomaterial with antimicrobial properties obtaining considerable antibacterial effect [37]. Thymol has also been used to modify polylactic acid (PLA) [38] and linear low-density polyethylene (LLDPE) [39] films using $scCO_2$ impregnation technique with the aim to prepare active materials for a wide range of applications such as food packaging and others. Furthermore, thymol was applied along with quercetin, a natural bioactive compound, to a film and foam-like structure N-carboxybutyl chitosan (CBC) film and agarose (AGR) using $scCO_2$ impregnation technique to fabricate wound dressing material. Impregnation was performed with the help of ethanol as a co-solvent, and higher impregnation yield was obtained at higher pressure and temperature. The obtained materials also exhibited a sustained release profile based on the release kinetic study [40]. Goñi et al. used Eugenol, a well-known natural antioxidant, and antimicrobial agent, to impregnate LLDPE films to fabricate active food packaging material [41]. The obtained film presented a good level of antioxidant property with some degree of heterogeneity and a decrease in crystallinity when higher treatment pressure was used. In another study, Eugenol was also used to impregnate polyamide fibers to fabricate antibacterial dental floss, and inhibition of more than 99.99% has been achieved [42]. Recently, Pajnik and co-workers impregnated pyrethrum extract to polypropylene, polyamide, and cellulose acetate in the form of films and beads using $scCO_2$ to fabricate functionalized materials with repellent properties [43]. In addition, chitosan and derivatives have been used to impregnate polyester in $scCO_2$ bath [44]. Results showed that low molecular weight chitosan and chitosan lactic acid salt were successfully impregnated whereas no chitin could be impregnated. More recently, very low molecular weight chitosan and chitosan lactate have also been successfully incorporated to polyester fabric using $scCO_2$ dyeing technique in our research group obtaining good antibacterial activity [45]. Overall, natural-based functional agents have shown a huge potential for the fabrication of various functional materials using $scCO_2$ impregnation technique.

4.4 Organometallic-based functional agents

In addition to the agents mentioned above, impregnation of organometallic compounds into polymer matrices using $scCO_2$ has also been widely studied for various functional applications. Antifungal textiles have been produced via $scCO_2$ impregnation of cotton with silver, Ag (hepta), and Ag (cod), demonstrating measurable inhibition [46]. Boggess et al. produced highly reflective polyimide films for aerospace application with silver-containing additive using $scCO_2$ infusion and subsequent curing at 300°C [47]. They have demonstrated that silver additive was incorporated into a polyimide film creating a reflective surface on both sides of the film. Chiu et al. [48] produced a wearable photocatalytic device via integration of Ni-P/TiO_2 onto silk fabric using $scCO_2$ impregnation technique. Co-deposition of photocatalytic TiO_2 and electrically conductive Ni-P metallization layer was achieved through $scCO_2$-assisted electroless plating and silk fabric with higher corrosion-resistant, and photocatalytic activity was achieved. Metallization of silk with platinum (Pt) was also conducted in $scCO_2$ medium obtaining a smooth and compact layer with improved adhesion promoted by $sccCO_2$ metallization [49]. The results demonstrated its applicability in medical and wearable devices. Cotton fabric has been impregnated with palladium (II) hexafluoroacetylacetonate to fabricate conductive fabrics [50]. Hematite nanoparticles were loaded to cellulosic fiber under $scCO_2$ to fabricate a water repellent composite fiber [51]. Peng et al. used silver nanoparticles to coat wool fabrics in $scCO_2$, and the coated fabric exhibited excellent catalytic, antistatic, and antibacterial activities [52]. Polycarbonate has been impregnated with silver nitrate in $scCO_2$, resulting up to 99.9% bacteria reduction [53]. Belmas and co-workers [54–56] have used $scCO_2$ process to impregnate a range of organometallic complexes in a synthetic polymer prior to electroless copper plating to improve the adhesion of copper to the polymer. The adhesion between the copper and polymer was much improved after $scCO_2$ impregnation of the organometallic complexes. Polyacrylate has been impregnated with copper (II) hexafluoroacetylacetonate in $scCO_2$ followed by thermal decomposition of the copper. The formation of copper oxide was evident ensuring improved wear resistance of polyacrylate [57]. In conclusion, owing to nanoscale metal microparticles, organometallic compounds have been successfully used to modify polymers in $scCO_2$ solvent for various functional applications and might be one potential area that needs further investigations in the future.

5. Conclusions

From the studies reviewed in this chapter, it has been shown that $scCO_2$ is a viable technique for the fabrication of various functional materials if appropriate agents suitable for the process are used. It can be an attractive alternative to traditional aqueous or organic solvents as it avoids toxic auxiliary chemicals and the use of water. Further studies are still required in selecting suitable functional agents which works best under $scCO_2$ solvent through investigation of their solubility, compatibility, and process optimizations. Due to its environmental advantages, the scientific community and the industrial compartment would expect an increase in research in this area as $scCO_2$ has the potential to replace the current water and solvent-based textile chemical processes in the future.

Acknowledgements

All the universities collaborating on the program are gratefully acknowledged.

Conflict of interest

The authors declare no conflict of interest.

Nomenclature

Ag	silver
AGR	agarose
CBC	carboxybutyl chitosan
CO_2	carbon dioxide
DMS	dimethyl siloxane
LLDPE	linear low-density polyethylene
m	meter
N	nitrogen
P	phosphorus
Pa	pascal
PCL	polycaprolactone
PCL-HA	polycaprolactone hydroxyapatite
PE	polyethylene
PET	polyester
PLA	polylactic acid
Pt	platinum
QAS	quaternary ammonium salt
$scCO_2$	supercritical carbon dioxide
T_g	glass transition temperature
TiO_2	titanium dioxide
TMP	2,2,6,6-tetramethyl-4-piperidinol

Author details

Molla Tadesse Abate[1,2,3]*, Ada Ferri[1], Jinping Guan[3], Guoqiang Chen[3] and Vincent Nierstrasz[2]

1 Department of Applied Science and Technology, Politecnico di Torino, Corso Duca degli Abruzzi, Torino, Italy

2 Textile Materials Technology, Department of Textile Technology, The Swedish School of Textiles, Faculty of Textiles, Engineering and Business, University of Borås, Borås, Sweden

3 College of Textile and Clothing Engineering, Soochow University, Suzhou, Jiangsu, China

*Address all correspondence to: molla.tadesse_abate@hb.se; molla.abate@polito.it

References

[1] Nalawade SP, Picchioni F, Janssen L. Supercritical carbon dioxide as a green solvent for processing polymer melts: Processing aspects and applications. Progress in Polymer Science. 2006;**31**:19-43. DOI: 10.1016/j.progpolymsci.2005.08.002

[2] Knez Ž, Markočič E, Leitgeb M, Primožič M, Knez Hrnčič M, Škerget M. Industrial applications of supercritical fluids: A review. Energy. 2014;**77**:235-243. DOI: 10.1016/j.energy.2014.07.044

[3] Ramsey E, Sun Q, Zhang Z, Zhang C, Gou W. Mini-review: Green sustainable processes using supercritical fluid carbon dioxide. Journal of Environmental Sciences. 2009;**21**:720-726. DOI: 10.1016/S1001-0742(08)62330-X

[4] Saus W, Knittel D, Schollmeyer E. Dyeing of textiles in supercritical carbon dioxide. Textile Research Journal. 1993;**63**:135-142. DOI: 10.1177/004051759306300302

[5] Weidner E. Impregnation via supercritical CO2–what we know and what we need to know. Journal of Supercritical Fluids. 2018;**134**:220-227. DOI: 10.1016/j.supflu.2017.12.024

[6] Kikic I, Vecchione F. Supercritical impregnation of polymers. Current Opinion in Solid State & Materials Science. 2003;7:399-405. DOI: 10.1016/j.cossms.2003.09.001

[7] Barros AA, Silva JM, Craveiro R, Paiva A, Reis RL, Duarte ARC. Green solvents for enhanced impregnation processes in biomedicine. Current Opinion in Green and Sustainable Chemistry. 2017;**5**:82-87. DOI: 10.1016/j.cogsc.2017.03.014

[8] Kazarian SG. Polymer processing with supercritical fluids. Polymer Science Series CC/C of vysokomolekuliarnye soedineniia. 2000;**42**:78-101. DOI: 10.1002/3527606726.ch10

[9] Zhong Z, Zheng S, Mi Y. High-pressure DSC study of thermal transitions of a poly(ethylene terephthalate)/carbon dioxide system. Polymer. 1999;**40**:3829-3834. DOI: 10.1016/S0032-3861(98)00594-1

[10] Chakraborty JN. Dyeing in supercritical carbon dioxide. In: Chakraborty JN, editor. Fundamentals and Practices in Colouration of Textiles. 2nd ed. New Delhi, India: Woodhead Publishing India Pvt. Ltd.; 2014. pp. 356-364. DOI: 10.1016/B978-93-80308-46-3.50028-5

[11] Champeau M, Thomassin JM, Tassaing T, Jérôme C. Drug loading of polymer implants by supercritical CO_2 assisted impregnation: A review. Journal of Controlled Release. 2015;**209**:248-259. DOI: 10.1016/j.jconrel.2015.05.002

[12] Gupta RB, Shim JJ. Solubility in Supercritical Carbon Dioxide. 1st ed. Boca Raton, Florida, USA: CRC press; 2006. 960 p. DOI: 10.1201/9781420005998

[13] Banchero M. Supercritical fluid dyeing of synthetic and natural textiles - a review. Coloration Technology. 2013;**129**:2-17. DOI: 10.1111/cote.12005

[14] Škerget M, Knez Ž, Knez-Hrnčič M. Solubility of solids in sub- and supercritical fluids: A review. Journal of Chemical & Engineering Data. 2011;**56**:694-719. DOI: 10.1021/je1011373

[15] Hyatt JA. Liquid and supercritical carbon dioxide as organic solvents. The Journal of Organic Chemistry. 1984;**49**:5097-5101. DOI: 10.1021/jo00200a016

[16] Draper SL, Montero GA, Smith B, Beck K. Solubility relationships for

disperse dyes in supercritical carbon dioxide. Dyes and Pigments. 2000;**45**:177-183. DOI: 10.1016/s0143-7208(00)00008-5

[17] Gebert B, Saus W, Knittel D, Buschmann H-J, Schollmeyer E. Dyeing natural fibers with disperse dyes in supercritical carbon dioxide. Textile Research Journal. 1994;**64**:371-374. DOI: 10.1177/004051759406400701

[18] Montero GA, Smith CB, Hendrix WA, Butcher DL. Supercritical fluid technology in textile processing: An overview. Industrial and Engineering Chemistry Research. 2000;**39**: 4806-4812. DOI: 10.1021/ie0002475

[19] McClain JB, Londono D, Combes JR, Romack TJ, Canelas DA, Betts DE, et al. Solution properties of a CO2-soluble fluoropolymer via small angle neutron scattering. Journal of the American Chemical Society. 1996;**118**:917-918. DOI: 10.1021/ja952750s

[20] Xiong XQ, Xu YY, Zheng LJ, Yan J, Zhao HJ, Zhang J, et al. Polyester fabric's fluorescent dyeing in supercritical carbon dioxide and its fluorescence imaging. Journal of Fluorescence. 2017;**27**:483-489. DOI: 10.1007/s10895-016-1975-0

[21] Abou Elmaaty T, Ma J, El-Taweel F, Abd El-Aziz E, Okubayashi S. Facile bifunctional dyeing of polyester under supercritical carbon dioxide medium with new antibacterial hydrazono propanenitrile dyes. Industrial and Engineering Chemistry Research. 2014;**53**:15566-15570. DOI: 10.1021/ie502088r

[22] Elmaaty T, El-Aziz E, Ma J, El-Taweel F, Okubayashi S. Eco-friendly disperse dyeing and functional finishing of nylon 6 using supercritical carbon dioxide. Fibers. 2015;**3**:309-322. DOI: 10.3390/fib3030309

[23] Herek L, Oliveira R, Rubira A, Pinheiro N. Impregnation of PET films and PHB granules with curcumin in supercritical CO2. Brazilian Journal of Chemical Engineering. 2006;**23**:227-234. DOI: 10.1590/S0104-66322006000200010

[24] Abate MT, Ferri A, Guan J, Chen G, Nierstrasz V. Colouration and bio-activation of polyester fabric with curcumin in supercritical CO2: Part I-investigating colouration properties. Journal of Supercritical Fluids. 2019;**152**:104548. DOI: 10.1016/j.supflu.2019.104548

[25] Mohamed AL, Er-Rafik M, Moller M. Supercritical carbon dioxide assisted silicon based finishing of cellulosic fabric: A novel approach. Carbohydrate Polymers. 2013;**98**: 1095-1107. DOI: 10.1016/j.carbpol.2013.06.027

[26] Chen Y, Niu MQ, Yuan S, Teng HN. Durable antimicrobial finishing of cellulose with QSA silicone by supercritical adsorption. Applied Surface Science. 2013;**264**:171-175. DOI: 10.1016/j.apsusc.2012.09.165

[27] Chen Y, Wang YY, Zhang Q, Yang CY, Han QX. Preparation of silicone containing 2,2,6,6-tetramethyl-4-piperidinol-based N-chloramine for antibacterial polyethylene via interpenetration in supercritical carbon dioxide. Journal of Applied Polymer Science. 2019;**136**:1-9. DOI: 10.1002/app.47614

[28] Chen Y, Zhang Q, Ma YJ, Han QX. Surface-oriented fluorinated pyridinium silicone with enhanced antibacterial activity on cotton via supercritical impregnation. Cellulose. 2018;**25**: 1499-1511. DOI: 10.1007/s10570-018-1657-y

[29] Prorokova NP, Kumeeva TY, Zavadskii AE, Nikitin LN. Modification of the surface of poly(ethylene terephthalate) fabrics by application of a water-repellent coating in

supercritical carbon dioxide medium. Fibre Chemistry. 2009;**41**:29-33. DOI: 10.1007/s10692-009-9121-2

[30] Prorokova NP, Kumeeva TY, Khorev AV, Buznik VM, Nikitin LN. Ensuring a high degree of water repellency of polyester textile materials by treating them with supercritical carbon dioxide. Fibre Chemistry. 2010;**42**:109-113. DOI: 10.1007/s10692-010-9233-8

[31] Xu Y-Y, Zheng L-J, Ye F, Qian Y-F, Yan J, Xiong X-Q. Water/oil repellent property of polyester fabrics after supercritical carbon dioxide finishing. Thermal Science. 2015;**19**:1273-1277. DOI: 10.2298/TSCI1504273X

[32] Zefirov VV, Lubimtsev NA, Stakhanov AI, Elmanovich IV, Kondratenko MS, Lokshin BV, et al. Durable crosslinked omniphobic coatings on textiles via supercritical carbon dioxide deposition. Journal of Supercritical Fluids. 2018;**133**:30-37. DOI: 10.1016/j.supflu.2017.09.020

[33] Milovanovic S, Stamenic M, Markovic D, Radetic M, Zizovic I. Solubility of thymol in supercritical carbon dioxide and its impregnation on cotton gauze. Journal of Supercritical Fluids. 2013;**84**:173-181. DOI: 10.1016/j.supflu.2013.10.003

[34] Milovanovic S, Stamenic M, Markovic D, Ivanovic J, Zizovic I. Supercritical impregnation of cellulose acetate with thymol. Journal of Supercritical Fluids. 2015;**97**:107-115. DOI: 10.1016/j.supflu.2014.11.011

[35] Milovanovic S, Markovic D, Aksentijevic K, Stojanovic DB, Ivanovic J, Zizovic I. Application of cellulose acetate for controlled release of thymol. Carbohydrate Polymers. 2016;**147**:344-353. DOI: 10.1016/j.carbpol.2016.03.093

[36] Markovic D, Milovanovic S, Radetic M, Jokic B, Zizovic I. Impregnation of corona modified polypropylene non-woven material with thymol in supercritical carbon dioxide for antimicrobial application. Journal of Supercritical Fluids. 2015;**101**:215-221. DOI: 10.1016/j.supflu.2015.03.022

[37] Milovanovic S, Adamovic T, Aksentijevic K, Misic D, Ivanovic J, Zizovic I. Cellulose acetate based material with antibacterial properties created by supercritical solvent impregnation. International Journal of Polymer Science. 2017;**2017**:9. DOI: 10.1155/2017/8762649

[38] Torres A, Ilabaca E, Rojas A, Rodríguez F, Galotto MJ, Guarda A, et al. Effect of processing conditions on the physical, chemical and transport properties of polylactic acid films containing thymol incorporated by supercritical impregnation. European Polymer Journal. 2017;**89**:195-210. DOI: 10.1016/j.eurpolymj.2017.01.019

[39] Torres A, Romero J, Macan A, Guarda A, Galotto MJ. Near critical and supercritical impregnation and kinetic release of thymol in LLDPE films used for food packaging. Journal of Supercritical Fluids. 2014;**85**:41-48. DOI: 10.1016/j.supflu.2013.10.011

[40] Dias AMA, Braga MEM, Seabra IJ, Ferreira P, Gil MH, de Sousa HC. Development of natural-based wound dressings impregnated with bioactive compounds and using supercritical carbon dioxide. International Journal of Pharmaceutics. 2011;**408**:9-19. DOI: 10.1016/j.ijpharm.2011.01.063

[41] Goñi ML, Gañán NA, Strumia MC, Martini RE. Eugenol-loaded LLDPE films with antioxidant activity by supercritical carbon dioxide impregnation. Journal of Supercritical Fluids. 2016;**111**:28-35. DOI: 10.1016/j.supflu.2016.01.012

[42] Mosquera JE, Goñi ML, Martini RE, Gañán NA. Supercritical carbon dioxide

assisted impregnation of eugenol into polyamide fibers for application as a dental floss. Journal of CO_2 Utilization. 2019;**32**:259-268. DOI: 10.1016/j.jcou.2019.04.016

[43] Pajnik J, Radetić M, Stojanovic DB, Jankovic-Častvan I, Tadic V, Stanković MV, et al. Functionalization of polypropylene, polyamide and cellulose acetate materials with pyrethrum extract as a natural repellent in supercritical carbon dioxide. Journal of Supercritical Fluids. 2018;**136**:70-81. DOI: 10.1016/j.supflu.2018.02.014

[44] Baba T, Hirogaki K, Tabata I, Okubayashi S, Hisada K, Hori T. Impregnation of chitin/chitosan into polyester fabric using supercritical carbon dioxide. Sen'i Gakkaishi (報文). 2010;**66**:63-69. DOI: 10.2115/fiber.66.63

[45] Abate MT, Ferri A, Guan J, Chen G, Ferreira JA, Nierstrasz V. Single-step disperse dyeing and antimicrobial functionalization of polyester fabric with chitosan and derivative in supercritical carbon dioxide. Journal of Supercritical Fluids. 2019;**147**:231-240. DOI: 10.1016/j.supflu.2018.11.002

[46] Gittard SD, Hojo D, Hyde GK, Scarel G, Narayan RJ, Parsons GN. Antifungal textiles formed using silver deposition in supercritical carbon dioxide. Journal of Materials Engineering and Performance. 2010;**19**:368-373. DOI: 10.1007/s11665-009-9514-7

[47] Boggess RK, Taylor LT, Stoakley DM, St. Clair AK. Highly reflective polyimide films created by supercritical fluid infusion of a silver additive. Journal of Applied Polymer Science. 1997;**64**: 1309-1317. DOI: 10.1002/(SICI)1097-4628(19970516)64:7<1309:AID-APP10>3.0.CO;2-S

[48] Chiu WT, Chen CY, Chang TFM, Hashimoto T, Kurosu H, Sone M. Ni-P and TiO2 codeposition on silk textile via supercritical CO2 promoted electroless plating for flexible and wearable photocatalytic devices. Electrochimica Acta. 2019;**294**:68-75. DOI: 10.1016/j.electacta.2018.10.076

[49] Chiu W-T, Chen C-Y, Chang T-FM, Tahara Y, Hashimoto T, Kurosu H, et al. Platinum coating on silk by a supercritical CO2 promoted metallization technique for applications of wearable devices. Surface and Coating Technology. 2018;**350**:1028-1035. DOI: 10.1016/j.surfcoat.2018.02.070

[50] Iwai Y, Sameshima S, Yonezawa S, Katayama S. Fabrication of conductive cotton by electroless plating method with supercritical carbon dioxide. Journal of Supercritical Fluids. 2015;**100**:46-51. DOI: 10.1016/j.supflu.2015.02.027

[51] Xu S, Shen D, Wu P. Fabrication of water-repellent cellulose fiber coated with magnetic nanoparticles under supercritical carbon dioxide. Journal of Nanoparticle Research. 2013;**15**:1577. DOI: 10.1007/s11051-013-1577-6

[52] Peng LH, Guo RH, Lan JW, Jiang SX, Li C, Zhang ZY. Synthesis of silver nanoparticles on wool fabric in supercritical carbon dioxide. Materials Express. 2017;**7**:405-410. DOI: 10.1166/mex.2017.1386

[53] Mölders N, Renner M, Errenst C, Weidner E. Incorporation of antibacterial active additives inside polycarbonate surfaces by using compressed carbon dioxide as transport aid. Journal of Supercritical Fluids. 2018;**132**:83-90. DOI: 10.1016/j.supflu.2017.02.009

[54] Belmas M, Tabata I, Hisada K, Hori T. Application of Dithiol compounds in supercritical carbon dioxide to improve the adhesive properties of copper-plated p-aramid fibers. Sen'i Gakkaishi (報文). 2010;**66**:229-235

[55] Belmas M, Tabata I, Hisada K, Hori T. Supercritical fluid-assisted Electroless copper plating of aramid film: The influence of surface treatment. Sen'i Gakkaishi (報文). 2010;**66**:215-221. DOI: 10.2115/fiber.66.215

[56] Belmas M, Tabata I, Hisada K, Hori T. Supercritical fluid-assisted Electroless metal plating onto aramid films: The influence of thermal treatment. Journal of Applied Polymer Science. 2011;**119**:2283-2291. DOI: 10.1002/app.32970

[57] Popov VK, Bagratashvili VN, Krasnov AP, Said-Galiyev EE, Nikitin LN, Afonicheva OV, et al. Modification of tribological properties of polyarylate by supercritical fluid impregnation of copper (II) hexafluoroacetylacetonate. Tribology Letters. 1998;**5**:297-301. DOI: 10.1023/a:1019110228703

Chapter 9

Foaming + Impregnation One-Step Process Using Supercritical CO_2

Antonio Montes, Clara Pereyra
and Enrique Martínez de la Ossa

Abstract

Polymers are widely used in everyday life due to their properties as toughness, viscoelasticity, and the possibility to form glasses and semicrystalline structures. For that reason, it is used in mostly drug delivery systems and tissue engineering and in pharmaceutical and biomedical investigations. Foaming process allows creating porous structure into the polymer leading to scaffolds. Scaffolds are the focus of many investigations as prolonged drug delivery systems and implants or injections which are used to deliver cells, drugs, and genes into the body. Particulate leaching, freeze-drying, thermally induced phase separation, rapid prototyping, powder compaction, sol–gel, and melt molding are the main techniques in front of supercritical fluid technology to prepare scaffolds. Supercritical foaming process using CO_2 presents advantages as a high dissolution in polymers and a green process because CO_2 is nontoxic, inexpensive, and reusable. Moreover, supercritical technology allows to do an impregnation with an active substance together with the foaming at the same time. Thus active substances entrapped into scaffolds could be fabricated in a one-step green process.

Keywords: supercritical, foaming, scaffold, polymer, impregnation

1. Introduction

For decades, drug delivery systems are the focus of many investigations [1–4] because it increases the effectiveness of formulations, avoiding the first-pass effect, and reduces the drug dosage of patients, producing delayed drug delivery to increase the patient comfort. Polymers are the coating agent frequently used in pharmaceutical technology due to their properties as toughness, viscoelasticity, and the possibility to form glasses and semicrystalline and porous structures.

Microencapsulation of active substance with polymers allows to produce drug delivery systems where the release phenomena is controlled by the diffusion of the active substance through the polymer and/or the erosion and degradation of the polymer at acid or basic media. Many authors have carried out investigations to prepare controlled drug delivery systems using supercritical technology [5–8]. In most of the cases, supercritical antisolvent process (SAS) has been the chosen technology because most of the active substances are insoluble in supercritical fluids. In this process an organic solution that contains the polymer and active substance is sprayed into a chamber filled with bulk supercritical fluid. The generated microdroplets improve the mass transfer between the supercritical fluid and the solution producing the dissolution into the solvent and the consequent solvent expansion and precipitation of particles of polymer and active substance by antisolvent effect. The result could be a coprecipitation of both compounds separately, the inclusion of particles into a matrix of polymer called composites, or the production of microcapsules with polymer coating as the active substance. In general, to avoid the separated precipitation, the ratio of polymer/active substance should be high. Moreover, operating conditions as pressure, temperature, concentration, flow rate ratios, and nozzle device have a relative influence on the final product characteristic.

However, some formulations require a long time drug delivery system as transdermal drug delivery where, for instance, hormone treatment could be carried out. Synthetic polymers, for example, polycaprolactone (PCL), polyvinyl alcohol (PVA), or polyvinylpyrrolidone (PVP), are good candidates to prepare for this kind of systems. An excellent alternative is the use of biopolymers, such as chitosan, alginate, starch, or hyaluronic acid [9].

Extended or long delay drug delivery systems are not often achieved in supercritical microencapsulation. If the active substance was placed into the pores of a polymeric porous structure, the drug release would be delayed most of the time. Thus supercritical impregnation into the pores of a polymer is an excellent alternative to prepare delay drug delivery systems.

In supercritical impregnation two processes could happen, the impregnation into the polymer pores and the foaming of polymer with the subsequent impregnated scaffold production. This fact will happen if the polymer structure is able to grow up in the depressurization step. For that many authors have carried out the foaming + impregnation one-step process and the other ones only the impregnation process. CO_2 is widely used as blowing agent because it presents properties that are nontoxic, inexpensive, and reusable and have a high dissolution in polymers.

When a polymer is put in contact with supercritical CO_2, in a first step the polymer is saturated with the gas above supercritical conditions. In a second step, the system is driven to a supersaturated state, usually decreasing the pressure or increasing temperature. This causes nucleation and relative growth of the porous cells within the polymeric matrix [10]. The fact that the polymer is under supercritical conditions alter physical properties as melting point and heat, glass transition and crystallization temperatures, crystallization rate, and swelling or foaming processes, among others, could be occasioned [11–13].

In general, as a solvent penetrated the polymer, it induced swelling and consequently facilitated the mobility of the chains, allowing reorientation of the chains to form the more thermodynamically favorable crystalline state and reducing the crystallization and melting temperatures [14].

Moreover, other authors conclude that crystallization rate of polymer-CO_2 depends only on the local degree of swelling inside the amorphous regions and the degrees of crystallinity itself [15].

Campardelli et al. [9] investigated the pore formation of PCL under CO_2 at 100–200 MPa of pressure and 35–40°C of temperature; due to a higher temperature, the polymer was melted. Process time was varied between 4 and 8 h. They concluded that formation of pores and thus the foaming of the polymer were only favored at 8 h when 100 MPa was used, but at higher pressures the foaming is produced independent of processing time. As pressure increases a regular pore structure was obtained with lower average pore diameter. However, as the temperature increases, the polymer swells more, forming a single structure, sticking polymer granules.

Thus polymer foaming could be achieved in some operating conditions producing scaffolds. The inclusion of active substance in these scaffolds is the focus of many new investigations. So when a polymer is going to be impregnated, it should be taken into account if the foaming process will carry out to produce scaffolds or the impregnation of the active substance happens mostly on the polymer surface.

2. Supercritical CO_2

A supercritical fluid is a substance above its critical temperature and pressure. A typical pressure-temperature phase diagram is shown in **Figure 1**. At this condition the fluid has unique properties as diffusivities that are two orders of magnitude larger than those of typical liquids, resulting in higher mass transfer rates. Moreover this state presents many exceptional characteristics, such as singularities in compressibility and viscosity and diminishing difference in liquid and vapor phases, among others. It is a good candidate to do extraction or impregnation processes because density can be adjusted continuously by altering the experimental conditions of temperature and pressure so solvent power and selectivity can be tuned.

The requirement that should fulfill the supercritical fluid is on the one hand low danger and on the other hand the relative low cost. In this sense CO_2 is GRAS solvent, noninflammable, nontoxic, and gaseous at room temperature which makes the separation process easy. Besides it does not present a high cost and presents relative mild conditions of its critical point (31.1°C and 71.8 bar), allowing the thermolabile solute processing.

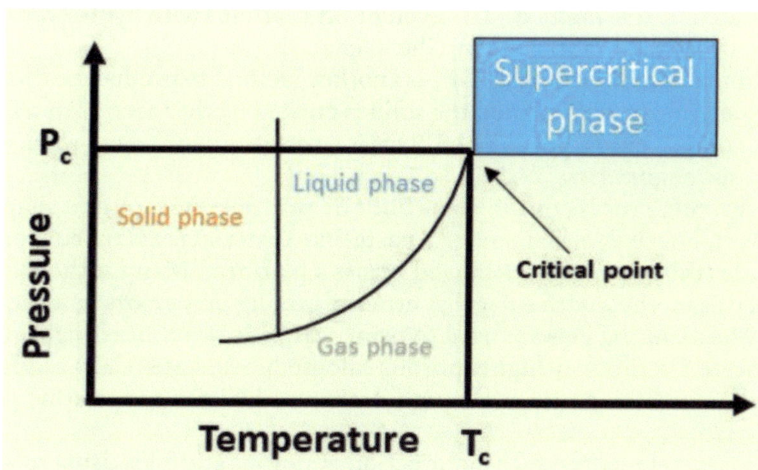

Figure 1.
Pressure-temperature phase diagram.

3. Scaffold fabrication

During the past two decades, biomedical research has advanced extensively to develop potentially applicable scaffolds. Several methods are used decades ago to manufacture these porous structures.

The solvent-casting particulate leaching (SCPL) technique is a standard method to produce polymer-based scaffolds. A polymer is dissolved in an organic solvent that contains mainly salts, with specific dimensions. Then, the mixture is shaped into a three-dimensional mold to produce a scaffold. Thus, when the solvent is removed by simple evaporation, it creates a structure of composite material consisting of the particles together with the polymer. At the end, particles are dissolved in a bath leaving behind a porous structure. In this way, Sola et al. fabricated innovative 3D porous structures to mimic the bone marrow niche in vitro using polymethyl methacrylate (PMMA) and a flexible polyurethane (PU) and NaCl, as an efficient porogen [16].

The preparation of porous structures from a thermoplastic polymer melt is a convenient route because of the production of scaffolds of many shapes and sizes reproducibly with the use of molds and without involving any solvents. These techniques typically include compression molding, extrusion, and injection molding. Scaffolds of any desired shape could be created by simply changing the mold to use for clinical applications. Moreover, various solid fillers as bioactive molecules could be employed as additives. However, the use of high molding temperatures could degrade and inactivate the biodegradable polymer or the impregnated bioactive molecules [17].

In freeze-gelation method, the porous structure is generated during the freeze of a polymer solution, and then the solvent is extracted by a non-solvent, or the polymer is gelled under the freezing condition. Thus, the porous structure destruction would be avoided during the subsequent drying stage. Porous PLLA, PLGA, chitosan, and alginate scaffolds were successfully fabricated with this method [18].

Various porous biodegradable scaffolds with these polymers have been also fabricated by thermally induced phase separation (TIPS) technique to be used in tissue engineering and drug delivery [19–20]. This technique is based on changes in Gibbs free energy to induce the demixing of a homogeneous polymer solution to obtain a multiphase system [21]. Highly porous scaffolds of biodegradable PCL have been fabricated by this method [22]. Even PLLA scaffolds with hydroxyapatite as filler were successfully fabricated by Ghersi et al. [23].

Three-dimensional printing (3DP) is another method to produce scaffolds for tissue engineering. In 3DP method the solid is created by the reaction of a liquid selectively sprayed onto a powder bed. This is a versatile method to produce scaffolds for tissue engineering [24].

In robocasting process the object is built by printing the required shape layer by layer. For that, a filament of a paste-like material is extruded from a small nozzle while the nozzle is moved across a platform. Many authors have combined this method with sol–gel synthesis, mixing precursors in an aqueous medium. The resulting gels are used to print scaffolds by robocasting. Houmard et al. fabricated in this way highly porous calcium phosphate (CaP) scaffolds for bone-tissue engineering using calcium nitrate tetrahydrate and triethyl phosphite precursors [25].

Besides, several reports on the fabrication of porous scaffolds using sol-gel technique are found in the literature [26–32].

Scaffold could be fabricated using supercritical CO_2 as blowing agent, avoiding the use of organic solvent, and thus the presence of solvent residue in the final product due to tradition processes tested until now does not allow the complete removal of the organic solvents involved in the starting solutions, avoiding high temperatures and long processing time (12–48 h) that can imply the stratification of the drug inside the scaffolds due to the separation of the loaded materials from the polymeric solutions. In this way, the efficiency of the generated devices sensibly decreases due to the inhomogeneous distribution of the drug [33].

In supercritical CO_2 foaming, the polymer is exposed to carbon dioxide, which plasticizes the polymer by reducing the glass transition temperature. Then in the depressurization step, thermodynamic instability causes supersaturation of the carbon dioxide dissolved in the polymer matrix, and hence, nucleation of cells occurs. Anyway gas foaming could not be used in polymers which have a very low affinity for CO_2 because the main requirement of process is that CO_2 can be dissolved in a sufficient amount in the polymer. So this technique is more commonly applied to amorphous polymers excluding polymers with high crystallinity or high glass transition temperatures [34].

In this way, for instance, the PCL scaffolds produced by supercritical fluid processing had a homogeneously interconnected porous structure, and they exhibited a narrow pore size distribution. Consequently, these results indicated that the PCL scaffolds made by supercritical fluid processing offer well-interconnected and nontoxic substrates for cell growth, avoiding problems associated with a solvent residue. This suggests that these elastic PCL scaffolds formed by supercritical fluid processing could be used for cartilage tissue engineering [35].

Other authors used supercritical CO_2 in the foaming process for the formation of polyvinylidene fluoride copolymerized with hexafluoropropylene loaded scaffolds which is a material that is semicrystalline and biocompatible with a good resistance to acid environments. They concluded that a higher pressure, a lower temperature, and a longer saturation time were more favorable for obtaining uniform foam. Moreover, the average pore cell diameter in low depressurization is larger than that in rapid depressurization. Lower crystallinity and higher melting temperature were induced in the formed scaffolds [36].

Supercritical CO_2 is used sometimes as dryer to prepare scaffolds. In this way polymer solution is prepared, and then this solution is put in contact with scCO_2. In that moment CO_2 solubilizes the organic solvent and the scaffold is formed.

In this sense a chitosan-based scaffold for tissue engineering applications has been prepared using supercritical CO_2 as dryer. The hydrogel fabricated was subsequently processed with supercritical CO_2. The highest porosity (87.03%) was achieved at 250 bar, 45°C, and 2 h of processing at 5 g/min CO_2 flow rate [37].

However other authors investigated about a new supercritical fluid-assisted technique for the formation of 3D scaffolds to overcome the main difficulty to obtain the coexistence of the macro- and microstructural characteristics necessary for a successful application. The process consists of the formation of a polymeric gel loaded with a solid porogen, then the drying of the gel using supercritical CO_2, and the washing with water to eliminate the porogen. In this way Reverchon et al. fabricated (PLLA) scaffolds with elevated porosity (>90%) and interconnectivity and with good mechanical properties [38]. Moreover they produced scaffolds with predetermined shape and size in a relatively short time (<30 h) and without an appreciable solvent residue (<5 ppm).

Tang et al. produced porous PCL scaffolds with open and interconnected architectures based on supercritical fluid-assisted hybrid processes of phase inversion and foaming. They achieved the encapsulation growth factor in these porous scaffolds, promoting the osteogenic differentiation and thus having also a significant potential in bone tissue engineering [39].

4. Impregnation

A scaffold where a bioactive substance can be incorporated that, for instance, can control proliferation and differentiation of cells is an excellent alternative to be used in tissue engineering. In this way the function of a scaffold is not limited only as a physical support but also as a bioactive element to control cell proliferation and differentiation. Anyway, scaffold impregnation process has been mostly studied for the preparation of long time drug delivery systems, with more or less delay depending on the final purpose of the delivery.

The conventional impregnation of scaffolds uses organic solvents that dissolve the drug which is going to be incorporated into the scaffolds, but this organic solution should swell and stretch the polymer to allow the diffusion of the drug at adequate depressurization rate. Posteriorly the organic solvent should be removed, leaving the impregnated scaffold. This method has several drawbacks, for example, residues of organic solvent in the final product need a last step to dry the scaffold, and the distribution of drug into scaffold is heterogeneous.

SCF impregnation removes all these drawbacks due to its properties as high diffusivity, low surface tension, and the ease of solvent recovery. Nevertheless, this methodology is limited by the solubility of the drug in the SCF, and the polymer can be swollen by the SCF. If these last requirements are fulfilled, a high-quality product free of residual solvents can be obtained, since no organic solvents are involved in process [40, 41]. In this process SCF is put in contact with the active substance that is going to be incorporated into the scaffolds. Then the SCF solubilizes this substance till saturation during the impregnation time. Later, in the depressurization step, the gas rapidly diffuses out of the polymer, deplasticizing it and avoiding the polymers and active substance exposition to high temperatures, which may degrade them.

CO_2 is the frequently used SCF because it is not dangerous, not toxic, not flammable, relatively cheap, and classified as a safe solvent. Anyway due to polarity limitations, it is often used as a cosolvent to increase the polarity and selectivity.

Duarte et al. prepared loaded chitosan scaffold able to sustain the release of dexamethasone using supercritical impregnation [42]. Sproule et al. [43] achieved the successful impregnation of a protein in PMMA scaffolds for biomedical applications holding unaltered the protein activity.

Campardelli et al. [9] prepared polycaprolactone/nimesulide patches using supercritical impregnation. In this work the authors achieved the foaming of the polymer in certain conditions at the same time that nimesulide is incorporated into its structure. Thus impregnated scaffolds are prepared in a one-step process for determined conditions. However in other operating conditions, foaming of the polymer is not favored, and scaffolds were not achieved.

Biodegradable PLA/PLGA foams impregnated with indomethacin in $scCO_2$ were studied by Cabezas et al. [44]. Authors observed that drug loading of foams was favored by high values of stirring rate. Moreover little pore sizes were obtained at slow depressurization rates. As it was expected, composition influenced the mechanical resistance, the PLA foams being more fragile.

Fanovich et al. studied the functionalized PCL scaffolds impregnated with natural compounds extracted from Patagonian *Usnea* lichen for tissue engineering [45]. An integrated process at high pressure for extraction/impregnation/foaming of PCL was developed. Authors concluded that the process is successful at 35°C and 15–17 MPa of CO_2 by foaming. The same researchers incorporated in a posterior work hydroxyapatite to these scaffolds, concluding that the scaffold obtained from PCL-HA with 20% of the HA shows the highest impregnation yield at 17 MPa and 35°C and subsequently also the best bactericidal effect on the tested *Staphylococcus aureus* strains [46].

The impregnation of chitosan with lactulose using supercritical fluids under various operating conditions, in order to improve the solubility of this natural polymer at neutral or basic pH, was carried out by Diaz et al. The highest impregnation yield was obtained using continuous process, 60-min contact time, 14% (v/v) of cosolvent ethanol/water (95:5), depressurization rate equal to 3.3 bar/min, 100 bar of pressure, and 100°C [47].

The impregnation of 5-fluorouracil, a chemotherapy agent, into a polymer based on D,L-lactide and glycolide was carried out at the same time to the foaming process in a one-step procedure. The possibility of regulating the rate of the scaffold degradation and the kinetics of drug release makes the usage of the copolymer more attractive for a further medical application. Venting rate is revealed to be the most important factor affecting the probes' pore size and their morphology. Thus, slow venting rates should be used to promote small pores in order to retard the drug release from the polymeric matrix. As it was expected, vigorous stirring rates favor the contact between supercritical CO_2 and the swelled polymer, improving the impregnation process. On the other hand, the presence of glycolide enhanced the mechanical strength of the foam preventing pore collapse [48].

In the same way, Salerno et al. prepared porous PCL scaffolds containing three different drugs: 5-fluorouracil, nicotinamide, and triflusal, in order to obtain systems with controlled drug delivery capabilities. ScCO_2 saturation and foaming conditions were optimized to create the porosity within the samples and demonstrated that the composition of the starting PCL/drug/solvent mixtures influenced polymer crystallization, scaffold morphology, and pore structure features. Moreover, it was found that drug loading depended on both initial solution composition and drug solubility in scCO_2. So, in the case of triflusal that is a highly scCO_2-soluble drug, loading efficiency was improved by adding a higher amount of free drug inside of the impregnation vessel. The drug delivery study, as it was expected, indicated that release profiles depended mainly on pore structure and scaffold composition [49].

However, the authors observed that the control on the pore interconnectivity and pore size with this technique still needs to be improved. They proposed the use of natural plasticizers as eugenol to overcome these limitations. Thus, eugenol-containing PCL scaffolds were prepared by supercritical foaming followed by a one- or a two-step depressurization profile. Moreover these scaffolds presented antimicrobial activity preventing the attachment of Gram-positive (*S. aureus*, *S. epidermidis*) bacteria and showed good tissue integration [50].

A hybrid porous scaffold of PLGA hydroxyapatite and collagen was prepared using a supercritical CO_2 saturation technique by Zhang et al. The results showed that the pore size and porosity of the scaffold could be controlled by manipulating these process conditions. The pore size and porosity can be regulated by supercritical CO_2 saturation temperature, saturation time, and saturation pressure [51].

5. One-step supercritical foaming + impregnation process: a particular case

In our facilities PCL scaffolds impregnated with quercetin were prepared using supercritical CO_2. PCL is a semicrystalline polyester with a melting point (Tm) of 329–334 K and a glass transition temperature (Tg) of 213 K [52]. Quercetin (Q) is a flavonoid present in many fruits and vegetables [53]. This flavonoid highlights its antioxidant action, but it has different benefits as antibacterial, cardiovascular health, anti-inflammatory, and anticancer effects [54, 55]. The study was supported by an experimental design to elucidate the influence of pressure (15–30 MPa), temperature (308–333 K), and depressurization rate (0.1–20) on foaming, melting temperature, melting heat, and amount of released quercetin.

The experiments were carried out in a plant RESS250 developed by Thar Technologies [56]. PCL and quercetin were mixed physically into an aluminum foil support (ratio 50:1 PCL/Q), and it was introduced into a stainless steel vessel. Then, CO_2 was pumped to the vessel till the desired operating pressure at the same time that the temperature was adjusted is reached. A determined impregnation time was awaited, and once finished, the output valve was opened to vent the CO_2 in a range of depressurization rate of 0.1–20 MPa min^{-1}. In the SEM image in **Figure 2**, it can be seen that PCL/quercetin foamed in our facilities [56].

The generated PCL/quercetin scaffold with higher pore density and smaller pore size was achieved for higher pressure and depressurization rate and lower temperatures (300 bar, 308 K, and 20 MPa min^{-1}). In general according to our results, the high level of temperature is recommended to obtain a pronounced effect of foaming to produce scaffold.

It was also observed that experiments done at lower pressure and temperature together with a higher depressurization rate led to a higher melting temperature. An increase of pressure and temperature leads to composite which released a higher amount of quercetin. However, depressurization rate has the opposite trend, so an increase of depressurization rate leads to a lower amount of released quercetin. These facts can be observed in the contour surface plots (**Figures 3** and **4**).

Release profiles showed that quercetin took five times longer to dissolve the same amount of quercetin into the first 8 h, demonstrating the efficacy of using PCL to control quercetin release and its possible use with other medical or pharmaceutical compounds (**Figure 5**).

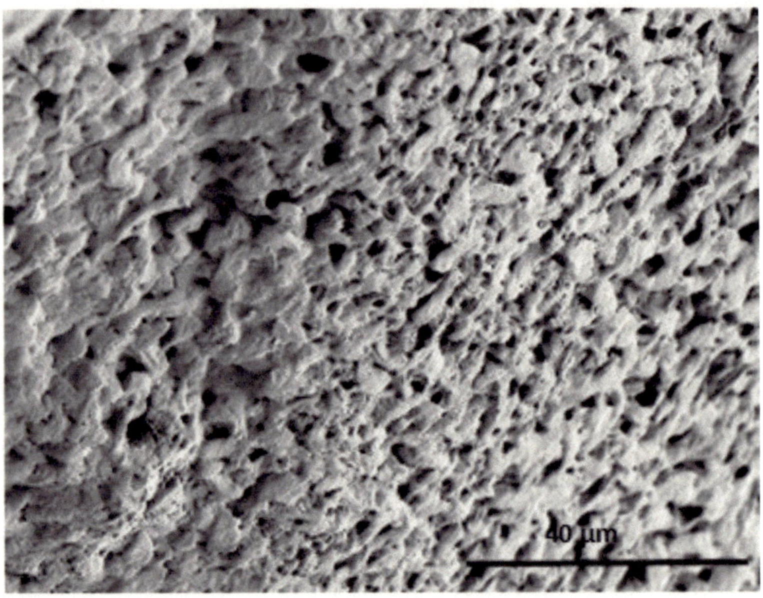

Figure 2.
SEM image of PCL/quercetin scaffold.

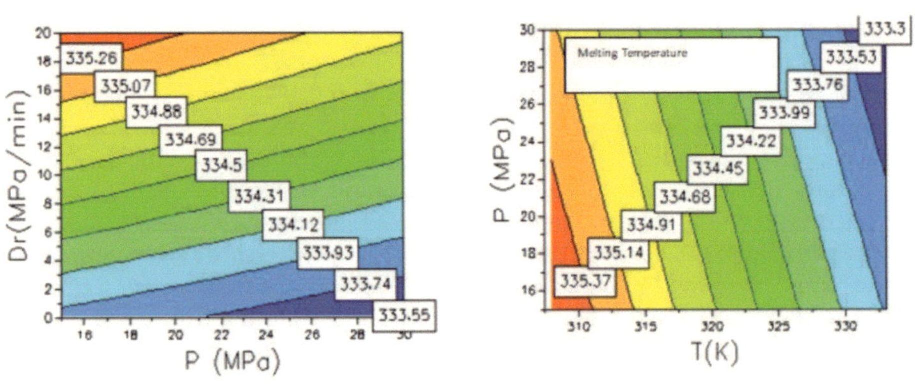

Figure 3.
Contour plot of melting temperature of PCL/quercetin scaffold.

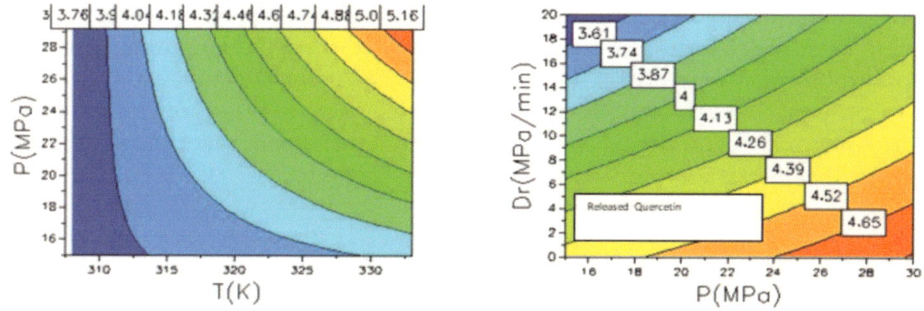

Figure 4.
Contour plot of release quercetin of PCL/quercetin scaffold.

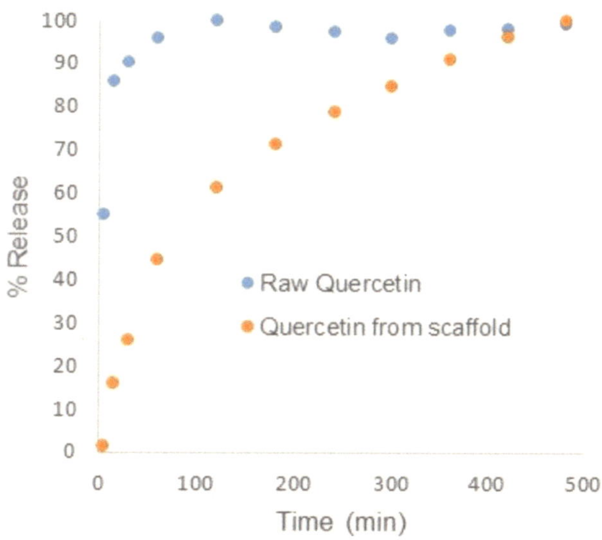

Figure 5.
Release profile from raw quercetin and PCL/quercetin scaffold.

6. Conclusions

Impregnated scaffolds are an interesting alternative to be used in pharmacology and biomedicine because scaffolds act not only as a physical support but also as a carrier of a bioactive substance with a controllable release. The possibility of regulating the rate of the scaffold degradation and the kinetics of drug release make it easy to fabricate particular drug release systems. Bone regeneration, implants, hormonal treatment, and tissue engineering applications are fields where scaffolds could be used. The way to create the porosity in the polymer originated a multiple scaffold fabrication methods based most of them in molding and removing the used organic solvent in a posterior step. Supercritical CO_2 has been used as dryer in many conventional methods as sol-gel where the solvent must be evaporated. However, in these methods supercritical CO_2 is able to remove almost the totality of used organic solvent, which requires several process steps. Another way to use supercritical CO_2 is in the gas foaming process. In this sense the porosity is created at the same time that the bioactive substance is incorporated, avoiding the use of organic solvent. Moreover these processes do not use high temperature, so the activity of the bioactive molecule would hold unaltered. Foaming process changes not only the polymer porosity but also other properties as melting, crystallization or glass transition temperature, melting heat, and so on and thus could produce more fragile or harder polymer depending on the foaming conditions. Particularly in our facilities, polycaprolactone/quercetin scaffolds were prepared using supercritical CO_2 foaming + impregnation one-step process in an efficient way. Release profiles showed that quercetin took five times longer to dissolve the same amount of quercetin into the first 8 h where it was placed into scaffold.

Acknowledgements

We gratefully acknowledge the Spanish Ministry of Science and Technology (Project CTQ2017-86661-R) and European Regional Development Funds (UNCA10-1E-1125 and 18INIA1103. 2011) for the financial support and the Central Services of Science and Technology of University of Cádiz for the analyses.

Conflict of interest

The authors declare no conflict of interest.

Author details

Antonio Montes*, Clara Pereyra and Enrique Martínez de la Ossa
Department of Chemical Engineering and Food Technology, Faculty of Science, University of Cádiz, International Excellence Agrifood Campus (CeiA3), Cádiz, Spain

*Address all correspondence to: antonio.montes@uca.es

References

[1] Rai M, Pandit R, Paralikar P, Nagaonkar D, Rehman F, Alves dos Santos C. Pharmaceutical applications of curcumin-loaded nanoparticles. In: Rai M, Alves dos Santos C, editors. Nanotechnology Applied to Pharmaceutical Technology. Cham: Springer; 2017. pp. 139-154

[2] Varshosaz J, Taymouri S, Jafari E, Jahanian-Najafabadi A, Taheri A. Formulation and characterization of cellulose acetate butyrate nanoparticles loaded with nevirapine for HIV treatment. Journal of Drug Delivery Science and Technology. 2018;**48**:9-20. DOI: 10.1016/j.jddst.2018.08.020

[3] Chaoqun Y, Hongshuai W, Minxing W, Zhiguo G, Baiwang S, Xiangyang Z. Synthesis and biological evaluation of redox/NIR dual stimulus-responsive polymeric nanoparticles for targeted delivery of cisplatin. Materials Science & Engineering C. 2018;**92**:453-462. DOI: 10.1016/j.msec.2018.06.044

[4] Liakos I, Iordache F, Carzino R, Scarpellini A, Oneto M, Bianchini P, et al. Cellulose acetate-essential oil nanocapsules with antimicrobial activity for biomedical applications. Colloids and Surfaces B: Biointerfaces. 2018;**172**:471-479. DOI: 10.1016/j.colsurfb.2018.08.069

[5] Zhong Q, Jin M, Davidson P, Zivanovic S. Sustained release of lysozyme from zein microcapsules produced by a supercritical anti-solvent process. Food Chemistry. 2009;**115**(2):697-700. DOI: 10.1016/j.foodchem.2008.12.063

[6] Soon Hong S, Lai Yeng L. Microencapsulation and nanoencapsulation using supercritical fluid (SCF) techniques. Pharmaceutics. 2019;**11**(1):21-31. DOI: 10.3390/pharmaceutics11010021

[7] Trucillo P, Campardelli R. Production of solid lipid nanoparticles with a supercritical fluid assisted process. The Journal of Supercritical Fluids. 2019;**143**:16-23. DOI: 10.1016/j.supflu.2018.08.001

[8] Olayinka J, Baloguna M, Swanepoel A, Bolaji R, Ahmed A, Labuschagne P. Microencapsulation of eucalyptol in polyethylene glycol and polycaprolactone using particles from gas-saturated solutions. RSC Advances. 2019;**9**:34039-34049. DOI: 10.1039/C9RA06419B

[9] Campardelli R, Franco P, Reverchon E, De Marco I. Polycaprolactone/nimesulide patches obtained by a one-step supercritical foaming + impregnation process. The Journal of Supercritical Fluids. 2019;**146**:47-54. DOI: 10.1016/j.supflu.2019.01.008

[10] Goel SK, Beckman EJ. Generation of microcellular polymeric foams using supercritical carbon dioxide. II: Cell growth and skin formation. Polymer Engineering and Science. 1994;**34**:1137-1147. DOI: 10.1002/pen.760341408

[11] Fanovich MA, Jaeger P. Sorption and diffusion of compressed carbon dioxide in polycaprolactone for the development of porous scaffolds. Materials Science and Engineering: C. 2012;**32**:961-968. DOI: 10.1016/j.msec.2012.02.021

[12] Lian Z, Epstein SA, Blenk CW, Shine AD. Carbon dioxide-induced melting point depression of biodegradable semicrystalline polymers. Journal of Supercritical Fluids. 2006;**39**:107-117. DOI: 10.1016/j.supflu.2006.02.001

[13] Yoganathan R, Mammucari R, Foster NR. Impregnation of ibuprofen into polycaprolactone using supercritical carbon dioxide. Journal of Physics Conference

Series. 2010;**215**:6-11. DOI: 10.1088/1742-6596/215/1/012087

[14] Zhang Z, Handa YP. CO_2-assisted melting of semicrystalline polymers. Macromolecules. 1997;**30**:8505-8507. DOI: 10.1021/ma9712211

[15] Kalospiros NS, Astarita G, Paulaitis ME. Coupled diffusion and morphological change in solid polymers. Chemical Engineering Science. 1993;**48**:23-40. DOI: 10.1016/0009-2509(93)80280-4

[16] Sola A, Bertacchini J, 'Avella D D, Anselmi L, Maraldi T, Marmiroli S, et al. Development of solvent-casting particulate leaching (SCPL) polymer scaffolds as improved three-dimensional supports to mimic the bone marrow niche. Materials Science & Engineering C. 2019;**96**:153-165. DOI: 10.1016/j.msec.2018.10.086

[17] Deng Y, Kuiper J. Functional 3D Tissue Engineering Scaffolds: Materials, Technologies, and Applications. 1st ed. Amsterdam, Netherlands: Elsevier; 2018. ISBN: 9780081009796

[18] Ming-Hua H, Pei-Yun K, Hsyue-Jen H, Tzu-Yang H, Lein-Tuan H, Juin-Yih L, et al. Preparation of porous scaffolds by using freeze-extraction and freeze-gelation methods. Biomaterials. 2004;**25**(1):129-138. DOI: 10.1016/S0142-9612(03)00483-6

[19] Yoon Sung N, Tae GP. Porous biodegradable polymeric scaffolds prepared by thermally induced phase separation. Journal of Biomedical Materials Research. 1999;**47**:8-17. DOI: 10.1002/ (SICI)1097-4636(199910)

[20] Fang Jun H, Jae Do N, Doo Sung L. Preparation of a macroporous poly(L-lactide) scaffold by liquid-liquid phase separation of a PLLA/1,4-dioxane/water ternary system in the presence of NaCl. Macromolecular Rapid Communications. 2001;**22**(13):1053-1057. DOI: 10.1002/jbm.10121

[21] Liumin H, Yanqing Z, Xiang Z, Daping Q, Susan L, Yuanshan Z, et al. Fabrication and characterization of poly(l-lactic acid) 3D nanofibrous scaffolds with controlled architecture by liquid-liquid phase separation from a ternary polymer–solvent system. Polymer. 2009;**50**(16):4128-4138. DOI: 10.1016/j.polymer.2009.06.025

[22] Guarino V, Guaccio A, Guarnieri D, Netti PA, Ambrosio L. Binary system thermodynamics to control pore architecture of PCL scaffold via temperature-driven phase separation process. Journal of Biomaterials Applications. 2011;**27**(3):241-254. DOI: 10.1177/0885328211401056

[23] Ghersi G, Carfì Pavia F, Conoscenti G, Mannella GA, Greco S, Rigogliuso S, et al. PLLA scaffold via TIPS for bone tissue engineering. Chemical Engineering Transactions. 2016;**49**:301-306. DOI: 10.3303/CET1649051

[24] Butschera A, Bohner M, Roth C, Ernstberger A, Heuberger R, Doebelin N, et al. Printability of calcium phosphate powders for three-dimensional printing of tissue engineering scaffolds. Acta Biomaterialia. 2012;**8**(1):373-385. DOI: 10.1016/j.actbio.2011.08.027

[25] Houmard M, Fu Q, Saiz E, Tomsia AP. Sol–gel method to fabricate CaP scaffolds by robocasting for tissue engineering. Journal of Materials Science. Materials in Medicine. 2012;**23**(4):921-930. DOI: 10.1007/s10856-012-4561-2

[26] Chen QZ, Thompson ID, Boccaccini AR. 45S5 bioglass®-derived glass-ceramic scaffolds for bone tissue engineering. Biomaterials. 2006;**27**:2414-2425. DOI: 10.1016/j.biomaterials.2005.11.025

[27] Rezwan K, Chen QZ, Blaker JJ, Boccaccini AR. Biodegradable and bioactive porous polymer/inorganic composite scaffolds for bone tissue engineering. Biomaterials. 2006;**27**:3413-3431. DOI: 10.1016/j.biomaterials.2006.01.039

[28] Jones JR, Ehrenfried LM, Hench LL. Optimising bioactive glass scaffolds for bone tissue engineering. Biomaterials. 2006;**27**:964-973. DOI: 10.1016/j.biomaterials.2005.07.017

[29] Jones JR, Tsigkou O, Coates EE, Stevens MM, Polak JM, Hench LL. Extracellular matrix formation and mineralization on a phosphate-free porous bioactive glass scaffold using primary human osteoblast (HOB) cells. Biomaterials. 2007;**28**:1653-1663. DOI: 10.1016/j.biomaterials.2006.11.022

[30] Almeida RM, Gama A, Vueva Y. Bioactive sol-gel scaffolds with dual porosity for tissue engineering. Journal of Sol-Gel Science and Technology. 2011;**57**:336-342. DOI: 10.1007/s10971-009-2134-8

[31] Marques AC, Jain H, Kiely C, Song K, Kiely CJ, Almeida RM. Nano/macroporous monolithic scaffolds prepared by the sol-gel method. Journal of Sol-Gel Science and Technology. 2009;**51**:42-47. DOI: 10.1007/s10971-009-1960-z

[32] Peña J, Román J, Cabañas MV, Vallet-Regí M. An alternative technique to shape scaffolds with hierarchical porosity at physiological temperature. Acta Biomaterialia. 2010;**6**:1288-1296. DOI: 10.1016/j.actbio.2009.10.049

[33] Cardea S, Reverchon E. 3-D loaded scaffolds obtained by supercritical CO_2 assisted process. In: 2014 Global Conference on Polymer and Composite Materials (PCM 2014), IOP Conf. Series: Materials Science and Engineering. 2014. DOI: 10.1088/1757-899X/62/1/012004

[34] Duarte ARC, Mano JF, Reis RL. Supercritical fluids in biomedical and tissue engineering applications: A review. International Materials Reviews. 2009;**54**:214-222. DOI: 10.1179/174328009X411181

[35] Kim SH, Jung Y, Kim SH. A biocompatible tissue scaffold produced by supercritical fluid processing for cartilage tissue engineering. Tissue Engineering. Part C, Methods. 2013;**19**(3):181-188. DOI: 10.1089/ten.TEC.2012.0170

[36] Muoi T, Tzu-Chi W. Foaming of poly (vinylidene fluoride-co-hexafluoropropylene) using supercritical carbon dioxide. Journal of the Taiwan Institute of Chemical Engineers. 2016;**7**(73):146-153. DOI: 10.1016/j.jtice.2016.09.038

[37] Ozdemir E, Sendemir-Urkmez A, Yesil-Celiktas O. Supercritical CO_2 processing of a chitosan-based scaffold: Can implantation of osteoblastic cells be enhanced? The Journal of Supercritical Fluids. 2013;**75**:120-127. DOI: 10.1016/j.supflu.2012.12.031

[38] Reverchon E, Cardea S, Rapuano C. A new supercritical fluid-based process to produce scaffolds for tissue replacement. Journal of Supercritical Fluids. 2008;**45**(3):365-373. DOI: 10.1016/j.supflu.2008.01.005

[39] Tang H, Kankala RK, Wang S. Supercritical fluid-assisted controllable fabrication of open and highly interconnected porous scaffolds for bone tissue engineering. Science China. Life Sciences. 2019;**62**:1-13. DOI: 10.1007/s11427-018-9393-8

[40] Kikic I, Vecchione F. Supercritical impregnation of polymers. Current Opinion in Solid State & Materials Science. 2003;**7**:399-405. DOI: 10.1016/j.cossms.2003.09.001

[41] Elvira C, Fanovich A, Fernandez M, Fraile J, Roman JS,

Domingo C. Evaluation of drug delivery characteristics of microspheres of PMMA-PCL-cholesterol obtained by supercritical-CO_2 impregnation and by dissolution-evaporation techniques. Journal of Controlled Release. 2004;**99**:231-240. DOI: 10.1016/j.jconrel.2004.06.020

[42] Duarte ARC, Mano JF, Reis RL. Preparation of chitosan scaffolds loaded with dexamethasone for tissue engineering applications using supercritical fluid technology. European Polymer Journal. 2009;**45**:141-148. DOI: 10.1016/j.eurpolymj.2008.10.004

[43] Sproule TL, Lee JA, Li H, Lannutti JJ, Tomasko DL. Bioactive polymer surfaces via supercritical fluids. Journal of Supercritical Fluids. 2004;**28**:241-248. DOI: 10.1016/S0896-8446(03)00044-5

[44] Cabezas LI, Fernández V, Mazarro R, Gracia I, de Lucas A, Rodríguez JF. Production of biodegradable porous scaffolds impregnated with indomethacin in supercritical CO_2. The Journal of Supercritical Fluids. 2012;**63**:155-160. DOI: 10.1016/j.supflu.2011.12.002

[45] Fanovich MA, Ivanovic J, Misic D, Alvarez MV, Jaeger P, Zizovic I, et al. Development of polycaprolactone scaffold with antibacterial activity by an integrated supercritical extraction and impregnation process. The Journal of Supercritical Fluids. 2013;**78**:42-53. DOI: 10.1016/j.supflu.2013.03.017

[46] Fanovich MA, Ivanovic J, Zizovic I, Misic D, Jaeger P. Functionalization of polycaprolactone/hydroxyapatite scaffolds with *Usnea lethariiformis* extract by using supercritical CO_2. Materials Science and Engineering: C. 2016;**58**:204-212. DOI: 10.1016/j.msec.2015.08.024

[47] Díez Municio M, Montilla A, Herrero M, Olano A, Ibáñez E. Supercritical CO_2 impregnation of lactulose on chitosan: A comparison between scaffolds and microspheres form. The Journal of Supercritical Fluids. 2011;**57**:73-79. DOI: 10.1016/j.supflu.2011.02.001

[48] Cabezas LI, Gracia I, García MT, de Lucas A, Rodríguez JF. Production of biodegradable porous scaffolds impregnated with 5-fluorouracil in supercritical CO_2. The Journal of Supercritical Fluids. 2013;**80**:1-8. DOI: 10.1016/j.supflu.2013.03.030

[49] Salerno A, Saurina J, Domingo C. Supercritical CO_2 foamed polycaprolactone scaffolds for controlled delivery of 5-fluorouracil, nicotinamide and triflusal. International Journal of Pharmaceutics. 2015;**496**(2):654-663. DOI: 10.1016/j.ijpharm.2015.11.012

[50] Salerno A, Diéguez S, Diaz-Gomez L, Gómez-Amoza JL, Magariños B, Concheiro A, et al. Synthetic scaffolds with full pore interconnectivity for bone regeneration prepared by supercritical foaming using advanced biofunctional plasticizers. Biofabrication. 2017;**9**(3):035002-035010. DOI: 10.1088/1758-5090/aa78c5

[51] Zhao-Li M, Li-Jun Z, Qin-An Z, Jing Z, Zhi-Qi Z. Preparation of porous PLGA/HA/collagen scaffolds with supercritical CO_2 and application in osteoblast cell culture. The Journal of Supercritical Fluids. 2011;**58**(3):398-406. DOI: 10.1016/j.supflu.2011.07.003

[52] Sinha VR, Bansal K, Kaushik R, Kumria R, Trehan A. Poly-ϵ-caprolactone microspheres and nanospheres: An overview. International Journal of Pharmaceutics. 2004;**278**:1-23. DOI: 10.1016/j.ijpharm.2004.01.044

[53] Formica JV, Regelson W. Review of the biology of quercetin and related bioflavonoids. Food and Chemical

Toxicology. 1995;**33**:1061-1080. DOI: 10.1016/0278-6915(95)00077-1

[54] Boots AW, Haenen GRM, Bast A. Health effects of quercetin: From antioxidant to nutraceutical. European Journal of Pharmacology. 2008;**585**:325-337. DOI: 10.1016/j.ejphar.2008.03.008

[55] Sahoo NG, Kakran M, Shaal LA, Li L, Müller RH, Pal M, et al. Preparation and characterization of quercetin nanocrystals. Journal of Pharmaceutical Sciences. 2011;**100**:2379-2390. DOI: 10.1002/jps.22446

[56] García-Casas I, Montes A, Valor D, Pereyra C, Martínez de la Ossa EJ. Foaming of polycaprolactone and its impregnation with quercetin using supercritical CO_2. Polymers. 2019;**11**:1390. DOI: 10.3390/polym11091390